AF847555

Gender and American Law

The Impact of the Law on the Lives of Women

Series Editor

Karen J. Maschke

Contents of the Series

1. Women and the American Legal Order
2. Reproduction, Sexuality, and the Family
3. The Employment Context
4. Educational Equity
5. The Legal Response to Violence Against Women
6. Pornography, Sex Work, and Hate Speech
7. Feminist Legal Theories

Reproduction, Sexuality, and the Family

Edited with introductions by

Karen J. Maschke

Routledge
Taylor & Francis Group
New York London

First published by Garland Publishing, Inc.

This edition published 2013 by Routledge

Routledge
Taylor & Francis Group
711 Third Avenue
New York, NY 10017

Routledge
Taylor & Francis Group
2 Park Square, Milton Park
Abingdon, Oxon OX14 4RN

Routledge is an imprint of the Taylor & Francis Group, an informa business

Introductions copyright © 1997 Karen J. Maschke. All rights reserved.

Library of Congress Cataloging-in-Publication Data

Reproduction, sexuality, and the family / edited with introductions by
Karen J. Maschke.
p. cm. — (Gender and American law ; 2)
Includes bibliographical references.
ISBN 0-8153-2516-9 (alk. paper)
1. Women—Legal status, laws, etc.—United States. 2. Women's rights—United States. 3. Abortion—Law and legislation—United States. 4. Custody of children—United States. 5. Human reproduction—Law and legislation—United States. I. Maschke, Karen J. II. Series.
KF478.A5R46 1997
346.7301'34—dc21 96-39903
CIP

Contents

vii Series Introduction

ix Volume Introduction

1 Who Gets the Child? Custody, Guardianship, and the Rise of a Judicial Patriarchy in Nineteenth-Century America
Michael Grossberg

27 Lesbian Mothers II: Long Night's Journey Into Day
Annamay T. Sheppard

55 Women as Fathers: Motherhood and Child Care Under a Modified Patriarchy
Barbara Katz Rothman

71 *Webster* v. *Reproductive Health Services*: Devaluing the Right to Choose
Gayle Binion

90 Images of Mothers in Poverty Discourses
Martha L. Fineman

113 Reproductive Freedoms and African American Women
Charlotte Rutherford

150 Reproductive Technologies and the U.S. Courts
Kim M. Blankenship et al.

175 Adolescent Pregnancy and Public Policy
Deborah L. Rhode

211 Keeping Abortion Clinics Open: The Importance of *Ragsdale* v. *Turnock* in the Post-*Casey* Era
Susan Gluck Mezey, Raymond Tatalovich, and Michael Walsh

227 The Unconstitutional Conditions Doctrine and Mandating Norplant for Women on Welfare Discourse
Laurence C. Nolan

251 Communitarianism and Feminism: The Case Against the Preference for the Two-Parent Family
Susan B. Apel

277 The Norplant Solution: Norplant and the Control of African-American Motherhood
Darci Elaine Burrell

321 "Employable Mothers" and "Suitable Work": A Re-Evaluation of Welfare and Wage-Earning for Women in the Twentieth-Century United States
Joanne L. Goodwin

344 Fathers' Rights, Mothers' Wrongs? Reflections on Unwed Fathers' Rights and Sex Equality
Mary L. Shanley

375 Acknowledgments

Series Introduction

From colonial times to the present, the law has been used to both expand and contract the rights of women. Over the last two decades, a body of literature has emerged that examines the ways in which the law has had an impact on the lives of women in the United States. The topics covered in this series include the historical development of women's legal rights, matters surrounding reproduction, sexuality and the family, equal employment opportunity, educational equity, violence against women, pornography, sex work, hate speech, and developments in feminist legal thought. The articles represent multidisciplinary approaches in examining women's experiences with the law and provide theoretical insights about the nature of gender equality.

A unifying theme in these articles is that women have been constructed historically as "different," and that this characterization "has had implications in regard to the way in which women are understood as objects and subjects of law" (Fineman 1992, p. 1). Biological differences between women and men and assumptions about the different "nature" of women and men have provided the basis for legal restrictions on women's ownership of property, guardianship of their children, ability to control their reproduction, and access to the workplace and educational institutions.

Even though women have more legal rights now than at any other time in history, many of the articles show that the law can both eradicate and reinforce women's subordination. Nineteenth-century child custody law is a telling example. While "judges granted women new legal power in family affairs," they also placed "severe limits on married women's overall legal rights" (Grossberg 1983, p. 235). On the one hand, by claiming the authority to determine which parent should be granted custody, judges dismantled the practice of fathers possessing unquestioned domain over matters of child custody and guardianship. Yet judges also "ensured that women's domestic powers did not translate into extensive external political and economic authority" (Grossberg 1983, p. 237).

The limitations of legal reform are also revealed by contemporary attempts to achieve gender equality. For example, in the 1970s many local governments adopted comparable worth wage policies. Such policies were "designed to correct historically discriminatory wages of female- and minority-dominated jobs" (Evans and Nelson 1989, p. 172). However, when some local communities were forced to raise women's wages to correct discriminatory practices, they eliminated or modified various jobs and social programs. Many of these programs, such as after school latchkey programs, were developed in response to the needs of two-paycheck and female-headed families. "In other words," note Evans and Nelson, "local governments are reacting to comparable worth with threats to renege on the emerging social commitment to policies addressing what have been traditionally defined as *women's* problems" (Evans and Nelson 1989, p. 183).

Other articles provide additional evidence of how legal reforms may actually disadvantage women in ways that are unanticipated. Furthermore, they reveal the problems in employing the model of equality as a basis for achieving gender justice. The emerging "difference theories" in feminist legal thought focus on the ways in which women and men live "gendered lives," and the ways in which legal and social institutions are shaped and operate according to gendered constructs (Fineman 1992).

The articles in this series also show how the law affects women in different ways. Women of color, poor women, and single mothers may experience the power of the law in ways that are different from the experiences of white, middle-class women. Poverty reform discourses are laden with images of single mothers as "bad" and often lump these women together "with drug addicts, criminals, and other socially-defined 'degenerates' in the newly-coined category of 'underclass'" (Fineman 1992, p. 283). Consequently, current welfare policies are not designed so much to help single mothers as they are to punish them for their bad behavior.

Several authors show how in the legal matters concerning male violence against women, "the experiences of women of color are frequently the product of intersecting patterns of racism and sexism" (Crenshaw 1991, p. 1243). These authors contend that theories of gender oppression must acknowledge the intersection of race and sex, the ways in which women have contributed to the social construction of race, and the oppression of black women. They also point out that the analyses of the law's role in women's oppression must take account of the intersectional identities of women and of how the law responds to, reinforces, and stigmatizes those identities.

The articles in this series bring together an outstanding selection from a growing body of work that examines how the law treats women and gender difference. They represent some of the most intriguing theoretical writing on the subject and reflect the strong multidisciplinary character of contemporary research on women and the legal order.

Notes

Cited works are contained in volumes five, three, seven, and two, respectively.

Crenshaw, Kimberle. 1991. Mapping the Margins: Intersectionality, Identity Politics, and Violence Against Women of Color. *Stanford Law Review* 43:1241–99.

Evans, Sara M. and Barbara J. Nelson. 1989. Comparable Worth: The Paradox of Technocratic Reform. *Feminist Studies* 15:171–90.

Fineman, Martha. 1992. Feminist Theory in Law: The Difference it Makes. *Columbia Journal of Gender and Law* 2:1–23.

Grossberg, Michael. 1983. Who Gets the Child? Custody, Guardianship, and the Rise of a Judicial Patriarchy in Nineteenth-Century America. *Feminist Studies* 9:235–60.

Volume Introduction

Women today have greater freedom to control their reproduction, express their sexuality, and choose nontraditional family arrangements than at any other time in history. Yet powerful social, political, legal, and religious forces continue to shape women's lives and the family arrangements under which they live. Informed by a patriarchal ideology of the family, these forces valorize the two-parent, heterosexual family. Women who stray from the traditional role of wife and mother are considered deviant.

Since colonial times, the family has been a site of social, legal, and political conflict. Under early American law, women were often denied the right to own property, to enter into contractual arrangements, or to be the legal guardians of their children. It was not until the 19th century that many of these barriers were removed by various legal and political reforms. Yet while these reforms gave women greater legal control over their property and family matters, they also gave the state a greater opportunity to intervene in family life and to supervise the actions of wives and mothers.

The "creation of a judicial patriarchy in the nineteenth century" is but one example of how the expansion of women's legal rights is often accompanied by state intervention that limits women's freedom concerning the family (Grossberg 1983, p. 254). Changes in family law recognized women's legal interests in family matters, particularly those concerning child custody. For example, the traditional doctrine of paternal rights in custody disputes was replaced by the notion of paternal fitness. Paternal rights were also diminished by the tender years rule, which "decreed that infants, children below puberty, and youngsters afflicted with health ailments be placed in a mother's care unless she proved unworthy of the responsibility" (Grossberg 1983, p. 247).

While these legal reforms appeared to expand the rights of mothers, in actuality they resulted in "a judicial tendency to use fault and gender-based social values to determine parental fitness" (Grossberg 1983, p. 250). Judges became the new patriarchs over custody decisions, and maternal preference did not mean that mothers were guaranteed permanent custody rights. On the contrary, mothers who "did not meet the standards of maternal conduct" decreed by the judicial patriarchs faced losing custody of their children at any time (Grossberg 1983, p. 250).

As we near the end of the 20th century, the standards of maternal conduct that determined parental fitness in the 19th century continue to shape legal and social policies surrounding family matters. At the core of these standards are gender-based social values that define the "good" and "bad" mother. Martha Fineman points out that "[M]otherhood has always been, and continues to be, a colonized concept—an event physically practiced and experienced by women, but occupied and defined, given content and value, by the core concepts of patriarchal ideology" (Fineman 1991, pp. 289–90). Good mothers are married, heterosexual women. Mothers who do not fall into these categories are bad mothers and classified as deviant.

In recent years, images of the bad mother have increasingly appeared in poverty discourses. These images portray a deviant form of motherhood, occupied by a single, poor, nonwhite woman. The imagery is often of a black "welfare queen"—constructed as an irresponsible, sexually promiscuous woman who bears multiple children in order to collect welfare. Her status as a "deviant" derives from two sources: the lack of attachment to a male and the fact that she is black (Fineman 1991). Contemporary welfare reform proposals that would limit women's access to public assistance are defended as measures designed to halt a "cycle of dependency" that afflicts families receiving welfare. Yet there is no conclusive evidence of such a generational cycle among all families receiving welfare. Nor is there conclusive evidence that women have children in order to qualify for welfare assistance. Nevertheless, mothers who are single, poor, and black are subjected to social control policies that not only limit their access to public assistance, but that also limit their reproductive freedom.

Dorothy Roberts points out that "[W]ithout the ability to determine their reproductive destinies, women will never achieve an equal role in social, economic and political life and will continue to be politically subordinate to and economically dependent on men" (Roberts 1990, p. 60). Traditionally, black women and poor women have had limited access to resources that would enable to them to assert control over their reproduction. They have also been "subjected to the patriarchal control of their reproduction" through the use of forced sterilizations and court-ordered interventions during pregnancy (Burrell 1995, pp. 406–7). In recent years the contraceptive device known as Norplant has emerged as the newest form of coerced contraception.

Norplant is a contraceptive device that a woman can have surgically inserted under the skin of her arm. Pregnancy is prevented for up to five years by the release into the bloodstream of the synthetic hormone tevnorgestill. While the goal of the contraceptive's developer was to create a "method to enhance reproductive freedom," Norplant has been used in ways that raise concerns about governmental forms of social control. It has been used as a condition for probation for some female offenders and as an inducement for welfare mothers to work. Norplant has also been offered to teenage girls in predominantly black schools, raising concerns that the contraceptive device is being used to control the pregnancy rates of black girls rather than to offer them greater reproductive freedom (Burrell 1995).

The controversy surrounding Norplant not only raises questions about coerced contraception, it also raises questions about reproductive technologies and the state's control over women's reproductive lives and status as mothers. It is important to keep in mind that the state determines which reproductive technologies will be available to women and under what conditions women may use those technologies. For example,

RU486, the so-called abortion pill, has not been approved by the Food and Drug Administration for use in the United States. Yet it has been widely used in France for many years (Blank and Merrick 1995).

Attempts to market RU486 in the United States have met with considerable resistance from antiabortion groups. The antiabortion movement has also succeeded in obtaining court rulings and legislative mandates that limit the availability of certain abortion services (Mezey, Tatalovich and Walsh 1994). Courts have also issued other rulings that limit women's reproductive freedom. In these cases, courts have tended 1) to promote fathers' rights over mothers' rights, 2) to ignore the "social relationships that constitute childbearing and child rearing," 3) to ignore the "extent to which social conditions shape women's reproductive choices," 4) to value men's over women's biological contribution to childbearing and childrearing, 5) to base their decisions on "certain assumptions about the proper roles of mothers and fathers," and 6) to exhibit an "unwavering commitment to the nuclear family structure" (Blankenship, Rushing, Onorato, and White 1993, p. 27; Shanley 1995).

It is the preference for the heterosexual, nuclear family that lies at the core of the patriarchal ideology of the family and that shapes U.S. social policy. For example, lesbian mothers face not only intolerance from the communities in which they live, but often are declared to be unfit mothers by courts and other legal authorities that refuse to grant them custody of their children. Although in recent years some lesbians have obtained legal custody of their children or have been permitted to adopt children, the legal rights of lesbian mothers varies from state to state.

As Susan Apel points out, "never, despite the continuing evolution of the family from the traditional patriarchy of the 1950's to more diverse forms of intimacy" have those who decry these developments been so adamant or in such agreement that society must return to "its traditional patriarchal, nuclear family structure" (Apel 1995, p. 2). Yet it is not only right-wing leaders and activists who extol the virtues of the two-parent, heterosexual family and contend that children must be raised within this familial structure. Indeed, liberals and conservatives waving the banner of communitarianism vie "energetically with one another to decry the present state of the American family" while promoting the primacy of the two-parent family (Apel 1995, p. 2).

As Marilyn Bowens notes, not everyone is best served by the traditional family structure the communitarians support. She reminds us that

> The long-standing policy of favoring the (so-called) 'traditional' family over others has operated to perpetuate a status quo in which white, middle- to upper-class heterosexual men (and, by association, the white women who 'appropriately' attach themselves to them, and their white children) are validated and extended privilege in this society at the expense of everyone else. It cannot be mere coincidence that those most negatively impacted by the pro-Cleaver [so named after the Cleaver family of the 1950's television] bias are women, African-Americans, and lesbians and gays (Bowens 1992–93, p. 58).

Whether we like it or not, there are more types of families in late 20th-century

America than one can find during the "family hour" on television. There are interracial families, interdenominational families, young and old families, lesbian and gay families, single-parent families, grandparent families, immigrant families, adoptive families, and stepparent families. Regardless of the type of family configuration women find themselves in, they continue to demand the freedom to define family on their own terms, to express their sexuality, and to control their reproductive lives.

Notes

* Articles marked with an asterisk (*) are contained in this volume.

* Apel, Susan B. 1995. Communitarianism and Feminism: The Case Against the Preference for the Two-Parent Family. *Wisconsin Women's Law Journal* 10:1–26.

Blank, Robert and Janna C. Merrick. 1995. *Human Reproduction, Emerging Technologies, and Conflicting Rights*, Washington, D.C.: CQ Press.

* Blankenship, Kim M., Beth Rushing, Suzanne A. Onorato, and Renee White. 1993. Reproductive Technologies and the U.S. Courts. *Gender and Society* 7:8–31.

Bowens, Marilyn. 1992–1993. The Cleaver Paradigm as an Anachronism in a Multi-Cultural Society. *Wisconsin Women's Law Journal* 7:25–65.

* Burrell, Darci Elaine. 1995. The Norplant Solution: Norplant and the Control of African American Motherhood. *UCLA Women's Law Journal* 5:401–44.

* Fineman, Martha L. 1991. Images of Mothers in Poverty Discourses. *Duke Law Journal* 1991:274–95.

* Grossberg, Michael. 1983. Who Gets the Child? Custody, Guardianship, and the Rise of a Judicial Patriarchy in Nineteenth-Century America. *Feminist Studies* 9:235–60.

* Mezey, Susan Gluck, Raymond Tatalovich, and Michael Walsh. 1994. Keeping Abortion Clinics Open: The Importance of *Ragsdale* v. *Turnock* in the Post-Casey Era. *Policy Studies Review* 13:110–26.

* Shanley, Mary L. 1995. Fathers' Rights, Mothers' Wrongs? Reflections on Unwed Fathers' Rights and Sex Equality. *Hypatia* 10:74–103.

Roberts, Dorothy. 1990. The Future of Reproductive Choice for Poor Women and Women of Color. *Women's Rights Law Reporter* 12:59–67.

WHO GETS THE CHILD? CUSTODY, GUARDIANSHIP, AND THE RISE OF A JUDICIAL PATRIARCHY IN NINETEENTH-CENTURY AMERICA

MICHAEL GROSSBERG

In the sprawling nineteenth-century American republic, law assumed a powerful role in defining status, rights, and appropriate behavior. A fundamental revision of the married woman's place in the legal order lay at the center of one of the law's branches, domestic relations.[1] Postrevolutionary changes in family life, ones which social historians label as "modern," such as child-centeredness, rigid age and sex divisions, companionate marital practices, and declining household production, required significant modifications in traditional gender-bound policies.[2] The state judiciary took the lead in determining the nature and extent of these changes. Their solutions institutionalized the era's gender beliefs and consequently were double-edged. Influenced by the society's growing glorification of motherhood and female domesticity, judges granted women new legal powers in family affairs; yet the courts used those same ideals to place severe limits on married women's overall legal rights. It was this critical distinction between dependent power based on judicial discretion and independent legal rights assertable by women themselves that proved crucial to feminine status in the emerging American law of domestic relations.[3]

The most telling example of these developments in the law comes from an area of nineteenth-century family life generally conceded to be the rightful domain of women—childrearing. Bourgeois, white married women mounted a campaign to attach superior legal rights to motherhood and thereby posed a direct challenge to traditional domestic governance. These family reformers used the domestic virtues newly ascribed to women as a wedge for extending the legal boundaries of the maternal

Feminist Studies 9, no. 2 (Summer 1983). © by Feminist Studies, Inc.

sphere. Their efforts engendered protracted and often bitter spousal, legislative, and judicial battles over child custody and guardianship, the unquestioned domain of fathers in previous legal practice. As a result of countless demands by women for their children, the traditional superiority given paternal custody and guardianship rights in England and colonial America became untenable. Significantly, however, instead of granting mothers those powers directly and independently, the courts usurped the role formerly held by fathers and claimed the authority to determine which parent would be granted custody. I will argue that this transfer of power from the male parent to the male jurist made judges a new kind of patriarch, one invested with a power over some families that came to rival their male predecessors.

Although patriarchy with its broad range of meanings must be carefully defined, it is the most accurate label for the new child placement powers of the nineteenth-century judiciary. As they did in so many areas of economic and social life, judges seized the power to determine what happened to children when one or both marital partners failed in their responsibilities. They did so first by wresting control of child placement from domestic patriarchs who presided over disrupted families, and then by using the broad discretionary authority sanctioned by equity law to reallocate custody and guardianship to mothers, fathers, and surrogate parents. In gaining control of child placement they created a distinctive set of doctrines and policies. Patriarchy identifies the content and context of those specific doctrines and policies. It describes the particular corpus of laws devised by nineteenth-century judges to decide which claimants triumphed in custody and guardianship contests.[4]

Courts acquired such authority because judicial power expanded greatly in postrevolutionary America. In the restructured republican polity, the judiciary aggressively assumed policymaking prerogatives. Judicial power increased as legislatures gradually conceded to the bench jurisdiction over matters as diverse as corporate charters and divorce, while professionally trained jurists ousted laypersons, and judges seized control of trials from juries. The judiciary also developed a more concrete conception of the common law as an instrument to shape and control society. As legal historian Mark DeWolfe Howe noted of the period, "it was clear to laymen as it was to lawyers, that the nature of American institutions, whether economic, social, or political, was largely to be determined by the judges. In such a period questions of private law were seen and considered as questions of social policy." Courts took the lead in devising policies to govern aspects of social and economic life ranging from railroads to families.[5]

Yet the sources of a judicial patriarchy extended beyond these institutional changes in the legal order. The assumption of patriarchal authority by the bench rested to a significant degree on the nineteenth century's increasingly rigid segregation of homebound female and worldly male functions. It represented a refined version of the distinction between the male authority to govern the home and the female responsibility to maintain it.[6] Indeed, the power of judges to define the legal abilities of married women helped perpetuate, albeit in altered form, patriarchal authority within republican society. Judges combined the traditional assumption that married women lacked economic and intellectual independence to act without male supervision (and thus needed special protection) with the new faith in separate and mutually exclusive sexual spheres so central to the organization of the modern family. The result was the double-edged character of married women's legal rights: judges remade the law to aid wives and mothers in performing their domestic chores while at the same time they used their powers to check radical alterations in the subordinate legal status of married women. Judicially inspired changes in custody and guardianship, as in most areas of domestic relations law, thus allowed for an expanded feminine presence in the legal order but in a way that ensured that women's domestic powers did not translate into extensive external political and economic authority.

The persistence of patriarchy in its new guise illustrates some of the basic changes and continuities in nineteenth-century family law. The rise of this potent judicial power must be clarified to reconstruct the changes in the relationship between married women and the state during the nineteenth century. However, the legal context of women's status has not received much attention.[7] Indeed, even the few historical studies of domestic relations law which touch on the topic have been marred by an uncritical reliance on modernization theory. Those works have been guided by implicit or explicit contentions that legal innovations such as adoption or married women's property acts reflected modern, and thus progressive, notions of family life. The changes are then classified more as parts of an onrushing, indeed inevitable, modernization of the entire society than as the products of distinctive historical situations. Such interpretations of legal change ignore or discount the complex character of new legal policies in which the family and gender acted as causal agents and where traditionalism not only persisted, but in some ways also gained new footholds in the law. Modernization theory, particularly the assumption that gender inequities must eventually disappear in modern society, unnecessarily reduces our understanding of the elements at work within nineteenth-century family law and offers an incomplete picture of the sources and results of this vital period in women's legal history.[8]

The following analysis will demonstrate that the legal rights of married women did not expand in a clear linear manner during the nineteenth century. Rather, alterations in child custody and guardianship took a distinctive course of legal development which blended innovation and traditionalism, and had their own chronology. Early in the century, judges took the lead in revamping traditional paternal rights and assumed control of child placement. Legislatures then began codifying the judicial creations. Around midcentury, under the press of increasingly insistent claims by mothers for custody and guardianship rights, the courts staked out the limits of the new maternal legal sphere. By the late nineteenth century, a distinctive American law of custody and guardianship had come to rest on the twin pillars of gender beliefs and a judicial patriarchy.

JUDGES TAKE CONTROL

Traditionally in Anglo-American law fathers had an almost unlimited right to the custody of their minor, legitimate children. The law assumed that the interests of children were best protected by making the father the natural guardian and by using a property-based standard of parental fitness. Custody law considered children to be dependent, subordinate beings. They were assets of estates in which fathers had a vested right. Their services, earnings, and so on became the property of their paternal masters in exchange for life and maintenance. This *quid pro quo* developed directly out of the settled conviction that fathers were best equipped to care for their offspring.[9]

The common law denied married women legal independence and relegated them to a status similar to that of children. Their lack of rights precluded them from maintaining custody and guardianship of their children. In the words of Sir William Blackstone, the great eighteenth-century compiler of the common law, a mother "was entitled to no power, but only to reverence and respect." Husbands not only had a paramount right to custody, but also the power to appoint a testamentary guardian and thus extend their authority beyond the grave. A man in a 1712 edition of *The Spectator,* an English journal with a wide provincial readership, described paternal domestic authority in stark terms, "in short, Sir, I look upon my family as a patriarchal sovereignty, in which I am both king and priest."[10]

Although equity jurisprudence often softened the harshness of the common law by expanding married women's right to legal redress, colonial law apparently made few inroads into the paternal bias of custody rules. Most likely many women gained the custody or guardianship of their children even though no specific legal justification for such legal deviations emerged. Rather, like much of colonial law, practices and customs that departed from English ways were not fully institutionalized until after the Revolution with the establishment of a formal and more self-consciously American legal culture.[11]

After the Revolution, state judges made a major contribution to that new legal culture by devising a new approach to child custody. They relied on two English legal innovations, the doctrine of *parens patriae* and the use of habeas corpus. In the seventeenth century, English chancery courts had assumed increasing jurisdiction over the welfare and property of minors by expanding the old doctrine of *parens patriae.* Under the doctrine, the courts claimed sovereign custodial power over children and other dependents. Although its sources are unclear, the development of judicial authority under the *parens patriae* rule accompanied changes in the legal standing of children and parental rights. Gradually the father's custody powers evolved from a property right to a trust tied to his responsibilities as a guardian. Armed with the authority of *parens patriae,* courts in England and America could, and did, dislodge the law's paternal bias.[12]

The eminent late-eighteenth-century English jurist Lord Mansfield provided the means. He issued a number of decisions authorizing the use of writs of habeas corpus to try custody controversies.[13] That procedure allowed judges to determine custody by examining a child's interests, rather than by merely assessing the legitimacy of parental claims. Judicial discretion, when linked to new attitudes toward the family and increasing demands by mothers for the custody of their children, enabled American judges to reshape the common law. In the first part of the nineteenth century they did so while resolving custody disputes triggered by separation, divorce, and death. At the center of their deliberations stood the issue of just how far to diminish domestic patriarchal authority.

In 1809 a South Carolina equity court heard Jennette Prather's demand for a marital separation and child custody. She charged her husband with living openly with another woman. The judges easily complied with her first request, but hesitated in granting the second. The chancellor, Henry Desaussure, was mindful, he said, of the father being the children's "natural guardian, invested by God and the law of the country with reasonable power over them. Unless his parental power has been monstrously and cruelly abused, this court would be very cautious in interfering with the execution of it." Even so, the court refused to let the adulterous man retain his full parental rights and ended up giving the custody of an infant daughter, though not the other older children, to Jennette. In doing so, the judges publicly acknowledged that they trod on new legal ground.[14]

The ambivalence of the South Carolina judges reveals the conflicting pressures custody cases placed on the early antebellum judiciary. Traditional male authority over the family remained a fundamental tenet of American family beliefs. Yet a new concern about child nurture and a rising conviction that women possessed special qualities for moral and religious leadership and for childrearing led to growing dissatisfaction over the primacy given paternal custody rights. The emergence of more clearly defined sexual spheres within antebellum society further reinforced that uneasiness. The Prather opinion illustrates how judicial discretion could be used to restrict paternal rights and make the law conform to new gender beliefs. Judges like those on the South Carolina court enlarged their authority to determine whether a father had so clearly abused his domestic authority that custody should be forfeited. These inquiries used the era's faith in the innate childrearing capacities of women as counterweights to paternal economic and political power and judicial policy-making prerogatives to counterbalance domestic patriarchy. Judicial innovations like *Prather v. Prather* gradually spread throughout the antebellum legal system by winning the support of more and more state judges. Custody rulings increasingly devalued paternally oriented, property-based welfare considerations and emphasized maternally biased child nurture ones.[15]

In his highly influential *Commentaries on American Law* published in the late 1820s, chancellor James Kent noted and accepted the changes under way in custody law. Maintaining that a "husband was the best judge of the wants of a family," he nevertheless cited Lord Mansfield and a number of American decisions to contend that paternal custody rights could be overruled when, as he put it, "the nature of the case appears to warrant it." The jurist took that position even though thirty years earlier he had revealed his own patriarchal sentiments by describing himself as "the independent. . .Lord of my own fireside." Justice Joseph Story reached similar conclusions in his equity treatise. Though he too genuflected to the traditional patriarchal ideal, Story urged judges to examine the fitness of each parent in selecting a custodian. He admitted an inclination to place a girl of "very tender years" with her mother.[16]

Legislators began to codify the judicial innovations as well. In an 1830s revision, New York formally recognized that husbands and wives were leaving one another without resort to the state's strict divorce laws. Apparently unwilling to leave child custody to informal arrangements, the new code authorized women to apply for writs of habeas corpus to let the courts settle child placement. In the first judicial test of the law, a vice-chancellor declared in 1840 that the legislature had "neutralized the rule of the common law," and placed parents "on an equality as to the future custody of the children, even if it does not create a presumption in favor of the wife." Massachusetts passed a similar statute in the same decade with an even more explicit charge to the courts that "the rights of the parents to their children, in the absence of misconduct, are equal and the happiness and welfare of the child are to determine its care and custody."[17]

Judges trimmed paternal custody rights not only by accepting maternal claims, but also by granting those of surrogate parents. The common law's allegiance to family integrity included a deep opposition to the informal relinquishment of custody. As a New Jersey chancery court phrased it in 1846, "the care and custody of minor children is a personal trust in the father, and he has no general power to dispose of them to another." Yet many antebellum fathers did just that. Judicial involvement commenced when a father changed his mind, demanded his children, and the surrogate parents refused.[18]

The inclination of antebellum judges to use child welfare as a means of monitoring such paternal claims appeared as early as 1796 in the Connecticut decision of *Nichols v. Giles.* The state supreme court dismissed a writ of habeas corpus issued on behalf of a man whose wife had left him and taken their three-year-old daughter to her family home. The judges concluded that because the child's "intemperate, propertyless father could not properly care for her" she should remain with her grandparents. They relied on the *parens patriae* doctrine which allowed courts to withhold custody from unfit fathers who had transferred or abandoned their children.[19]

But this negative power became a more positive and potent discretionary weapon under the influence of the newly developed standard of child welfare. In the first decades of the nineteenth century, judges broadened their custody authority by enlarging their definition of the child's interests. A New York court took a step along that path in 1816, when it was confronted with the case of a father attempting to regain his daughter after his financial setbacks had led the child and her mother to seek refuge with the woman's parents. Although at first the father visited his daughter frequently, he stopped doing so. The girl had been reared and educated by her grandparents after her mother had died. Chief Justice Smith Thompson seized upon rulings by Lord Mansfield and the Pennsylvania bench to decide the child's fate: "From the affidavits which have been laid before the court, little doubt can be entertained that it will be more for the benefit of the child to remain with her grandparents than to be put under the care of her father; and if this court has any discretion in such case, it will no doubt be discretely exercised by permitting the child to remain where she is."[20]

Eight years later Justice Story faced an almost identical case, except that before her death the wife extracted a promise from her husband to leave their daughter with her maternal grandparents. Denying his pledge, Elisha Williams had a writ issued to retrieve the girl. The paramount right of the father, Story insisted, "is not on account of any absolute right to the father, but for the benefit of the infant, the law presuming it to be for his interest to be under the nurture and care of his natural protector, both for maintenance and education." Story told Williams that courts settled custody disputes with regard to the "real, permanent interests of the infant." He added that Williams had no "absolute, vested right in the custody." Williams arrived at a private settlement with his father-in-law, spurred no doubt by a realization of the limits the courts imposed upon paternal rights.[21]

Contrasting Georgia and Ohio decisions illustrate the new standards of child nurture and parental fitness judges used to resolve custody disputes. The Georgia case involved the 1836 claim of a well-respected physician to his three-month-old child. The mother had died at childbirth and on her deathbed made her husband agree to let her parents raise the child. But the doctor soon disagreed with his in-laws' childrearing methods and demanded the baby. Superior Court Judge Robert M. Charlton issued the writ and summarized the bench's authority to settle custody disputes. "All legal rights, even those of personal security and liberty, may be forfeited by improper conduct," he maintained, "and so this legal right of the father to the possession of his child must be made subservient to the true interests and safety of the child, and to the duty of the State to protect its citizens of whatever age." Deciding that both parties appeared to be responsible custodians, Charlton deferred to the father's superior claim.[22]

The Cincinnati Superior Court treated Stephen Ball's assertion of paternal rights much less respectfully. Ball had abandoned his daughters and their mother when he and his father quit their carpentry trade to join the millennial Millerite movement. Without support, Balls' family had been taken in by his wife's parents. Ball's wife died, he joined the Shakers, and then sought custody in order to have the girls raised by the sect.

Ball's conduct and objective outraged the Cincinnati tribunal. Judge William Johnston argued that the right to custody flowed from the parental responsibility to clothe, feed, and educate a child. "Separate from the duty of providing," he lectured, "the right to custody does not exist." The Shaker's communal childrearing practices, which demanded that parental ties be broken and that all children be raised by caretakers, repelled the judge. Under such a system, he contended, the Shakers, not Ball, would actually receive custody, something which neither the grandmother nor the judge thought to be in their best interests. Johnston derided Ball's attempt to "sever them from the bosom of their grandmother and from his own bosom, and plant them in the cold ascetic bosoms of the 'female caretakers,' and transfer all his rights, title and interest in the child which God has given him, to total strangers." Such plans proved Ball's unfitness, Johnston decided, and could not counter the grandmother's "moral lien."[23]

By midcentury a distinctive judicial doctrine governed the

custodial rights of fathers who allowed their children to join other families. The judiciary used its newly assumed patriarchal authority to clothe these new family bonds with legitimacy. While paternal rights to custody remained superior, they became subject to much greater judicial supervision. Judges claimed the right to weigh assertions of paternal rights against their evaluation of the character and benefits of the new family ties.[24]

Judges and legislators began to alter guardianship much as they did custody. Fathers continued to be considered the proper legal guardian of children. Most states codified paternal guardianship with few alterations from the English tradition. Even Thomas Jefferson's liberal revision of Virginia statutes took paternal testamentary power for granted. Many men apparently appointed their wives as guardians, but the law's underlying premise remained: children should be distributed as men saw fit.[25]

Even when a mother secured the guardianship of her children, the law restricted her authority much more than it did a father's. Mothers had less claim to children's services, less control of their offspring's property, fewer defenses to removal from office, and inferior custody rights. The law made custody dependent on support, and the general assumption that widows lacked financial independence undermined their demands for guardianship as it did many other claims. Diminished guardianship echoed the social status in antebellum America of widows who, more often than not, were viewed as objects of pity and feared as potential drains on community resources. Consequently, widows, to an even greater extent than other middle-class women, were seen as persons in need of protection rather than as individuals with valid claims for legal rights.[26]

Yet the new approaches to child placement evident in custody disputes compelled judges and legislators to reformulate maternal guardianship. In 1835 when a New Jersey mother challenged the school selection her son's testamentary guardian had made, a chancery judge pointedly reminded the woman of her legal powerlessness. After a father appoints a testamentary guardian, he lectured, "the natural right of the mother must yield to the will of the father. It is paramount and [testamentary guardianship] is considered a continuation of the father's authority." But the jurists themselves no longer felt so constrained by paternal rights. The chancellor felt free to express his distrust of the paternal fiat and to use his discretion to assess the boy's best interests. Upon doing so, the court complied with the mother's wishes.[27]

The bench rewrote guardianship law not only by circumscribing paternal power, but also by enlarging maternal authority. For example, the traditional common law rule that remarriage extinguished maternal guardianship fell afoul of new judicial sentiments.[28] The bench's faith in women's innate childrearing instincts led jurists to question the old assumption that a woman became so dominated by her new husband that marital deference and affection overwhelmed maternal responsibility. In 1852, the Virginia Court of Appeals upheld the guardianship claim of a newly remarried woman against her ex-father-in-law. State law allowed a mother to request the guardianship of her children after her husband's death. The justices decided that the "right was not lost by her remarriage, there being no legal guardian to the child; and the facts and circumstances disclose nothing which would induce the court, in the exercise of its discretion, to deprive her of the custody." In issuing a similar ruling an Alabama judge explained the change: "it is safe to presume. . .that a mother. . .would be more careful of the moral, intellectual, and physical well-being of her children than any other person in the world." Such instances of judicial rule making narrowed the gulf between maternal and paternal guardianship, but did so by placing both firmly within the bench's domain.[29]

A widely publicized custody fight in the 1840s between the daughter of a wealthy Boston manufacturer and her husband, a Swiss nobleman, spelled out the implications of these legal changes. The pair parted because of the wife's dissatisfaction with her husband's filial dependence and his refusal to spend part of each year in America with her family. Fearing the baron might try to seize his heir, a son born in Boston, the distraught mother searched the East Coast for a maternal custody haven. Eventually she chose Philadelphia because of a series of appellate rulings by the state courts in favor of mothers. The judiciary vindicated her choice when it rebuffed her husband's challenge of its jurisdiction over the case and then refused his argument that paternal custody rights be considered paramount in the absence of clear proof of unfitness.

In a careful weaving of English and American decisions, the state supreme court argued "the reputation of a father may be stainless as crystal, he may not be afflicted with the slightest mental, moral, or physical disqualification from superintending the general welfare of the infant. . .and yet the interest of the child may imperatively demand the denial of the father's right, and its continuance with the mother." Such was the situation with the two-year-old boy before them, the judges asserted. They concluded by offering a paean to motherhood. "Not doubting that parental anxiety would seek for and obtain the best substitute which could be procured, every instinct of humanity unerringly proclaims that no substitute can supply the place of her, whose watchfulness over the sleeping cradle or waking moments of her offspring is prompted by deeper and holier feelings than the most liberal allowance of a nurse's wages could possibly stimulate." The court ordered the child to stay with his mother. John Cadwaller, the victorious mother's lawyer, added a coda to his argument that aptly summarized the growing gender orientation of custody law and its sources in the economic and social changes undermining earlier allocations of sexual responsibilities: "Everyone knows that a father is unfit to take care of an infant; physically unfit and unfit by reason of his avocations."[30]

The decision led a Massachusetts lawyer to publish an anonymous pamphlet in protest. He savagely attacked the contention that a wife could be granted custody without proving her husband had violated his spousal or fatherly duties. The unity of the husband and wife, he claimed, blocked such an outcome. Married women had no separate custody rights, just as they had none to sue or make a contract. Directly attacking the judiciary's growing patriarchal authority, the Bay State attorney charged that the Pennsylvania decision represented "nothing less than an *assumption of power by a court* . . . to determine the domestic arrangements of *a man's family*."[31]

The lawyer's tirade exposed the central tenets of an emerging American law of custody: each family member had distinct legal interests and the judiciary had the duty of resolving conflicts between them. This adversary view of the family and the patriarchal judicial powers it encouraged upset traditional family law. Legal rules had rested on a conception of the family as a community of interests governed by a patriarch. The displacement of that ideal initiated the major legal changes of the era. Among other things, by the 1860s it ensured that parental fitness not paternal rights would be the focus of custody disputes and that judges would assume the mantle of patriarchs.[32]

THE LIMITS OF CHANGE

Despite appearances, the judicially inspired changes in custody and guardianship shifted the locus of patriarchal authority much more than they challenged the subordinate status of women. They were symptomatic of the denial of full legal and political equality to wives and mothers in all areas of the law. Postrevolutionary egalitarianism, popular democracy, and capitalist individualism significantly enlarged the legal abilities of most white males. Relegation to a special sphere excluded women from those benefits. Married women acquired a distinct legal identity, but the greatest expansion of their legal capacities came in the form of familial privileges like the maternal custody preference. Exclusion from the polity and the economy thus helped women gain a foothold in the law in the form of protections for their domestic responsibilities. Yet that very homebound role imposed severe limitations on the extent of married women's overall legal rights and enhanced judicial power over family life. This double-edged character of antebellum family law is evident in the application of custody law by midcentury judges and in the persistent opposition to formal maternal rights.

In an 1858 self-help manual, *Every Woman Her Own Lawyer,* attorney George Bishop advised women that the law empowered courts to "give custody to the party that, in the court's judgment, is most competent to bring them up with advantage to the children themselves and benefit to society." As Bishop's statement indicates, the decline of paternal rights did not automatically increase maternal ones. On the contrary, the law reduced the rights of parenthood itself. Courts applied judicially created standards of child welfare and parental fitness to take the ultimate decision of child placement out of the hands of both parents. The standards paid heed to the child's best interest, tender years, and established ties, and gave priority to the innocent party in separation and divorce proceedings. Drawn from the bench's new commitment to nurture-based child welfare policies, they often biased custody determinations toward mothers. Yet they were so rooted in judicial sexual biases, that mothers could claim that preference only when they conformed to the bench's ideal of womanhood.[33]

The most sexually biased doctrine, the tender years rule, aptly illustrates the judicial approach. It decreed that infants, children below puberty, and youngsters afflicted with health ailments be placed in a mother's care unless she proved unworthy of the responsibility. The doctrine acted as a powerful leveler of paternal rights. Under it, mothers gained a presumptive claim to their young children, as an 1860 New Jersey codification of the rule reveals: "The mother is entitled to the custody of her children under the age of seven unless it affirmatively appears that, in her custody, they would be exposed to either neglect, cruelty, or the acquisition of immoral habits and principles." Nine years after the act's passage it won the hearty endorsement of the state court of errors and appeals. The justices declared that "it is not the dry, technical right of the father, but the welfare of the child which will form the substantial basis of judgment."[34]

The tender years doctrine thus institutionalized antebellum gender ideology. As courts began to refine their policies, they extended the doctrine by insisting that daughters of all ages were best cared for by mothers. An 1876 Alabama divorce decision explained the judicial rationale for broadening the rule. A woman, whose name the court refused to divulge, sued for divorce, charging cruelty. The court refused her petition, but accepted the fact that the pair would live apart. Agreeing that the woman had sufficient cause to do so, and to have custody of her daughter, the justices proclaimed: "All must feel, that no greater calamity can befall an infant daughter, than a deprivation of a mother's care, vigilant precept, and example. A mother's sympathy and culture exerts an influence on her life and character, perceptible only in its results. Therefore, courts are reluctant to deprive her of the custody of her infant daughter and but seldom if ever, do so, unless misconduct is imputable to her."[35]

Following Solomon-like logic, courts often used the tender years doctrine to split custody by giving fathers the custody of older sons and mothers that of girls and younger children. The Virginia Supreme Court considered such an action necessary in one case because, after the "tender nursing period has passed," it became time for the masculine duty of training a boy for life outside the home. Although the policy led to the separation of siblings, it relied on the antebellum insistence that specific gender skills and responsibilities be transferred from mother to daughter and father to son.[36]

Judges also used their authority to ensure that mothers who violated judicial expectations received no legal assistance. A Pennsylvania couple discovered the subtleties of judicial family governance early in the nineteenth century. Joseph Lee petitioned for the custody of his children after obtaining a divorce from his wife Barbara for adultery. He contended that her misdeeds and subsequent marriage to her paramour disqualified Barbara obtaining custody of the couple's daughters, ages seven and ten. Chief

Justice William Tilghman registered the court's "disapprobation of the mother's conduct," but noted that her care of the children had been faultless. "Our anxiety is principally directed," he asserted, "to the children. It appears to us, that considering their tender years, they stand in need of the kind of assistance which can be afforded by none so well as a mother." Consequently, "It is on their account . . . that exercising the discretion with which the law has invested us, we think it best at present not to take them from her."

But the court monitored the situation and three years later Joseph triumphed. Tilghman again based his decision on the needs of the children, now ruling that age had rendered them less in need of maternal nurture. Invoking his power to place children, the judge argued that Barbara's conduct disqualified her from continued custody because she had immoral ideas about "the obligation of the marriage contract." The court could not allow two potential wives to be reared by a mother who thought a marital vow could be so easily broken. "At the present they may not reflect upon it," he sagely concluded, "but soon they will, and when they inquire why it was that they were separated from their mother, they will be taught, as far as our opinion can teach them, that in good fortune or bad, sickness or in health, in happiness or in misery, the marriage contract, unless dissolved by the law of the country, is sacred and inviolable."[37]

The Supreme Court of Georgia issued an even more emphatic declaration on maternal fitness in 1854 when it refused the custody request of another adulterous mother. In response to her spouse's petition for a writ on habeas corpus, she claimed that his cruelty and lack of financial support had forced her to take her daughter and seek refuge with another man. The justices sustained a lower court ruling that the child's welfare would be best served by being placed with her father. They observed that "there may be no difference in the sins of the man and woman who violate the laws of chastity," but "in the opinion of society it is otherwise." Accordingly, when a man committed adultery, he did not automatically lose the respect of the community and his children would not necessarily be excluded from association with "decent people" and "may be educated to become good and useful members of society." Rigorously invoking the sexual double standard, the court reminded its listeners that with "the frail female" the outcome became quite different. Having violated the marriage contract, the wife inevitably found herself reduced to "utter and irredeemable ruin, where her associations are with the vulgar, the vile, and the depraved. If her children be with her, their characters must be, more or less, influenced and formed by the circumstances which surround them." In other words, the court decided that because the social stigma of adultery fell more heavily on women than men it negated maternal preferences.[38]

The Pennsylvania and Georgia verdicts illustrate a judicial tendency to use fault and gender-based social values to determine parental fitness. Although custom conceded a wife the right to sue for divorce or to file for separation, the ideological strictures of doctrines like the tender years policy and the fault principle ensured that the bench had the ultimate authority in determining the fate of a child when its parents parted. A firm belief in sexually endowed traits, the accepted division of the home from the world outside, and the priority accorded child nurture worked to replace paternal rights with maternal preferences in custody disputes. Crucially, however, the newly instituted maternal preference remained a discretionary policy, not a statutory prerogative. As a result, it could be easily revoked any time a mother did not meet the standards of maternal conduct decreed by judicial patriarchs.[39]

Frustration over the limited gains of the early antebellum period helped make equal custody and guardianship central goals of the women's rights movement that emerged in the 1840s. The "Declaration of Sentiments" issued at the first women's rights convention in 1848 assailed men for framing the law of custody after separations and of guardianship in disregard of the "happiness of women—the law in all such cases going upon the false supposition of the supremacy of man, and giving all powers into his hands."[40]

From the 1840s on, custody and guardianship figured prominently in every women's rights meeting and political campaign. An 1854 address to the New York legislature discloses the basic arguments advanced in support of equal rights, particularly the use of the same gender beliefs seized upon by judges to reformulate family law. The petition pleaded for the special nurturing qualities of women—"There is no human love so generous, strong, and steadfast as that of the mother for her child." It assailed the law for being "cruel and ruthless" because man "in his inordinante love of power" used it to defy nature's command to give children to women. Arguing from the worst possible case, and reflecting a widespread antebellum distrust of the domestic responsibility of worldly males, the address charged that gamblers and rum sellers had more secure legal rights to their children than did mothers. "By your laws," the petition concluded, "the child is the absolute property of the father, wholly at his disposal in life and at death." Its authors demanded that the legislature give mothers a superior right to their children.[41]

Unlike the incremental judicial adjustments in custody and guardianship, the mass campaign for maternal rights generated a mixed, often hostile response. Two examples reveal these divergent responses and suggest why judicial control won greater popular and professional support than did statutory reforms.

As women's rights campaigner Clarina I. Howard Nichols traveled from Vermont to Massachusetts on an 1852 lecture tour, an elderly man and a sheriff boarded her train a few miles from the Bay State border. The pair tried to seize the young children of a female passenger. Nichols immediately stood up and told the startled passengers: "It means my friends that a woman has no legal right to her own babies; that the law-givers of this Christian country (!) have given the custody of the babies to the father, drunken or sober, and he may send the sheriff . . . to arrest and rob her of her little ones! You have heard sneers at Women's Rights. This is one of the rights—a mother's right to the care and custody of her little ones." Learning that the husband had transferred his custody rights to his own father, Nichols quickly explained that a recent Massachusetts court decision held that only a father could take children away from their mother. The passengers then threw the pair out, and after the train crossed the border the mother got a favorable custody ruling from a local magistrate.[42]

Demands for formal custody rights, secured by statutes and not by judicial discretion, could receive a far different public response as was evident in an 1854 *New York Tribune* account of a women's rights rally. When a woman stood up and demanded custody and guardianship rights, the paper reported, male hecklers greeted her with cries of "Oh, Dry Up!" "Bow-Wow!" "Waugh!" "Hiss-s-s-s!" "Get Out!"[43] A perhaps more reasoned expression of the same sentiments came two decades later when lawyer and historian James Schouler published his 1870 *Treatise on Domestic Relations.* Discussing legislative changes in married women's legal status, he argued: "The danger to be apprehended from all legislation of this sort is that it will weaken the ties of marriage by forcing both sexes into an unnatural antagonism;

teaching them to be independent of one another, and to earn their own living apart; whereas God's law points to the family and the mutual intercourse of man and woman as among the strongest safeguards of human happiness." Given the choice of using the law for securing greater independence for women or for maintaining female subordination, Schouler declared that the best path would be to ensure that the law provided, as he put it, "honorably, faithfully, and generously against all possible misfortune," and taught a woman to "lean upon the stronger arm of her husband, and to look to man for guidance."[44]

Stubborn opposition to formal custody and guardianship rights for married women offers graphic evidence that many opponents of women's rights resisted all significant changes in the legal status of women, not just those, such as suffrage, which would clearly extend feminine authority outside the home. Opposition was intense even to changes within the home. The pleas for rights stirred up deep fears within turbulent mid-nineteenth-century America that legal reforms might upset a needed sexual balance of power within the home based on male sovereignty and thus contribute to the larger social dislocations of a society which considered the family its primary anchor. Critics of the women's rights movement pictured demands for domestic rights, such as child custody, as threats to the family. Massachusetts women's rights advocate Samuel Sewall identified these concerns in an 1869 tract: "The greatest difficulty to overcome is not that most men are unwilling to do complete justice to the sex, or that the majority of women care nothing for this object; but it is simply a superstitious dread lest a change so radical could unsettle all the foundations of society and bring down the whole fabric in ruins." In other words, aiding a defenseless mother on a train was one thing, giving her equal or superior statutory rights proved to be quite another.[45]

Popular resistance to women's rights reinforced the judicial assumption of power over the home and placed sturdy obstacles in the path of radical legislative changes in custody and guardianship. The fate of the few statutory victories of the women's movement indicates the lure of a judicial patriarchy. Most statutes merely granted women the right to petition the courts for custody of their children in the event of a separation or divorce. The most far-reaching acts provided for equal guardianship. In 1860, New York, the seat of the most organized women's rights effort, passed the first major act which enlarged married women's property rights and declared a wife to be the "joint guardian of her children along with her husband, with equal powers, rights, and duties in regard to them, with her husband." Even that was too much; in 1862 the lawmakers retreated and amended the act to dictate only that husbands obtain their wives' consent before appointing testamentary guardians or indenturing their children.[46]

Nevertheless, New York judges easily integrated the legislation into their custody determinations so as to preserve, indeed augment, their patriarchal powers. A state tribunal rejected a mother's contention that the 1860 act gave her independent custody rights. The justices held that her rights could only be exercised in conjunction with her husband unless she proved him an unfit father and spouse. "The common law," they declared, "remains, except as modified by the joining of the wife with him." By leaving her husband without a judicially acceptable cause, this mother lost her newly won custody rights. Two years later in *People v. Boice,* a widow challenged her husband's appointment of a testamentary guardian without her consent. The court construed the intent of the new legislation to be an enlargement of the "sphere of maternal authority" and ruled that unilateral paternal action nullified the appointment. Reverting back to the best interests of the child doctrine, they used their discretion to award custody to the widow.[47]

At the turn of the century, only the District of Columbia and nine states formally granted women equal custody and guardianship. The rest invested judges with the patriarchal power to place children.[48]

CONCLUSION

The reformation of custody and guardianship begun in antebellum America came to dominate the law after the Civil War. Legal consolidation and refinement occurred, but the basic principles and priorities of the earlier era remained in place. These included the use of child welfare custody standards, class and sexually biased rules, and especially a maternal preference. As a result, child custody and guardianship remained under the purview of judicial patriarchs. Their power represented, in part, a persistent refusal to place superior legal rights in the hands of women. It was as well a refusal issued by male institutions which excluded women as legislators, lawyers, judges, and jurors and granted them access only as spectators, witnesses, and supplicants. The law retained, in the words of the 1883 *American Legal Register,* "a sort of guardianship over the power and property of women."[49] Antebellum legal changes transferred that guardianship from domestic to judicial patriarchs.

The persistence of patriarchy underlines the importance of continuities in American domestic relations. The law's purpose, the traditional one of producing stable families, and male governance proved inseparable. That pairing and the feminist resistance it stirred are critical to an understanding of nineteenth-century family law. As an 1880 women's rights convention declared: "The theory of a masculine head to rule the family, the church, or the State is contrary to Republican principles and the fruitful source of rebellion and corruption."[50] Rebels remained on the sexual battlefield because women refused to mute their demands for maternal rights and accept the limitations sanctioned by law. These conflicts are better understood in their historical context as part of the nineteenth-century attempt to accommodate new family practices and sentiments than as examples of the modernization of the law. The character of legal change, such as alterations in custody and guardianship, can only be determined by locating their complex sources and actual impact on the lives of nineteenth-century families and not by trying to find a place for them on an ahistorical conveyor belt of Western social evolution.

The creation of a judicial patriarchy in the nineteenth century reveals part of that historical environment. A new relationship emerged between the family and the state; one described quite vividly by a lawyer in 1882. "While in ruder forms of government more power was left with the Family itself for its own protection, the State now has taken to itself most of the old *patria potestas* and other forms of family autonomy and assumed the maintenance of all rights pertaining to the family."[51] The assumption of patriarchal authority by the state, and its delegation to the courts, created an antagonistic relationship between the two. The liberal ideal of emancipating households from governmental supervision often clashed with the courts' growing insistence that families should perform their vital functions in a socially beneficial manner. The inherent tension between the two goals led to divergent legal policies. Family law placed restrictions on the right of the state to intervene in what the society considered normally functioning families; and yet, conversely, it augmented state power to intervene in homes labeled troubled or abnormal. Judges seized the power to make these determinations.

Perhaps the best way to summarize the new level of judicial responsibility for family governance produced by nineteenth-century family law is to let a judge describe his authority. Robert Grant served as a probate judge in Boston for over thirty years. Known in his time mainly for the popular romantic novels he penned, he is perhaps best remembered today as a member of the commission which upheld the conviction of Sacco and Vanzetti.

Grant learned the law in the 1870s when the changes of the antebellum era had become legal commonplaces. In 1919, at the end of his career, he published a volume on family law. In it the judge acknowledged the sexual biases of custody law: "The attitude of the courts where parents battle over children has inclined so steadily toward the mother that, unless she has shown herself wanton or exceptionally recreant or heartless, she is not likely to be separated from them." Yet Grant felt trapped by this judicial creation. He feared that the "pendulum has swung so far in the opposite direction and the theory of paternal ownership been so completely discredited that the boot is sometimes upon the other leg, and women are heard asserting that they own their children because they bore them, and ought under no circumstances to be deprived of them—a complete reversal of the original injustice." To let women know their place, Grant asserted the judicial ideal of child placement:

> Mother-love, though set upon a pinnacle in the conscience of modern courts, must yield to a higher consideration, the well-being of her offspring. Where the custody of children is concerned the only enemy which the modern woman has to fear is her own unfitness. This is more apt to be challenged by the social workers and charitable societies, who might be called liaison officers of the courts of domestic relations, than by masculine ill-will. The beneficent body-guards, who probe into and bring to the attention of the court the conditions which menace the child, serve as a buffer between it and maternal Bolshevism.[52]

Thirteen years earlier Grant had expressed similar sentiments a bit less ponderously, when he offered a local bar association his poetic description of the patriarchal powers he so confidently exercised.

A Probate Judge who outlives you
May break your will—yes tax it too.
. . . Concerning other things,
His power outrivals that of kings,
Your children, when you prove unfit,

Are whisked away by sovereign writ.
In short, it may truly be said,
He has you living, he has you dead.
The moral is, as on you trudge,
Propitiate the Probate Judge.[53]

NOTES

I would like to thank several colleagues for their comments on various drafts of this essay: Stephen Presser, Jamil Zainaldin, Peter Garlock, Bertram Wyatt-Brown, Carl Ubbelohde, Brian Gatton, Ann Warren, and Roberta Wollons. Earlier versions of this article were presented at the 1982 meeting of the American Society for the Study of Legal History, 23 October, 1982, Washington D.C., and the Interdisciplinary Faculty Seminar of the Charles Armington Foundation at Case Western Reserve University, Cleveland, Ohio, 17 February, 1983. Research for this article was made possible by Crown Fellowships from Brandeis University, Waltham, Massachusetts, a National Endowment for the Humanities Summer Stipend, and a grant from the Armington Foundation.

1. For a more detailed discussion of these issues see Michael Grossberg, "Law and the Family in Nineteenth-Century America" (Ph.D. diss., Brandeis University, 1979).
2. See, for example, Edward Shorter, *The Making of the Modern Family* (New York: Basic, 1975); Lawrence Stone, *The Family, Sex, and Marriage, 1500-1800* (New York: Harper & Row, 1977).
3. In *At Odds: Women and the Family in America, from the Revolution to the Present* (New York: Oxford University Press, 1980), Carl N. Degler argues that the changing status of women is the key issue in all areas of family history.
4. For discussions of patriarchy, see Gordon J. Schochet, *Patriarchalism in Political Thought* (New York, Basic Books, 1975); Janet Rifkin, "Toward a Theory of Law and Patriarchy," *Harvard Women's Law Journal* 3 (1980): 83-95.
5. Mark DeWolfe Howe, "The Creative Period in the Law of Massachusetts," *Proceedings of the Massachusetts Historical Society* 69 (1947-50): 237; and see Morton J. Horwitz, *The Transformation of American Law, 1780-1860* (Cambridge: Harvard University Press, 1977), esp. chap. 1; William E. Nelson, *The Americanization of the Common Law: The Impact of Legal Change in Massachusetts Society, 1780-1860* (Cambridge: Harvard University Press, 1975); Leonard W. Levy, *The Law of the Commonwealth and Chief Justice Shaw* (Cambridge: Harvard University Press, 1957).
6. See for example, Eleanor M. Boatright, "The Political and Civil Status of Women in Georgia, 1783-1860," *Georgia Historical Quarterly* 25 (1941): 301-24.
7. For descriptions of restrictive legal controls on women, see James C. Mohr, *Abortion in America: The Origins and Evolution of National Policy* (New York: Oxford University Press, 1978); Ellen C. Dubois, *Feminism and Suffrage: The Emergence of an Independent Women's Movement in America, 1848-1869* (Ithaca, N.Y.: Cornell University Press, 1978); Suzanne D. Lebsock, "Radical Reconstruction and the Property Rights of Southern Women," *Journal of Southern History* 43 (1977): 195-216. For a general discussion of the paternalistic orientation of antebellum law, see Linda Kerber, *Women*

of the Republic: Intellect and Ideology in Revolutionary America (Chapel Hill: University of North Carolina Press, 1980), chap. 5; Norma Basch, "Invisible Women: The Legal Fiction of Marital Unity in Nineteenth-Century America," *Feminist Studies* 5 (Summer 1979): 346-66; Albie Sachs and Joan Hoff Wilson, *Sexism and the Law: Male Beliefs and Legal Bias in Britain and the United States* (New York: Free Press, 1978).

8. See for example Maxwell Bloomfield, *American Lawyers in a Changing Society, 1776-1876* (Cambridge: Howard University Press, 1976), chap. 4; Jamil Zainaldin, "The Emergence of a Modern American Family Law: Child Custody, Adoption, and the Courts, 1796-1851," *Northwestern Law Review* 73 (1979): 1038, especially 1046-47; Stephen B. Presser, "The Historical Background of the American Law of Adoption," *Journal of Family Law* 11 (1971): 443-516; and for a full discussion of the problems and possibilities of modernization theory, see Raymond Grew, "Modernization and Its Discontents," *American Behavioral Scientist* 21 (1977): 289-312.

9. P.H. Pettit, "Parental Control and Guardianship," in *A Century of Family Law* ed. Ronald H. Graveson and F.R. Crane (Cambridge, England: Cambridge University Press, 1957), 56-70; Sir Frederick Pollack and Frederic W. Maitland, *The History of English Law,* 2 vols. (London, 1898), 1: 436-47; William Blackstone, *Commentaries on the Law of England,* 4 vols. (London, 1765-69), 1: 438, 456, 462.

10. William Blackstone, *Commentaries on the Laws of England: First American Edition, Philadelphia, 1771-1772,* ed. George Sharwood, (Philadelphia, 1860), 2: 452; Spectator quote in Julia C. Spruill, *Women's Life and Work in the Southern Colonies* (Chapel Hill, University of North Carolina Press, 1938), 44; and see Arthur Calhoun, *A Social History of the American Family* (Cleveland, 1917), 1: 77, 255; Mary S. Benson, *Women in Eighteenth-Century America: A Study of Opinion and Social Usage* (1935; reprint, Port Washington, N.Y.: Kenitkatt, 1966); Bloomfield, 97-99.

11. Lorena S. Walsh, "Child Custody in the Early Colonial Chesapeake: A Case Study" (Paper presented at the Berkshire Conference on Women's History, Vassar College, June 1981); Richard B. Morris, *Studies in the History of Early American Law* (New York: Columbia University Press, 1930), 126-200; Mary Beth Norton, *Liberty's Daughters: The Revolutionary Experience of American Women, 1750-1800* (Boston: Little, Brown & Co., 1980), 46-57.

12. Douglas T. Rendleman, "Parens Patriae: From Chancery to the Juvenile Court," *South Carolina Law Review* 23 (1971): 205-23; Howard Neil Cozen, "The English Background: Origins of Parens Patriae," *South Carolina Law Review* 22 (1970): 147-51.

13. Rex V. Delaval, 3 Burr. 1435 (K.B. 1763); and see Rollin C. Hurd, *A Treatise on the Right of Personal Liberty, and on the Writ of Habeas Corpus and the Practice Connected with It* (Albany, 1858), 453-72.

14. Prather v. Prather (S.C. 1809); 4 Desau. 33, 34, 44; and for the later acceptance of this power see *Ex Parte* Hewitt, 11 Rich. Law 326 (S.C. 1858); see also Anonymous, 4 Desau. 94 (S.C. 1820); Williams v. Williams, 4 Desau. 183 (S.C. 1811).

15. See for example, Commonwealth v. Nutt, 1 Browne 143 (Pa. 1810); Bedell v. Bedell, 1 Johns. Chan. 605 (N.Y. 1815); Richmond v. Richmond, 1 Green Chan. 90 (N.J. 1838); Clark v. Clark, Wright 226 (Ohio, 1833); and Grossberg, 262-68.,

16. James Kent, *Commentaries on America Law,* 4 vols. (New York, 1826-30), 2: 193, 194; Norton, 5; Joseph Story, *Commentaries on Equity Jurisprudence* (Boston, 1839), 2: 596-97; and see also Francis Hillard, *Elements of Law* (New York, 1835), 24.

17. Volume 2 Revised New York Statutes 562, art. 2; People v. Chegaray, 18 Wend. 637, 641 (N.Y. 1836); Ahrenfeldt v. Ahrenfeldt, 1 Hoffm. Chan. 497, 502 (N.Y. 1840); Bloomfield, 118-119; and see People v. Nickerson, 19 Wend. 16 (N.Y. 1837); State v. King, 1 Ga. Dec. 93 (1841).

18. For a general discussion of these issues see Grossberg, 281-85.
19. Nichols v. Giles, 2 Root 461 (Conn. 1796).
20. Matter of Waldron, 13 Johns. 417 (N.Y. 1816).
21. U.S. v. Green, 1 Mason 482 (U.S. C.C., R.I., 1824, Fed. Case #15,256).
22. In the Matter of Mitchell, 3 Charlton 489, 494-95 (Ga. 1836).
23. State v. Hand, 1 Ohio Dec. 238, 242, 244 (1848).
24. The American position was summarized in the New Hampshire decision of State v. Libby: "It is quite apparent that there may be cases where the father's conduct is such, as by permitting tacitly or by express agreement, another to assume and discharge for many years the duties of parent to his child, that he could not afterward attempt to reclaim his child in good faith, nor without subjecting to serious hazards its interests and happiness. In such a case the award of the custody of the child would not be consistent with the exercise of sound judicial discretion." State v. Libby, 44 Me. 469 (1858); and see *In Re* Kottman, 2 Hill 363 (S.C. 1834); Hutson v. Townsend, 6 Rich. Eq. 249 (S.C. 1854); People v. Cooper, 8 How. Pr. 288 (N.Y. 1853); *Ex parte* Williams, 11 Rich. 452 (S.C. 1838); *Ex parte* Schumpert, 6 Rich. 244 (S.C. 1953); Rollin C. Hurd, *A Treatise on the Right of Personal Liberty* (Albany, 1858), 540-48.
25. Tapping Reeve, *The Law of Baron and Femme* (New York, 1814), 295, 311-337; Kent, 2: 219-31; Hillard, *Elements*, 24-25; Boatright, 116; Grossberg, 274-75, 285-88.
26. Ibid.; and see Eldridge v. Lippincott, 1 Coxe's Law, 397 (N.J. 1795); Fields v. Law, 2 Root 320 (Conn. 1796); Kline v. Beebe, 6 Conn. 494 (1827).
27. In the Matter of Van Houten, 2 Green Chan. 220, 226 (N.J. 1835); and see Copp v. Copp, 20 N.H. 284 (1850); Williamson v. Jordan, Busb. Dq. 40 (N.C. 1852); Fullerton v. Jackson, 5 John. Chan. 277 (1821); Corrigan v. Kiernan, 1 Bradf. 203 (N.Y. 1850).
28. For invocations of the rule see, for example, Stenbrook v. McColm, 7 Halst. Law 97 (N.J. 1830); State v. Sheeseman, 2 South. 445 (N.J. 1819); Huie v. Nixon, 6 Porter 77 (Ala. 1837); Holley v. Chamberlain, 1 Redf. 333 (N.Y. 1860); Webb v. Webb, 5 La. An. (1850); Hovey v. Morris, 7 Blackf. 559 (Ind. 1812); Ransom Tyler, *The Law of Infancy* (Chicago, 1888), 253-54.
29. Armstrong v. Stone, 9 Gratt. 102, 107 (Va. 1852); Striplin v. Ware, 36 Ala. 87, 90 (1860); and see Lea v. Richardson, 8 La. An. 94 (1853); People v. Wilcox, 22 Barb. 178 (N.Y. 1854); Level v. Bettis, 3 Bush, 74 (Ky. 1867); Carlisle v. Tuttle, 30 Ala. 613 (1857); Porch v. Fries, 3 Green 204 (N.J. 1867).
30. *Report of the D'Hauteville Case* (Philadelphia, 1840), 292-93, 234, and see 47-50.
31. A Member of the Boston Bar, *Review of the D'Hauteville Case* (Boston, 1841), 25, 37, 41 (emphasis added).
32. An 1865 article by Vermont Chief Justice Isaac Redfield summarized the new judicial view. "The rule of the common law which recognizes the superior rights of the husband as to all matters pertaining to the control of the family, while the parents continue to live together, has not the same reasons of necessity of convenience in its favor after they become separated. The court may then fix the custody of the child as they think will most induce to their careful training, education, and comfort, without violation of any rule affecting quiet and good order in families; a right which is now generously secured in American states, either by custom or statute, but not in all, we are sorry to say." See "Recent Developments in English Jurisprudence," *American Law Register*, n.s., 4 (1865): 137; and see Grossberg, 597-612.
33. George Bishop, *Every Woman Her Own Lawyer* (New York, 1858), 181; Joel P. Bishop, *Commentaries on the Law of Marriage and Divorce* (Boston, 1852), 517-18,

525-26; Grossberg, 273-4, 598-612.
34. State v. Baird and Torrey, 21 N.J. Eq. 384, 388 (1869), and 18 N.J. Eq. 184 (1867); Landis v. Landis, 39 N.J. Law 274 (1877); State v. Kirkpatrick, 54 Ia. 373 (1880); Matter of Pray, 60 Howard Pr. 194 (1881); McKim v. McKim, 12 R.I. 462 (1879); Latham v. Latham, 30 Gratt. 364 (1878); Matter of Watson, 10 Abb. N.C. 215 (1882); Greenwood v. Greenwood, 84 Minn. 203 (1901).
35. Anonymous, 55 Ala. 428, 432-433 (1876).
36. Carr v. Carr, 22 Gratt. 168, 174 (Va. 1872); and see Bennett v. Bennett, 43 Conn. 313 (1876).
37. Commonwealth v. Addicks, 5 Binn. 520 (Pa. 1813); 2 Serge and Rawle, 174 (Pa. 1816).
38. Lindsey v. Lindsey, 14 Ga. 657 (1854); and see Matter of Viele, 44 How, Pr. 14 (1872).
39. People v. Olmstead, 27 Barb. 9, 31, 32; and see State v. Sitgall, 2 Zabr. 286 (N.J. 1849); Magee v. Holland, 3 Dutch. 86 (N.J. 1858); People v. Rhodes, 24 Barb. 521, (N.Y. 1857); Bryan v. Bryan, 34 Ala. 516 (1859); Lawrence M. Friedman and Robert V. Percival, "Who Sues for Divorce? From Fault through Fiction to Freedom," *Journal of Legal Studies* 5 (1976): 61-82.
40. Stanton et al., *History of Woman Suffrage,* 6 vols. (1881-1922), 1:71. This demand was repeated in women's rights platforms and addresses for the rest of the century, see 105, 194, 241, 254-55, 260-61, 562-63, 579-89, 776-79.
41. E.C. Strauder, *Address to the Legislature of New York* (Albany, 1854); Caroline H. Dall, *Woman's Rights under the Law* (Boston, 1861), 60; Caroline H. Dall, *Women, College, and Court* (Boston, 1878), 266-360.
42. Stanton et al., 1: 174-78.
43. Cited in Bell Squire, *The Woman Movement in America* (Chicago, 1911), 46.
44. James Schouler, *A Treatise on Domestic Relations* (Boston, 1870), 17, 18. For similar expressions of concern see Stanton et al., 3: 152.
45. Samuel Sewall, *The Legal Condition of Women* (Boston, 1869), 14-15; and for continued feminist agitation for custody rights see Stanton et al., 2: 102, 799, 3: 419; Dubois, 188; Rheta Childe Dorr, *What Eight Million Women Want* (Boston, 1910); Lucy Stone, "The Progress of Fifty Years," in *The Congress of Women,* ed. Mary K. Oldham Eagle (Chicago, 1894), 60; Alice Paul, *Outline of the Legal Position of Women in Pennsylvania* (Philadelphia, 1911), 31-32.
46. Stanton et al., 1: 686, 747-78 and see in general, chap. 14. For discussions of the property reforms that accompanied these custody changes, see Peggy A. Rabkin, *Fathers to Daughters: The Legal Foundations of Female Emancipation* (Westport, Conn.: Greenwood Press, 1980).
47. People v. Brooks, 35 Barb. 85, 91 (N.Y. 1861); People v. Boice, 39 Barb. 307, 309-10 (N.Y. 1862).
48. Stanton et al., 4: chaps. 24-72. For a complete assessment of the status of the law in turn-of-the-century America, see Jeannie C. Wilson, *The Legal Rights and Duties of Women* (New York, 1913); John Norton Pomeroy, *A Treatise on Equity Jurisprudence,* 2d. ed. (San Francisco, 1892), 3: 2012-14; Annie G. Porritt, *Laws Affecting Women and Children* (New York, National Woman Suffrage Publishing Co., 1916), 137; United States Bureau of Education, "The Legal Rights of Children" (Washington, D.C., 1880), 16-27; Roscoe Pound, "Individual Interests in the Domestic Relations," *Michigan Law Review* 14 (1916): 182-85.

49. Quoted in Morton Keller, *Affairs of State: Public Life in Late Nineteenth-Century America* (Cambridge: Howard University Press, 1977), 470. For a contemporary endorsement of this policy which reveals its paternalistic foundations, see Russell H. Conwell, *Women and the Law* (Boston, 1876), esp. 16-21, 28-29, 35-55.
50. Stanton et al., 2: 152. For an endorsement of judicial control of custody by a female attorney, see Mary A. Greene, "The American Mother's Right to Her Child," *American Law Review* 52 (1918): 371-82; George A.O. Ernst, *The Legal Status of Married Women in Massachusetts* (Boston, 1895), 24-31; Calhoun, 3: 108, 394.
51. M.H. Buckham, "Habeas Corpus: Custody of Infants," *Central Law Journal* 15 (1882): 384; and see Pound.
52. Robert Grant, *Law and the Family* (New York, 1919), 181-82, 183.
53. Ibid., v-vi.

Lesbian Mothers II: Long Night's Journey Into Day

Annamay T. Sheppard

"Women are making progress, but the pace is much too slow. To get full equality, we'll have to change some basic social structures, for example our child care system. I hope that women now entering the legal profession will help to create that change," says Annamay Sheppard.

Professor Sheppard earned her bachelor's and master's degrees from Cornell in 1958 and her law degree from Rutgers Law School. After more than a decade as a litigator, she returned to Rutgers in the Fall of 1970 to set up the Urban Legal Clinic of which she was director from 1970 to 1976. She has been a full professor since 1975, and although she now spends most of her teaching time in the classroom, she returns to clinic work occasionally, most recently to the Women's Rights Litigation Clinic in which she taught during the 1984-85 school year.

In addition to her academic responsibilities (at this writing, she is hard at work on a New Jersey Practice casebook), Sheppard finds time to contribute her services to many organizations. Currently, she is an associate editor of the New Jersey Law Journal, a member of the Committee to Support Legal Services, a member of the New Jersey Supreme Court Task Force on Women in the Courts and co-chair of its Subcommittee on Matrimonial Law, an at-large member of the Board and cooperating counsel of the New Jersey ACLU and a trustee of the Education Law Center.

While on academic leave from Rutgers, Professor Sheppard, who has litigated several important cases on behalf of women, was asked to help a lesbian couple in a custody battle. Preparation and trial of the case were arduous, but her reward went beyond the court's ruling for joint custody. "My clients taught me a lot about human feelings and relationships. They renewed my faith in the human spirit."

CONTENTS — PAGE

I. INTRODUCTION 220
II. THE STATE OF THE ART: MEDICAL DEVELOPMENTS 220
III. THE STATE OF THE LAW: CONSTITUTION AND CASES 224
A. The Constitutional Framework 224
1. The Constitutional Right of Privacy 225
2. The Right to Parent 227
B. The Cases 228
1. Custody Contenders and Children (Chart I) 232
2. Mother's Domestic Arrangements (Chart II) 234
3. Negative Allegations About Mothers (Chart III) 236
a. Lesbian Personality and Parenting Behaviors 236
b. The Effects of Bad Conduct Allegations on Case Outcomes 236
4. Investigators and Experts (Chart IV) 238
a. Best Interest Investigators 238

[Women's Rights Law Reporter, Volume 8, Number 4, Fall 1985]
© 1985 by Women's Rights Law Reporter, Rutgers—The State University
0085-8269/80/0908

b. Experts 238
5. The Law and the Outcomes (Chart V) 241
6. Summary of the Cases 241
IV. PROSPECTS 243

I. INTRODUCTION

More than ten years ago, these pages carried an article in which the author canvassed the multiple strands of religious, psychiatric, and legal doctrine that conjoin to threaten lesbian mothers with deprivation of the custody of their children. In a frankly polemical challenge to the homophobic core of religious, medical and legal assumptions about homosexuality, the author argued (1) that Judeo-Christian condemnation of homosexuality had its origin not in some unreviewable moral truth but in the practical necessity to insure population, and later, in the reflexive religious equation of all non-procreative and non-marital sex with sin; (2) that psychiatry recast biblical opprobrium in contemporary terms by inaccurately and unscientifically labelling homosexuality as mental illness; and (3) that American law folded the dichotomous sin/illness characterizations of homosexuality into an undifferentiated judgment that homosexual conduct is properly punishable as deviant (ill and/or sinful) and therefore wrong (sinful and/or ill). Using a group of representative cases, the article analyzed the ways in which the label "deviant" adversely affects a lesbian mother's chance to retain or regain custody of her children.[1]

A substantial body of legal literature has since been published urging that lesbian mothers should, in justice and rationality, be accorded the same custodial rights as heterosexual mothers.[2] The literature has been produced by, and has in all likelihood produced, a significant number of cases in which lesbian mothers have challenged the legitimacy of their assumed custodial disentitlement.[3] Parallel with that challenge, mainstream psychiatry has broken stride with religion and the law.[4]

This article surveys the present state of psychiatric thinking about homosexuality in general and lesbian mothers in particular, reviews and comments upon the law's emerging responses to the custodial claims of lesbian mothers and argues that the rights of all women are linked to securing like rights for lesbian women.

II. THE STATE OF THE ART: MEDICAL DEVELOPMENTS

Until about fifteen years ago, there appeared to be near medical consensus that homosexuality was a manifestation of mental illness. Varying and contradictory theories were offered to explain its origins, particularly in males, but dominant voices rejected any view that placed it within the parameters of psychological normalcy.[5]

In 1973, after extended debate, the American

1. Basile, *Lesbian Mothers I*, WOMEN'S RTS. L. REP., Dec. 1974 at 3.

2. *See, e.g.*, Armanno, *The Lesbian Mother: Her Right to Child Custody*, 4 GOLDEN GATE 1 (1973); Davis, *Representing the Lesbian Mother*, 1 FAM. ADVOC. 21 (Winter 1979) [hereinafter cited as Davis]; Goodman, *Homosexuality of a Parent: A New Issue in Custody Disputes*, 5 MONASH U.L. REV. 305 (1979); Hitchens & Price, *Trial Strategy in Lesbian Mother Custody Cases: The Use of Expert Testimony*, 9 GOLDEN GATE 451 (1978-79); Hitchens, Martin & Morgan, *Child Custody and the Homosexual Parent*, 18 JUDGES J. 33 (Fall 1979); Hunter & Polikoff, *Custody Rights of Lesbian Mothers: Legal Theory and Litigation Strategy*, 25 BUFFALO L. REV. 691 (1976); Riley, *The Avowed Lesbian Mother and Her Right to Child Custody: A Constitutional Challenge That Can No Longer Be Denied*, 12 SAN DIEGO L. REV. 799 (1975).

3. This kind of challenge is quite new. The gay and lesbian rights movement did not fully emerge as a political force until the racial civil rights and feminist movement of the 1960's had taken firm hold and the citadels of racial and gender discrimination had begun to crumble. For gays and lesbians, the social upheavals created by these movements produced a climate of hope that, if they risked coming out despite the pervasive homophobia that confronted them, they, too, might achieve rights theretofore denied to them. *See* R. BAYER, HOMOSEXUALITY AND AMERICAN PSYCHIATRY: THE POLITICS OF DIAGNOSIS, 88-100 (1981) [hereinafter cited as BAYER].

For lesbian mothers, coming out carries both promise and peril. For discussions of the need for and consequences of openness, see Abbitt & Bennett, *Being a Lesbian Mother*, in POSITIVELY GAY 123 (B. Berzon & R. Leighton eds. 1979); Berzon, *Sharing Your Lesbian Identity With Your Children: A Case for Openness*, in OUR RIGHT TO LOVE: A LESBIAN RESOURCE BOOK (G. Vida ed. 1978); Hitchens, *Social Attitudes, Legal Standards and Personal Trauma in Child Custody Cases*, 5 J. HOMOSEXUALITY 89 (1979/80).

4. *See infra* for a discussion of the reasons for and process by which this occurred.

5. Early psychological literature on homosexuality contains very little information about lesbians. *See* M. Kirkpatrick, R. Roy & K. Smith, Studies of a New Population: The Lesbian Mother, 2 (unpublished galley pages 1980) [hereinafter cited as KIRKPATRICK, ROY & SMITH].

Psychiatric Association ("APA") announced a departure from illness theory, removing homosexuality from its mental illness nomenclature and substituting "sexual orientation disturbance" to describe individuals "who are either disturbed by, in conflict with or wish to change their sexual orientation." It added: "This diagnostic category is distinguished from homosexuality which itself does not necessarily constitute a psychiatric disorder."[6]

Several factors contributed to the APA's changed perception. First, illness theorists had produced a conceptual Tower of Babel. Freudians insisted that homosexuality is the consequence of unresolved oedipal conflict.[7] Neo-Freudians and non-Freudians offered a variety of alternative interactional hypotheses, some attributing homosexual outcomes to infantile fears of engulfment by a dominant mother;[8] some finding the roots of homosexuality in unsatisfactory relationships with a same-sex parent or other family member;[9] some concluding that it is the product of imitation or modeling the behavior of some homosexual person who plays a significant role in the child's life;[10] some hypothesizing that it is a response to fear of heterosexual functioning[11] or, alternatively, a form of rebellion against authority.[12] Some members of the profession rejected interactional hypotheses altogether, suggesting instead that homosexuality might well have genetic[13] or endocrinological[14] origins. None of these theories was supported by solid, objective

6. American Psychiatric Association, press release, December 15, 1973, *cited in* BAYER, *supra* note 3, at 137 & n.82.

7. Freud himself had a less categorical perspective, only one part of which was ultimately embraced by his inheritors. *See* HOMOSEXUALITY: SOCIAL, PSYCHOLOGICAL AND BIOLOGICAL ISSUES (W. Paul, J. Weinrich, J. Gonsiorek, and M. Hotvedt eds. 1982) [hereinafter cited as HOMOSEXUALITY] where Gonsiorek notes:

> Freud (1905 [1953]) suggested . . . that a basic bisexuality is inherent in all persons. Early learned experiences and the ensuing personality characteristics then determine which individuals as adults were homosexual, heterosexual, or bisexual. Specifically, Freud suggested that the manner in which a child handled the Oedipal conflict was a primary determinant of later sexual orientation (Freud, 1920 [1971]). However, many psychoanalytic thinkers, especially in the United States, have rejected the concept of inherent bisexuality while retaining the idea of the Oedipal conflict as central in the determination of sexual orientation.
>
> Therefore, despite Freud's reconceptualization that the potential for all varieties of sexual expression was inherent in all individuals, in the mid-twentieth century the prevailing opinion held that sexual orientation was essentially a dichotomous phenomenon, with most individuals being heterosexual and a few homosexual.

Id. at 57-58.

8. Socarides, *Psychoanalytic Therapy of a Male Homosexual*, 38 PSYCHOANALYTIC Q. 173 (1969). In his book, THE OVERT HOMOSEXUAL, Socarides offers the following:

> Homosexuality is based on the fear of the mother, the aggressive attack against the father, and is filled with aggression, destruction, and self deceit. It is a masquerade of life in which certain psychic energies are neutralized and held in a somewhat quiescent state. However, the unconscious manifestations of hate, destructiveness, incest and fear are always threatening to break through.

SOCARIDES, THE OVERT HOMOSEXUAL 8, (1968) [hereinafter cited as SOCARIDES].

9. SOCIETY OF MEDICAL PSYCHOANALYSTS, SEXUALITY: A PSYCHOANALYTIC STUDY 79-84 (1962) [hereinafter cited as SEXUALITY]. Their conclusions were drawn from a study of 106 homosexual and 100 heterosexual men. *Id.* at 24. Information about them was gathered from 77 psychiatrists. *Id.* at 31. The society's analysis of the data includes the conclusion that a poor bond between father and son causes the son to turn to men, seeking the affection his father never provided. *Id.* at 114-17. They also point to the failure of detached fathers to provide their sons with masculine role models, and to protect their sons from consuming mothers. *Id.* at 115.

10. Green, *The Best Interests of the Child with a Lesbian Mother*, 10 BULL. AM. ACAD. PSYCHIATRY & L. 1, at 8-9 (1982) [hereinafter cited as Green]. For a discussion of this and other role model theories, see E. MACCOBY & C. JACKLIN, THE PSYCHOLOGY OF SEX DIFFERENCES (1974).

11. S. RADO, PSYCHOANALYSIS OF BEHAVIOR II 206 (1962). Rado theorizes that homosexuality is a "reparative" attempt at sexual pleasure, needed when heterosexual functioning presents a threat; *see also* Bieber, *Sexual Deviations II: Homosexuality*, in COMPREHENSIVE TEXTBOOK OF PSYCHIATRY 963, 970, 974-75 (A. Freedman & H. Kaplan eds. 1967).

12. Gould, *What About Homosexuality*?, N.Y. Times Magazine, Feb. 24, 1974 at 13 [hereinafter cited as Gould].

13. Kallmann, *Comparative Twin Study on the Genetic Aspects of Male Homosexuality*, 115 J. NERVOUS MENTAL DISEASE 283 (1952). This study compared the concordance for homosexuality in monozygotic ("identical") and dizygotic ("fraternal") twins, and found 100 percent concordance in the former as opposed to 12 percent in the latter group. *Id.* at 296-97. Using a smaller sample, in 1968, Heston and Shields, *Homosexuality in Twins*, 18 ARCHIVES OF GEN. PSYCHIATRY 149 (1968), found 40 percent concordance in identical twins and 14 percent in fraternals. The most obvious flaw in this method of analyzing genetic origins is that it does not separate environmental and psychological factors. The author obviously relies on these factors to cancel each other out where mono and dizygotic twins are compared. However, it cannot be said that twins raised knowing (and seeing) that they are "identical" have the same experience as those who are fraternal.

14. Dorner, Rohde, Stahl, Krell, & Masius, *A Neuroendocrine Predisposition for Homosexuality in Men*, 4 ARCHIVES SEXUAL BEHAV. 1 (1975). The authors find their tests to suggest "that homosexual men possess a predominantly female-differentiated brain which may be activated to homosexual behavior by normal or approximately normal androgen levels in adulthood." *Id.* at 1; *see also* Kolodny, Masters, Hendryx & Toro, *Plasma Testosterone and Semen Analysis in Male Homosexuals*, 285 NEW ENG. J. MED. 1170

evidence, and none could be satisfactorily harmonized with any of the others. As one commentator put it:

> The only point that the "illness" theorists agree upon is that homosexuality is a disorder "in function." Beyond this, there are widely divergent views about when and how the "disfunction" occurs and what causes it. Freud says the problem develops in the Oedipal stage; Bergler and Socarides say it is pre-oedipal, and Bieber says the timing is irrelevant—only the pattern of family interaction is important. But what pattern? Again, the "illness" theorists disagree.[15]

Moreover, the data supporting illness theory were suspect. The conclusions of illness theorists had been drawn from work with a subset of homosexuals who had voluntarily sought or had been pressed into seeking psychiatric help because of personal unhappiness or social maladjustment. General conclusions about the mental health of all homosexuals had been extrapolated from this relatively small and not necessarily representative population.[16] Nor did findings of ongoing studies support important factual premises of some illness theory. For example, the theory that identifiable patterns of family interaction account for homosexual outcomes was not borne out by medical experience. Rather, experience taught that homosexuals often emerge from families where supposed causative family interactions do not exist and, conversely, that heterosexuals often emerge from families where such interactions do exist.[17] It also became clear that, since homosexuals are usually the offspring of functioning heterosexuals, homosexuality cannot have its genesis in teaching by or modeling upon the behaviors of a homosexual parent.[18] Finally, sexuality studies commenced in the 1940's revealed that the line between heterosexual and homosexual orientation is not nearly as sharp as had been supposed. Data from these studies showed that a substantial percent of self-defined, practicing heterosexuals had had some consensual homosexual experience and that many homosexuals, in like fashion, had had heterosexual experience.[19] The conjunction of these factors with a gay rights political initiative[20] led the APA to abandon illness theory as lacking in evidential foundation.[21]

While the APA's change of position was a significant breakthrough for lesbian mothers enmeshed in custody disputes, standing alone it represented a less than adequate weapon with which to defend against the negative images supplied from other sources. In the view of many heterosexuals, "lesbian" and "mother" were mutually exclusive constructs. "Mother" was keeper of order, provider of safety, moral guide. "Lesbian" conjured up images of darkness, disorder, and moral, perceptual, and behavioral deformation—the antithesis of "mother."[22] To counteract the impact of that damaging imagery, much more was needed than a declaration that homosexuality is not pathology. That "more" is now increasingly available. Studies of lesbian women begun in the late 1950's have substantially eroded long

---

(1971) (suggesting a correlation between male homosexuality and sperm count); Loraine, Ismail, Adamopoulos & Dove, *Endocrine Function in Male and Female Homosexuals*, 4 BRIT. MED. J. 406 (1970); Margolese, *Homosexuality: A New Endocrine Correlate*, 1 HORMONES & BEHAV. 151 (1970).

15. Gould, *supra* note 12, at 57.

16. Most notably, SEXUALITY, *supra* note 9, studied 106 homosexual and 100 heterosexual men who sought psychoanalysis. *Id.* at 24. Moreover, the study consisted of a questionnaire completed by the psychiatrist. *Id.* at 21. Jones, *The Phallic Phase*, 14 INT'L J. PSYCHO-ANALYSIS 1 (1933) (no subjects); SOCARIDES, *supra* note 8, at 105-207 (six patients); Stoller, *The Intersexed Patient—Counsel and Management*, in SEXUAL PROBLEMS: DIAGNOSIS AND TREATMENT IN MEDICAL PRACTICE (C. Wahl ed. 1967).

For a study which asserts a tendency for those who volunteer for psychological research to be less well adjusted than those who do not, see Burdick & Stewart, *Differences Between "Show" and "No-show" Volunteers in a Homosexual Population*, 92 J. SOC. PSYCHOLOGY 159 (1974).

17. Siegelman, *Parental Backgrounds of Homosexual and Heterosexual Men: A Cross National Replication*, 10 ARCHIVES SEXUAL BEHAV. 505 (1981); A. KINSEY, W. POMEROY and C. MARTIN, SEXUAL BEHAVIOR IN THE HUMAN MALE 661 (1948) [hereinafter cited as KINSEY, POMEROY & MARTIN].

18. Green, *supra* note 10, at 9; *see also* Green, *Sexual Identity of 37 Children Raised by Homosexual or Transsexual Parents*, 135 AM. J. PSYCHIATRY 692 (1978).

19. *See* KINSEY, POMEROY & MARTIN, *supra* note 17, at 623-26, 660; *see also* C. FORD & F. BEACH, PATTERNS OF SEXUAL BEHAVIOR 125-29 (1951).

20. *See* BAYER, *supra* note 3, for a full scale account of the effect of gay rights protest on the mental illness nomenclature debate that led to the declassification of homosexuality.

21. The American Psychological Association followed suit. *See* Riddle, *Relating To Children: Gays as Role Models*, J. SOC. ISSUES 38 (1978).

22. KIRKPATRICK, ROY & SMITH, *supra* note 5, at 1; *see also* Kirkpatrick, *Lesbian Mother Families*, 12 PSYCHIATRIC ANNALS 842 (1982).

held beliefs that, in terms of life content, personal adjustment and social functioning, lesbian women are significantly different from heterosexual women.[23] In fact, if a difference exists between the two groups, it appears to lie in the greater self-reliance and better self-image displayed by avowed lesbians, and this despite the extraordinary social pressures to which they are subject.[24] Recent studies also show that the difference between lesbian parenting behaviors and those of divorced heterosexual mothers is negligible[25] and that the gender identity and social adjustment of children being raised by lesbian mothers parallels that of children being raised by divorced heterosexual mothers.[26] It is true, of course, that longitudinal studies tracking long term psychological and social outcomes for the custodial children of such mothers are not yet available, but there is nothing in the present data to suggest that, in the absence of a persecutorial environment, the adult adjustment of such children should be less satisfactory than that of their generational peers. In light of present knowledge, the lesbian sexual orientation of mothers, without more, should therefore play a diminished role in deciding their custody claims. In fact, sexuality is still the fulcrum of most custody cases involving such mothers.[27] Judicial unawareness or rejection of present medical perspectives accounts, in part, for the continuing centrality of the parental sexuality issue.[28] Illness theorists, dissenting from the contemporary views of the majority of their peers, play some part in preserving the importance of the sexuality issue.[29] And despite its changed stance on the illness question, mainstream psychiatry itself contributes to the difficulties faced by lesbian mothers. Although most psychiatrists appear now to be satisfied that homosexual orientation is not psychologically aberrational, they are quite unclear about what it is. Their continuing quest for a unique homosexual etiology[30] reinforces fears that parental homosexuality, mysterious in origin, may indeed create unknown and unknowable dangers for children.

---

23. Adelman, *A Comparison of Professionally Employed Lesbians and Heterosexual Women on the MMPI*, 6 ARCHIVES OF SEXUAL BEHAV. 193 (1977) (study compared non-patient lesbians and heterosexual women); Armon, *Some Personality Variables in Overt Female Homosexuality*, 24 J. PROJECTIVE TECH. 292 (1960); Gundlach & Riess, *Self and Sexual Identity in Women: A Study of Female Homosexuals*, in 1 NEW DIRECTIONS IN MENTAL HEALTH 205 (B. Riess ed. 1968) (used a non-patient sample); Hogan, Fox and Kirchner, *Attitudes, Opinions, and Sexual Development of 205 Homosexual Women*, 3 J. HOMOSEXUALITY 123 (1977) (used a questionnaire, without comparing results to a heterosexual control group); Hopkins, *Lesbian Signs on the Rorschach*, 15 BRIT. J. PROJECTIVE PSYCHOLOGY & PERSONALITY STUDY 7 (1970); Kingdon, *Lesbians*, 8 THE COUNS. PSYCHOLOGIST 44 (1979); Miller, *Characteristics of Homosexually Involved Incarcerated Females*, 30 J. CONSULTING PSYCHOLOGY 193 (1966); D. ROSEN, LESBIANISM: A STUDY OF FEMALE HOMOSEXUALITY 65-73 (1974) (studying non-patient lesbians, author found no correlation between lesbianism and poor mental health, and that a majority of lesbians do not practice "butch/femme" roles). Thompson, McCandless & Strickland, *Personal Adjustment of Male and Female Homosexuals and Heterosexuals*, 78 J. ABNORMAL PSCYHOLOGY 237 (1971).

24. Ferguson & Finkler, *An Involvement and Overtness Measure for Lesbians: Its Development and Relation to Anxiety and Social Zeitgeist*, 7 ARCHIVES SEXUAL BEHAV. 211 (1978) (indicating that lesbians tested as less anxious and better adjusted than the norm); M. FREEDMAN, HOMOSEXUALITY AND PSYCHOLOGICAL FUNCTIONING 77-78 (1971) (found no difference in overall psychological adjustment but found that lesbians are more independent, more sensitive to their own needs, more oriented toward the present, more candid, less defensive, and show higher levels of work satisfaction); Oberstone & Sukoneck, *Psychological Adjustment and Life Style of Single Lesbians and Single Heterosexual Women*, 1 PSYCHOLOGY WOMEN Q. 172 (1976) (lesbians scored higher in quality of interpersonal relationships, were less apt to compromise vocational goals, and demonstrated higher levels of independence and self determination); Siegelman, *Adjustment of Homosexual and Heterosexual Women*, 120 BRIT. J. PSYCHIATRY 477 (1972) (lesbians found higher on goal directedness and self acceptance and lower in depression scale); Van Cleave, *Self Identification, Self Identification Discrepancy and Environmental Perspectives of Women with a Same-Sex Sexual Preference*, 38 DISSERTATION ABSTRACTS INT'L 5932-A (1978); Wilson & Greene, *Personality Characteristics of Female Homosexuals*, 28 PSYCHOLOGICAL REP. 407 (1971) (lesbians found higher on dominance, endurance, capacity for status and intellectual efficiency.)

25. Miller, Jacobsen and Bigner, *The Child's Home Environment for Lesbian vs. Heterosexual Mothers: A Neglected Area of Research*, 7 J. HOMOSEXUALITY 49 (1981); Green, *supra* note 10, at 13-14. For a summary of many unpublished studies that come to the same conclusion, see KIRKPATRICK, ROY & SMITH, *supra* note 5, at 6.

26. Hoeffer, *Children's Acquisition of Sex Role Behavior in Lesbian-Mother Families*, 51 AM. J. ORTHOPSYCHIATRY 536 (1981); Kirkpatrick, *Lesbian Mother Families*, 12 PSYCHIATRIC ANNALS 842 (1982); Kirkpatrick, Smith & Roy, *Lesbian Mothers and Their Children: A Comparative Survey*, 51 AM. J. ORTHOPSYCHIATRY 545 (1981); and Green, *supra* note 10, at 13-14.

27. *See* Davis, *supra* note 2, at 22-23.

28. Rand, Graham & Rawlings, *Psychological Health and Factors the Court Seeks to Control in Lesbian Mother Custody Trials*, 8 J. HOMOSEXUALITY 27 (1982) (study found a positive correlation between psychological health and openness to employer, ex-husband, children, and a lesbian community).

29. Bieber and Socarides continue to publish their views.

30. *See, e.g.*, Hencken, *Homosexuality and Psychoanalysis: Toward a Mutual Understanding* in HOMOSEXUALITY, *supra* note 7, at 121, 134-47.

In sum, although psychiatry is no longer the condemnatory partner of religion and the law, it is not yet a formidable advocate for lesbian mothers in their struggles for the custody of their children; nor, in all likelihood, will it become such an advocate until it is satisfied either that homosexuality has physiological origins or that it is a manifestation of the inherent pansexuality of all human beings.[31]

III. THE STATE OF THE LAW: CONSTITUTION AND CASES

A. The Constitutional Framework

Lesbians and gay males are pervasively socially stigmatized. Although their lifestyles and self-presentations are in fact as varied as those of heterosexuals,[32] caricaturing street level imagery portrays their conduct as uniformly sexually charged, clandestine, predatory, promiscuous, violent, compulsively self-gratifying, and now diseased, and the manifestations of their sexuality as counterfeit representations of opposite sex roles.[33] Fundamentalist religion continues to condemn them.[34] The state may constitutionally exact criminal penalties for their commission of private adult, consensual homosexual acts;[35] and it is constitutionally permissible to bar them from the armed forces,[36] from public and private employment,[37] from entry into[38] and citizenship[39] in the United States, and from marriage to one another.[40] The underlying premise for their exclusion from the protections of the Constitution seems to be that, unlike race, gender or birth status, homosexuality is the product of free and, by definition, wrongful choice. For lesbian mothers in custody battles, the impact of that assumption

31. As stated in note 7 *supra*, this is not a new theory. Sigmund Freud long ago postulated that human beings are inherently bisexual. The theory is still at least tentatively advanced by some psychiatrists, although sharply criticized by others. *See* Green, *Homosexuality as a Mental Illness*, 10 INT'L J. PSYCHIATRY 77 (1972) and the immediately following responses of some of his peers where the debate is well captured. *Id.* at 99-125. (Bell, Hatterer, Hoffman, Karlen, Marmor and Socarides); *see also* Poole, *The Etiology of Gender Identity and the Lesbian*, 87 J. SOC. PSYCHOLOGY 51 (1972); A. FREUD, A CHILD IS BEING BEATEN, SEXUALITY AND PSYCHOLOGY OF LOVE 131 (1970).

32. Paul and Weinreich, *Whom and What We Study: Definition and Scope of Sexual Orientation* in HOMOSEXUALITY, *supra* note 7, at 23.

33. Paul, *Social Issues and Homosexual Behavior, A Taxonomy of Categories and Themes in Anti-Gay Argument* in HOMOSEXUALITY, *supra* note 7, at 29. This chapter examines quite thoroughly research on the prominent stereotypes and those who hold them.

34. *See* Barrett, *Legal Homophobia and the Christian Church*, 30 HASTINGS L.J. 1019 (1979).

35. Doe v. Commonwealth's Attorney, 403 F. Supp. 1199 (E.D. Va. 1975) (wherein a three-judge court sustained the constitutionality of Virginia's statute criminalizing adult, consensual, private homosexual acts against a challenge based upon the First, Fifth, Eighth, Ninth and Fourteenth Amendments) *aff'd without opinion*, 425 U.S. 901 (1976). This is not to say that lower courts have universally rejected constitutional challenges to such statutes. *See, e.g.*, People v. Onofre, 51 N.Y.2d 476, 415 N.E.2d 936, 434 N.Y.S.2d 947 (1980) (striking down New York's anti-sodomy statute as a violation of privacy); Hardwick v. Bowers, 760 F.2d 1202, *reh'g en banc denied*, 765 F.2d 1123 (11th Cir. 1985), *cert. granted*, 106 S. Ct. 342 (1985). In *Hardwick*, the Eleventh Circuit, declining to read *Doe v. Commonwealth's Attn'y* as foreclosing precedent, reversed the district court's Rule 12(b)(6) dismissal of a constitutional challenge to Georgia's sodomy statute. The challenge was brought, inter alia, by a gay man who had been targeted for prosecution for a consensual sexual act with an adult male in his own home. The appellate court, relying on Griswold v. Connecticut, 381 U.S. 479 (1965); Eisenstadt v. Baird, 405 U.S. 438 (1972); and Stanley v. Georgia, 394 U.S. 557 (1969), concluded that there is a fundamental constitutional right to sexual privacy which can only be abridged to serve a compelling state interest. The Circuit ordered a remand for proceedings in which the State of Georgia would be held to that standard. However, on November 4, 1985, the United States Supreme Court granted the State's petition for certification. At this writing, the matter is pending before the High Court. Obviously its outcome will profoundly affect the quest of lesbians as well as gay men for a right to be themselves.

36. Matlovich v. Secretary of the Air Force, 591 F.2d 852 (D.C. Cir. 1978); Berg v. Clayton, 591 F.2d 849 (D.C. Cir. 1978); *see* Rivera, *Our Straight-Laced Judges: The Legal Position of Homosexual Persons in the United States*, 30 HASTINGS L.J. 799, 837-855 (1979) [hereinafter cited as *Our Straight-Laced Judges*]; Rivera, *Recent Developments in Sexual Preference Law*, 30 DRAKE L. REV. 311 (1980-81) [hereinafter cited as *Recent Developments*].

37. For an exhaustive analysis of the cases in the area of federal, state, and local government employment, as well as those declining to extend the protection of Title VII to private sector employees, *see Our Straight-Laced Judges*, *supra* note 36, at 805-837, 855-875; *Recent Developments*, *supra* note 36, at 312-24; and E. BOGGAN, M. HAFT, C. LISTER, and J. RUPPS, THE RIGHTS OF GAY PEOPLE 21-73 (1975) [hereinafter cited as THE RIGHTS OF GAY PEOPLE]; see also Richter, ANTI-GAY LEGISLATION: AN ATTEMPT TO SANCTION INEQUALITY? (1982) (published by LAMBDA LEGAL DEFENSE).

38. *See Our Straight-Laced Judges*, *supra* note 36, at 934-42; *Recent Developments*, *supra* note 36, at 343-46; THE RIGHTS OF GAY PEOPLE, note 37, at 75-83.

39. *Id.*

40. Singer v. Hara, 11 Wash. App. 24, 522 P.2d 1187 (1974); *see Our Straight-Laced Judges*, *supra* note 36, at 874-78; *Recent Developments*, *supra* note 36, at 324-25; *see also* THE RIGHTS OF GAY PEOPLE, *supra* note 37, at 103-35.

is potentially disastrous. Since these women have generally had heterosexual marriages and the heterosexual relations that produce offspring, they are in particular danger of being perceived as willfully and immorally adopting an interdicted mode of sexual expression, to the detriment of society and their children. Neither privacy nor parenting rights protect them from the consequences of that perception when their custodial entitlement is challenged by family adversaries.

1. *The Constitutional Right of Privacy*

Griswold v. Connecticut[41] marked the Supreme Court's first significant limitation of the state's power to define sexual morality. In *Griswold*, the Court concluded that Connecticut's statute barring access to contraceptive information violated the constitutional right of marital privacy.[42] *Eisenstadt v. Baird*,[43] decided seven years later, identified a broader right to contraceptives, one that attached to all persons regardless of marital status.[44] In like fashion, *Roe v. Wade*[45] and its companion case[46] announced the pre-viability right of all women, married and unmarried, to terminate unwanted pre-viability pregnancies, subject only to the requirements of medical safety.[47] Since then, the Court has repeatedly bridled state power to limit procreative choice.[48]

Although logic would suggest that the right of unmarried persons to have access to contraceptives and to choose abortion over childbirth must, sub silentio, encompass a right to have non-marital sexual intercourse, it would be a mistake to conclude that the Court has so held. Statutes criminalizing nonmarital and extramarital sex remain on the books of twenty-five states.[49] In the years since *Eisenstadt* and *Roe* were decided, the Court has not entertained a challenge to any of them. *Jarrett v. Jarrett*,[50] a decision that came well after the Court's announcement of the right of procreative privacy, aptly illustrates their continuing impact. In *Jarrett*, plaintiff wife was awarded custody of three marital children at the time of divorce. Seven months later, her former husband sought to take the children from her on the sole ground that she was living in the marital home, in a monogamous and apparently loving relationship, with a man to whom she was not married. The former husband offered no proof of any specific harm to his children flowing from their mother's conduct. Nonetheless, the trial court divested her of custody.[51] Its decision was affirmed by the Illinois Supreme Court.[52] Rejecting the mother's Fourteenth Amendment claims, the court found it entirely proper to construct the requisite harm from her open violation of the Illinois anti-fornication law.

In the eyes of the Illinois court, teaching one's children by example to ignore the state's

41. 381 U.S. 479 (1965).
42. *Id.* at 485.
43. 405 U.S. 438 (1972).
44. *Id.* at 453.
45. 410 U.S. 113 (1973).
46. Doe v. Bolton, 410 U.S. 179 (1973).
47. *Id.* at 195.
48. *See, e.g.*, Planned Parenthood of Central Missouri v. Danforth, 428 U.S. 52 (1976) (statute requiring parental or spousal consent to abortion unconstitutional); Carey v. Population Serv. Int'l, 431 U.S. 678 (1977) (invalidating restriction on distribution and advertising of non-prescription contraceptives); Bellotti v. Baird, 443 U.S. 622 (1979) (parent consent); Colautti v. Franklin, 439 U.S. 379 (1979) (invalidating imposition of criminal sanctions upon doctors failing to follow a statutorily prescribed procedure); Akron v. Akron Center for Reproductive Health, 462 U.S. 416 (1983) (restrictions on abortions not done in hospitals unconstitutional). *But see* Harris v. McRae, 448 U.S. 297 (1980) (statutory bar on use of medicaid funds for abortion constitutional).
49. States with statutes criminalizing adultery: ALA. CODE § 13A-13-2 (1975) & (Supp. 1985); CAL. PENAL CODE § 269a (West 1970); COLO. REV. STAT. § 18-6-501 (1973); CONN. GEN. STAT. ANN. § 53a-81 (West 1972); FLA. STAT. ANN. § 798.01 (West 1976); GA. CODE ANN. § 16-6-19 (1985); IDAHO CODE § 18-6601 (1979 & Supp. 1985); ILL. ANN. STAT. ch. 38 § 11-8 (Smith-Hurd 1985); KAN. STAT. ANN. § 21-3-507 (1981); MD. ANN. CODE ART. 27 § 4 (1957); MASS. ANN. LAWS ch. 272 § 17 (Michie/Law. Co-op. 1982); MICH. COMP. LAWS § 750-30 (1979); NEB. REV. STAT. § 28-704 (1979); N.H. REV. STAT. ANN. § 645.3 (1974); N.Y. PENAL LAW § 255:17 (McKinney 1975); N.D. CENT. CODE § 12.1-20-09 (1976); OK. STAT. ANN. tit. 21 § 872 (West 1983); R.I. GEN. LAWS § 11-6-2 (1981); UTAH CODE ANN. § 76-9-507 (1978); VA. CODE § 18.2.-365 (1982 & Supp. 1985); WIS. STAT. ANN. § 944:16 (West 1982 & Supp. 1985).

States with anti-fornication statutes: FLA. STAT. ANN. § 798.03 (West 1976); GA. CODE ANN. § 16-6-18 (1985); IDAHO CODE § 18-6603 (1979 & Supp. 1985); ILL. ANN. STAT. ch. 38 § 11-8 (Smith-Hurd 1985); MASS. ANN. LAWS ch. 272 § 18 (Michie/Law Co-op 1979); R. I. GEN. LAWS § 11-6-3 (1981); UTAH CODE ANN. § 76-7-104 (1978); VT. STAT. ANN. tit. 13 § 205 (1974 & Supp. 1984); VA. CODE § 18.2-344 (1982); WIS. STAT. ANN. § 944:15 (West 1982 & West Supp. 1985).

States with a combined statute, proscribing both adultery and fornication: MISS. CODE ANN. § 97-291 (1973 & Supp. 1985); N.C. GEN. STAT. §§ 14-184 (1981); S.C. CODE ANN. § 16-15-60 (1976) W. VA. CODE § 61-83 (1984).
50. 78 Ill.2d 337, 400 N.E.2d 421 (1979), *cert. denied*, 449 U.S. 927, *reh'g denied*, 449 U.S. 1067 (1980).
51. 449 U.S. at 928.
52. 78 Ill.2d 337, 350, 400 N.E.2d 421, 422 (1979).

proscription of nonmarital sex was sufficient, without more, to justify loss of their custody.[53] Ms. Jarrett's petition for certification to the United States Supreme Court was denied.[54] Justice Brennan, with whom Justice Marshall concurred,[55] filed this vigorous dissent:

> The decision of the Illinois Supreme Court that, in effect, a divorced woman's ostensible violation of the Illinois fornication statute presumptively harmed the best interests of the children and that this was conclusive for purposes of custody presents a serious question under the Fourteenth Amendment. Giving conclusive effect to such a violation would appear to contravene the teaching of *Stanley v. Illinois*. . . . Stanley established the proposition that "the interest of a parent in the companionship, care, custody and management of his or her children," [citation omitted] cannot be determined by the evidentiary shortcut of a conclusive presumption. Thus, for purposes of this case, Stanley would seem to foreclose custody modification on the basis of a similar conclusive presumption of serious adverse effect on the children's best interests despite whatever contrary evidence may have been or might be adduced. This is particularly true since there is no rational basis for the conclusive presumption actually utilized, whether Jacqueline is viewed as having violated the fornication statute only or as being a lawbreaker generally.
>
> Nothing in the record or in logic supports a conclusion that divorced parents who fornicate, for that reason alone, are unfit or adversely affect the well-being and development of their children in any degree over and above whatever adverse effect separation and divorce may already have had on the children. Illinois seldom, if ever, enforces its fornication statute and therefore can hardly contend that there is a rational correlation between divorced parents who fornicate and divorced parents who impair the healthy development of their children.
>
> Nor can Jacqueline be presumed to have an adverse effect on her children's well-being because she is a lawbreaker, for surely such a presumption would be irrationally overbroad. It would make no sense to treat murder, fornication and traffic violations similarly for purposes of custody. . . .
>
> Moreover, not only is there no basis for conclusively presuming that Jacqueline's cohabitation would adversely affect her children sufficiently to justify modification, but any such conclusion is unequivocally rejected by the record which affirmatively shows that the "children were healthy, well adjusted, and well cared for." [citations omitted] There was no evidence of actual harm; nor was there evidence, statistical or otherwise, to suggest that the children's current exposure to their mother's cohabitation might result in harm to them that might become manifest only in the future. Surely, in any event, it is no more likely that divorced mothers who fornicate are unfit than are unwed fathers. Thus, this case squarely presents the question whether the Due Process Clause entitles Jacqueline to a meaningful hearing at which the trial judge determines, without use of a conclusive presumption, whether violation of the fornication statute adversely affects the well-being of the children.[56]

The majority, however, was apparently constitutionally content with the state court's equation of non-marital sex with criminality and therefore with custodial disentitlement. Denial of certification in *Jarrett* must therefore mean that the Court found no present necessity, in the light of *Eisenstadt, Roe* and their progeny, to extend the sexual privacy right beyond the four corners of procreative choice.

The constitutional claims of lesbians and gay males in general and lesbian mothers in particular are hostage to that outcome. If the state can proscribe and penalize heterosexual intercourse outside of marriage, it can proscribe and penalize homosexual intercourse outside of marriage. For homosexuals, who are nowhere granted the right to marry, that amounts to the power to prohibit all sexual expression. For lesbian mothers, it

53. *Id.*
54. 449 U.S. 297 (1980).
55. Justice Blackmun would also have set the case for argument. *Id.* at 931.
56. *Id.* at 927-31.

amounts to the power to take their children from them if they are sexual in the ways dictated by their sexual orientation or, indeed, at all outside of heterosexual marriage.

If the Supreme Court ultimately recognizes a constitutionally based privacy right to engage in nonmarital heterosexual intercourse, the multiple criminal and civil penalties that are presently imposed on homosexuals will be vulnerable to attack on establishment, privacy and equal protection grounds. Constitutional recognition of a right to be heterosexually active will certainly signal the Court's acknowledgement that sexual relations are a valuable form of personal expression independent of their procreative potential. In that posture, it will be quite impossible to justify the continued interdiction of homosexual acts on the ground that such acts lack a procreative mission. And since, in light of present medical knowledge, the proposition that homosexuality is a disorder requiring treatment and control has lost credibility,[57] religious interdiction will inexorably emerge as the linchpin supporting its suppression.[58] In our tradition, the commands of religion have never been thought sufficient, without more, to justify state limitations on personal autonomy. It is true, of course, that religious teaching lies at the base of much of our law.[59] The concordance of legal with religious proscriptions gives no cause for constitutional complaint when the conduct that they jointly condemn creates identifiable social harm. But in the absence of such harm, religion may not invoke the powers of the state for the fulfillment of its mandates.[60] Arguably, private, adult, consensual homosexual relations create no more inherent social harm than do like nonmarital heterosexual relations.[61] That being the case, a fully developed right of heterosexual privacy would position homosexuals to assert that they are constitutionally entitled to a comparable right. Ironically, it is for want of a full heterosexual right that they remain unable to claim any solid constitutional ground for themselves in the nation's highest court.

2. *The Right to Parent*

We begin with the settled principle that there is a right, lodged in the penumbrae of the First, Fifth, Ninth and Fourteenth Amendments, to the care, custody and society of one's children.[62] Because of the fundamental nature of the right, the state may not interfere in the parent-child relationship absent a compelling need that touches the child's health, safety, welfare or morals.[63] And even then, it may not interfere to a greater degree than the need requires.[64]

In guardianship proceedings and proceedings to terminate parental rights, the identity of the intruder is clear; it is the state itself that seeks to

57. This is not to suggest that mental health professionals unanimously agree with the APA's removal of homosexuality from its mental illness nomenclature. It is clear, however, that a majority of those professionals, at least as their views are made known through their professional organizations, supported and continues to support declassification.

See BAYER, *supra* note 3, at 162-68, (where the author discusses the present status of homosexuality within psychiatry).

58. For a brief discussion of the Judeo-Christian roots of anti-homosexual condemnation, see BAYER, *supra* note 3, at 15-18.

59. The principle of separation of church and state is firmly rooted in the United States Constitution, the First Amendment of which prohibits government from establishing or interfering with the free exercise of religion.

60. Stone v. Graham, 449 U.S. 39 (1980) (required posting of 10 commandments in school unconstitutional); Engel v. Vitale, 370 U.S. 421 (1962) (invalidating use of a state composed "non-demoninational prayer"); Abington School Dist. v. Schempp, 374 U.S. 203 (1963) (required bible reading or prayer in school unconstitutional); McLean v. Arkansas Bd. of Educ., 529 F. Supp. 1255 (E.D. Ark. 1982) (not constitutional to require equal time be spent on creation and evolution theories).

61. Linkage of Acquired Immune Deficiency Syndrome ("AIDS") with gay male sexual activity should not, over the long term, erode the legitimacy of this perspective. In the first place, AIDS is not uniquely a disease of homosexuals. Although gay men are the largest fraction of people with AIDS, the disease also affects, and is transmittable by and to heterosexuals. Anal intercourse is not required for its transmission. *See* R. Redfield, P. Markham, S. Salahuddin, C. Wright, M. Serngadharan, & R. Gallo, *Heterosexually Acquired HTLV-III/LAV Disease (AIDS Related Complex and AIDS), Epidemiologic Evidence for Female-to-Male Transmission*, 254 J. A.M.A. 2094-96 (1985); MORBIDITY AND MORTALITY WEEKLY REPORT, CENTERS FOR DISEASE CONTROL, ATLANTA, *Heterosexual Transmission of Human T-Lymphotropic Virus Type III/Lymphadenopathy - Associated Virus*, 254 J. A.M.A. 2051-54 (1985).

Moreover, there is nothing to suggest that, with sufficient scientific effort, a mode of prevention or a cure will not be found for AIDS as it has been for other sexually transmitted diseases such as syphilis.

62. Stanley v. Illinois, 405 U.S. 645, 651 (1972); Little v. Streater, 452 U.S. 1 (1981).

63. Mathews v. Eldridge, 424 U.S. 319, 335 (1976).

64. Santowsky v. Kramer, 45 U.S. 745 (1982).

displace the parent. To do so, the state must establish that its actions are rooted in provable necessity.[65] Parental custody disputes, in contrast, facially involve only private parties neither of whom has a greater or more basic right than the other. In such disputes, the state, acting through the court, appears in the role of dispassionate adjustor rather than active contender. In that role, the state's judgments are insulated from constitutional challenge because of its apparent lack of direct investment in the outcome. Thus insulated, it is free to disfavor the custodial claims of a presumptively fit parent to the extent thought necessary for the advancement of the child's "best interests and welfare."[66]

The best interests standard teaches only that the well-being of the child must be the court's preeminent concern. It makes no substantive contribution to resolving the question, "What risks and what benefits will this child experience as the result of a decision in favor of one or the other custody contender?" In attempting to answer that question, a custody court must look to variables that lie outside and are independent of the standard itself. In so doing, it ordinarily explores the past and existing relationship of each of the parents with the child, the character of the home and care offered by each and, if the child is old enough, the desires of the child.[67] It gathers information from court-appointed investigators, from the testimony of the parties and their witnesses, from meetings with the child, and in some cases, from expert evaluators.[68] Legal presumptions, such as the fast-fading presumption favoring maternal custody of young children[69] or the presumption that a child should remain with the primary parental caretaker,[70] may be decisional ingredients. Judicially adopted psychological doctrines, such as the notion that the parent to whom the child has "bonded", is the child's psychological parent and therefore the proper custodian,[71] may operate to tip the scales in one direction or the other. But the calculus by which the court decides is, in the end, a highly subjective response to a mix of unweighted factors. A parental custody decision is no more than an intuitive, prediction that the child will be better off with the winning party.[72] Such a prediction is essentially unreviewable. Unless a trial court has committed gross and palpable error, no appellate tribunal will inpugn its net judgment.[73] The decisional setting thus presents extraordinary risks for a lesbian mother. In states that retain statutes criminalizing homosexual conduct, she is in no better position than was Ms. Jarrett. Her sexual activity in defiance of the law may raise the irrebuttable presumption of which Justice Brennan so eloquently complained in *Jarrett*. And even where there is no proscriptive statute, the court need only find the environment that she offers less "normal," or less protective of the child's social and psychological well-being or generally less desirable than the environment offered by her heterosexual opponent to deny her custody.[74] Given the constitutionally unarmed position of such a mother, it would be a reasonable assumption that she will generally lose a custody struggle with her heterosexual parental adversary. Surprisingly, she wins in almost half of the cases that we now examine.

B. The Cases

The charts that follow summarize a decade

65. *Id.*

66. The best interests standard was formulated by Cardozo in Finlay v. Finlay, 240 N.Y. 429, 148 N.E. 624 (1925); *see also* J. NOBLE & W. NOBLE, THE CUSTODY TRAP 138-9 (1975).

67. For a good overview of the standards used to determine child custody, see S. KRAM & N. FRANK, THE LAW OF CHILD CUSTODY: DEVELOPMENT OF THE SUBSTANTIVE LAW 43-77 (1982) [hereinafter cited as KRAM & FRANK]; *see also* Freed & Walker, *Family Law in the Fifty States: An Overview*, 18 FAM. L.Q. 369, 431-437 (1985).

68. *Id.* at 431-37.

69. For a description of the tender years presumption see CLARK, DOMESTIC RELATION § 17.4 (1968).

70. Garsky v. McCoy, 278 S.E.2d 357 (W. Va. 1981).

71. J. GOLDSTEIN, A. FREUD and A. SOLNIT, BEYOND THE BEST INTERESTS OF THE CHILD (1979); Polikoff, *Why are Mothers Losing: A Brief Analysis of Criteria Used in Child Custody Decisions*, 7 WOMEN'S RTS. L. REP. 235 (1982) (author argues that this standard can be sex neutral).

72. *See supra* note 67; *see also* Jacobson v. Jacobson 314 N.W.2d 78 (N.D. 1981) (where the appellate court removed custody from the mother primarily because of her lesbian relationship). The court found that where all other factors were equal, the mother's lesbianism was "the overriding factor." *Id.* at 80.

73. *See, e.g.*, Adams v. Adams, 357 So.2d 881, 883 (La. Ct. App. 1978) (declining to disturb a finding upon no "abuse of discretion"); Chaffin v. Frye, 45 Cal. App.3d 39, 119 Cal. Rptr. 22, 26 (1975) (no abuse of discretion); D.H. v. J.H., 418 N.E.2d 286, 296 (Ind. Ct. App. 1981) (requiring manifest abuse of discretion); DiStefano v. DiStefano, 60 A.D.2d 976, 977, 401 N.Y.2d 636, 638 (1978) (sole issue whether "the trial court's award is supported in the record").

74. *See In re* Jane B., 85 Misc.2d 515, 380 N.Y.S.2d 848, (Sup. Ct. 1976).

of reported and unreported custody decisions involving lesbian mothers and familial opponents.[75] In reviewing them, the reader should keep these reliability discounters in mind: First, the raw material out of which custody decisions flow is highly individualized. Each case involves nonfungible actors with unique histories, personalities, capacities and interactive patterns. The idiosyncratic characteristics of the contestants and their children may influence the outcome of a custody case in powerful but unarticulated ways. Second, because most events involving family members occur out of public view, the factual evidence offered by witnesses is insulated from credibility challenge and may be heavily overlaid with personal interest or bias. These impediments make it difficult to assess the reliability of the facts identified by a court as critical to its decision. Third, medical-psychological views on the nature of homosexuality and its development in or impact on children are in process of change. When the testimony of experts is in conflict, the court has the option of choosing one expert view over another, a task for which it is ill-equipped by training or experience, or falling back on its own perceptions and intuitions. The way in which a court has worked this problem out is not always clear from its written decision. Fourth, the essentially content-less best interests standard invites the court to assess the past and predict the future based on a wide variety of factors none of which, alone or in specific combinations with others, has binding legal significance. A written decision may not always reveal what permutations of factors were considered crucial; nor, usually, will it set out the writer's own religious, emotional or moralistic attitudes toward homosexuality although these may have had a profound, if masked, influence on the result. Fifth, the texture of the evidence as a whole is frequently obscured by the trial court's selective recitation and will almost certainly be lost in the truncated recitation of a reviewing court. Analysis through the milk glass of a written decision must, therefore, by definition, be less than fully revelatory. Finally, it is important to bear in mind that reported and unreported full scale decisions represent an indeterminable fraction of all litigated cases involving lesbian mothers. While available written decisions offer a window through which to observe, it is one that provides only a partial view.[76]

Nonetheless, parsing the available decisions is a useful exercise. It offers some insights into lesbian personalities and parenting behaviors as perceived by children, investigators, custodial opponents and courts; it provides an opportunity to measure the extent to which professionals reflect, testimonally, the official views of their organizations; it sheds some light on the process by which judges possessed of extremely broad discretion arrive at results; and, at least in gross fashion, it tells us how far lesbian mothers, only now emerging from the secrecy imposed by a condemnatory society, have progressed in their quest for parental rights.[77]

The method here used was to chart the winning and losing cases separately, each with respect to a variety of factors. Chart I sets out data

75. Alphabetically arranged by case name, the decision surveys are: Adams v. Adams, 357 So.2d 881 (La. App. 1978); Ashling v. Ashling, 42 Or. App. 47, 599 P.2d 475 (1979); Belmont v. Belmont, No. M-16937-74 (N.J. Super., Ch. Div. July 22, 1980); Bennett v. Clemens, 230 Ga. 317, 196 S.E.2d 842 (1973); Buck v. Buck, 238 Ga. 540, 233 S.E.2d 792 (1977); Chaffin v. Frye, 45 Cal. App.3d 39, 119 Cal. Rptr. 22 (1975); Dailey v. Dailey, 635 S.W.2d 391 (Tenn. Ct. App. 1981), *permission to appeal denied* Feb. 16, 1982; D.H. v. J.H., 418 N.E.2d 286 (Ind. Ct. App. 1981); DiStefano v. DiStefano, 60 A.D.2d 976, 401 N.Y.S.2d 636 (1978); Doe v. Doe, 9 Fam. L. Rep. (BNA) 2662 (Mass. App. Ct. Aug. 9, 1983); Gay v. Gay, 149 Ga. App. 173, 253 S.E.2d 846 (1979); Gerdes v. Gerdes, No. D-21615 (Ariz. Super. Ct. Sept. 10, 1979); Hall v. Hall, 95 Mich. App. 614, 291 N.W.2d 143 (1980); *In re* Jane B., 85 Misc.2d 515, 380 N.Y.S.2d 848 (Sup. Ct. 1976); *In re* L.J.R. and J.A.R., No. 98812 (Wis. County Ct. Oct. 18, 1976); Jacobson v. Jacobson, 314 N.W.2d 78 (N.D. 1981); L. v. D., 630 S.W.2d 240 (Mo. Ct. App. 1982); M.J.P. v. J.G.P., 640 P.2d 966 (Okla. 1982); M.P. v. S.P., 169 N.J. Super. 425, 404 A.2d 1256 (App. Div. 1979); Mueller v. Mueller, No. 79 DR 1246 (Colo. App. Nov. 19, 1981); N.K.M. v. L.E.M., 606 S.W.2d 179 (Mo. Ct. App. 1980); Peterson v. Peterson, No. D-66634 (Colo. Dist. Ct. Apr. 26, 1978); Roberts v. Roberts, 25 N.C. App. 198, 212 S.E.2d 410 (1975); S. v. S., 608 S.W.2d 64 (Ky. Ct. App. 1980); Schneider v. Schneider, No. 77-1622 (Mich. Ct. App. Oct. 19, 1978); Schuster v. Schuster and Isaacson v. Isaacson (consolidated), 90 Wash. 2d 626, 585 P.2d 130 (1978); Spence v. Durham, 283 N.C. 671, 198 S.E.2d 537 (1973), *cert. denied*, 415 U.S. 918 (1974); Unglaub v. Unglaub, No. 62380 (Minn. Dist. Ct. June 12, 1978); Whitehead v. Black, No. CV-76-422 and CV-76-426 (Maine Super. Ct. June 14, 1976).

In the charts that follow, these cases are rostered without repetition of citation.

76. In preparing this paper, an attempt was made to pierce the decisional veil by sending detailed questionnaires to the attorneys who participated in the cases. Unfortunately, the responses were too few, and too sketchy when made, to be useful.

77. *See supra* notes 70 & 71.

on the identity of opponents, custodial arrangements when the litigation began, and the ages, sexes and custodial preferences of the involved children. The objective was to determine, as nearly as possible, the extent to which usual concerns for continuity of care, the tendency of courts to award custody of young and same sex children to the mother[78] and the traditional deference given by courts to the wishes of mature children[79] are displaced when the mother is a lesbian. Chart II identifies the domestic arrangements between the mother and her lover. This factor was separated out for the purpose of determining whether courts react more negatively to open relationships to which children are privy than to discreet liaisons. Chart III marshalls "bad parent" allegations not addressed directly to the mother's sexuality. Here the objectives were twofold. First, an opponent in a custody struggle may be expected to produce every ounce of accusatory information at his disposal. Therefore, the frequency with which such accusatory information is offered, and its nature, tell us something about the social and parenting behaviors of a lesbian mother. In addition, the data supply some insights about the influence of such information on case outcomes. Chart IV deals with the roles and influence of mental health professionals and court appointed investigators in the decision making process. Chart V focuses on the law applied in these cases. Specifically, it sets out anti-gay criminal proscriptions in an attempt to discover whether custody courts in jurisdictions having such proscriptions are influenced by the state's criminal law; it notes the frequency with which constitutional arguments are made by mothers and the responses of courts; and it attempts to identify the standards applied by decision makers in answering the question, "How are the child's best interests affected by the mother's sexuality?"

78. *See supra* note 69; *see also* M.P. v. S.P., 169 N.J. Super. 425, 404 A.2d 1256 (App. Div. 1979).

79. *See supra* note 68; *see also* Unglaub v. Unglaub, No. 62380 (Minn. Dist. Ct. June 12, 1978); MO. REV. STAT. § 452.375(2) (Supp. 1985).

© JEB (Joan E. Biren) 1985

I. CUSTODY CONTENDERS AND CHILDREN

	Custodian when case started			Relationship of Opponent to Child(ren)			Age(s) of Child(ren)			Preference(s) of Child(ren)			Sex(es) of Child(ren)		
	Mother	Opponent	Other	Father	Grand-parents	Other	Under 10	Over 10	Not Stated	Mother	Opponent	Not Stated	Male	Female	Not Stated
WINNING CASES (13) (ARRANGED BY JURISDICTION)															
Gerdes v. Gerdes (Arizona, 1979)	■			■			■	■		■ (1)	■ (1)			■	
Peterson v. Peterson (Colorado, 1978)	■			■					■			■	■	■	
Gay v. Gay (Georgia, 1979)	■	■		■	■				■			■			■
Buck v. Buck (Georgia, 1977)	■			■			■					■			■
Whitehead v. Black (Maine, 1976)	■			■			■	■				■	■	■	
Doe v. Doe (Massachusetts, 1983)	■ (Joint Custody)	■		■			■			■ (Both)	■		■		
Schneider v. Schneider (Michigan, 1978)	■			■					■			■			■
Unglaub v. Unglaub (Minnesota, 1978)	■			■			■	■		■			■		
M.P. v. S.P. (New Jersey, 1979)	■			■				■			■ (1)	■ (1)		■	
Belmont v. Belmont (New Jersey, 1980)	■			■			■	■		■			■	■	
Spence v. Durham (No.Carolina, 1973)		■	■		■		■	■				■		■	
Schuster v. Schuster and Isaacson v. Isaacson (Washington, 1978)	■			■					■			■			■
In re L.J.R. v. J.A.R. (Wisconsin, 1976)	■			■			■					■			■
LOSING CASES (16) (ARRANGED BY JURISDICTION)															
Chaffin v. Frye (California, 1975)		■			■			■		■				■	
Mueller v. Mueller (Colorado, 1981)	■			■			■				■		■	■	
Bennett v. Clemens (Georgia, 1973)	■				■		■					■		■	
D.H. v. J.H. (Indiana, 1981)		■		■			■	■				■	■	■	
S. v. S. (Kentucky, 1980)	■			■			■					■		■	
Adams v. Adams (Lousiana, 1978)		■		■					■			■		■	
Hall v. Hall (Michigan, 1980)		■		■					■			■			■
N.K.M. v. L.E.M. (Missouri, 1980)	■			■			■			■				■	
L. v. D. (Missouri, 1982)		■		■				■		■			■		
In re Jane B. (New York, 1976)	■			■			■			■				■	
DiStefano v. DiStefano (New York, 1978)		■		■			■					■			■
Roberts v. Roberts (No.Carolina, 1975)	■					■	■					■			■
Jacobson v. Jacobson (No.Dakota, 1981)	■			■			■					■			■
M.J.P. v. J.G.P. (Oklahoma, 1982)	■			■			■					■	■		
Ashling v. Ashling (Oregon, 1969)		■		■					■			■			■
Dailey v. Dailey (Tennessee, 1981)	■			■			■				■		■		

1. *Custody Contenders and Children (Chart I)*

In almost all of these cases the contenders were mothers and fathers, each having a presumptively equal right to custody.[80] However, in most of the winning cases actual custody was in mothers when the litigation began. It may therefore be assumed, at least inferentially, that judicial concern for continuity of care influenced outcomes in these cases. In the losing cases, only slightly more than half of mothers had preexisting custody. How far continuity concerns affected outcomes in these cases is unclear. Certainly, for those mothers who lost preexisting custody, such concerns were displaced by other considerations.

The ages of children did not have any demonstrable effect in the winning cases, where children under and over ten years were about evenly distributed. In the losing cases, the majority of children were ten years old or younger. Obviously, the tender years of these children did not persuade decision makers to lodge custody with mothers.

To the extent noted, the sexes of children were also about evenly divided in the winning cases. In the losing cases, female children predominated. Here it may be legitimate to infer that sometimes unarticulated preoccupations with the dangers of "modeling" may have influenced judges against awarding custody to mothers.

The custodial preferences of children were noted in only twelve of the thirty decisions. Where noted, two thirds of the children expressed a preference to live with their mothers. Their preference was ignored in the losing cases, perhaps because of the young age of most of them, and had little articulated influence on the outcomes in the cases where mothers won.

Overall, continuity concerns appear to have had a subsurface influence in the winning cases and an ambiguous role in the losing cases. There was no discernible linkage between the ages, sexes or preferences of children and case outcomes.

80. Until the mid-nineteenth century Anglo-American common law gave the father a virtually absolute right to custody. *See* W. Forsyth, Custody of Infants § 6 (1850). The tender years presumption, which shifted the law's preference from fathers to mothers, was an expression of the Victorian Era's iconization of women as breeders and nurturers. *See, e.g.*, Jenkins v. Jenkins, 173 Wisc. 592, 181 N.W. 826 (1921). In recent years, and especially since the advent of the modern women's movement the law has again shifted, this time toward a paternal equality model. *See* Kram & Frank, *supra* note 67, at 19-35.

© JEB (Joan E. Biren) 1986

II. MOTHER'S DOMESTIC ARRANGEMENTS

	Mother's Relationships			
WINNING CASES (13) (ARRANGED BY JURISDICTION)	Living with children	Living with Lover	Non live-in Lover	Past Lover
Gerdes v. Gerdes (Arizona, 1979)	■	■		
Peterson v. Peterson (Colorado, 1978)	■	■		
Gay v. Gay (Georgia, 1979)			■	
Buck v. Buck (Georgia, 1977)	■			■
Whitehead v. Black (Maine, 1976)	■	■		
Doe v. Doe (Massachusetts, 1983)	■	■		
Schneider v. Schneider (Michigan, 1978)	■	■		
Unglaub v. Unglaub (Minnesota, 1978)	■			■
M.P. v. S.P. (New Jersey, 1979)	■		■	
Belmont v. Belmont (New Jersey, 1980)	■	■		
Spence v. Durham (No.Carolina, 1973)				■
Schuster v. Schuster and Isaacson v. Isaacson (Washington, 1978)	■	■		
In re L.J.R. v. J.A.R. (Wisconsin, 1976)	■	■		
LOSING CASES (16) (ARRANGED BY JURISDICTION)				
Chaffin v. Frye (California, 1975)		■		
Mueller v. Mueller (Colorado, 1981)	■		■	
Bennett v. Clemens (Georgia, 1973)	■			■
D.H. v. J.H. (Indiana, 1981)			■	
S. v. S. (Kentucky, 1980)	■	■		
Adams v. Adams (Lousiana, 1978)				■
Hall v. Hall (Michigan, 1980)			■	
N.K.M. v. L.E.M. (Missouri, 1980)	■		■	
L. v. D. (Missouri, 1982)		■		
In re Jane B. (New York, 1976)	■	■		
DiStefano v. DiStefano (New York, 1978)		■		
Roberts v. Roberts (No.Carolina, 1975)	■		■	
Jacobson v. Jacobson (No.Dakota, 1981)	■	■		
M.J.P. v. J.G.P. (Oklahoma, 1982)	■	■		
Ashling v. Ashling (Oregon, 1969)			■	
Dailey v. Dailey (Tennessee, 1981)	■	■		

2. *Mother's Domestic Arrangements (Chart II)*

Given traditional aversions to homosexuality, it would be reasonable to assume that lesbian mothers who keep their sexual relationships separate from their lives with children will be more likely to win custody struggles than those who involve their children by establishing domestic relationships with lovers. As this chart shows, that assumption is not correct. Eleven out of fourteen winning mothers were living with lovers. Nine of these had domestic arrangements wherein mother, lover and children lived in one household. Only half of mothers who lost custody were living with lovers, and of these, only five were living with lover and children.

From this, it seems clear that domestic arrangements are not pivotal to outcomes. Indeed, the very openness and monogamous stability of a lesbian household may be a positive factor for judges who do not bring a fundamental anti-gay bias to their determinations. For those judges who bring such a bias, keeping lesbian sexuality separate from the maternal household provides no assurance that the mother will win. Intuition in fact suggests that separateness may subliminally reinforce stereotypes about the sinister, underground lifestyles of homosexuals.

LESBIAN AND GAY PRIDE MARCH, NEW YORK CITY, JUNE 1979

© JEB (Joan E. Biren) 1986

III: NEGATIVE ALLEGATIONS ABOUT MOTHERS

	AS PARENTS				OTHERWISE
WINNING CASES (13) (ARRANGED BY JURISDICTION)	Abuse	Neglect	Proselytizing	Sexual Conduct In Children's Present	Self-destructive or anti-social conduct
Gerdes v. Gerdes (Arizona, 1979)					
Peterson v. Peterson (Colorado, 1978)					[shaded]
Gay v. Gay (Georgia, 1979)					
Buck v. Buck (Georgia, 1977)					
Whitehead v. Black (Maine, 1976)					
Doe v. Doe (Massachusetts, 1983)					
Schneider v. Schneider (Michigan, 1978)					
Unglaub v. Unglaub (Minnesota, 1978)					
M.P. v. S.P. (New Jersey, 1979)					
Belmont v. Belmont (New Jersey, 1980)					
Spence v. Durham (No Carolina, 1973)					
Schuster v. Schuster and Isaacson v. Isaacson (Washington, 1978)			[shaded]		
In re L.J.R. v. J.A.R. (Wisconsin, 1976)					
LOSING CASES (16) (ARRANGED BY JURISDICTION)					
Chaffin v. Frye (California, 1975)	[shaded]	[shaded]			[shaded]
Mueller v. Mueller (Colorado, 1981)					
Bennett v. Clemens (Georgia, 1973)	[shaded]	[shaded]			[shaded]
D.H. v. J.H. (Indiana, 1981)		[shaded]		[shaded]	
S. v. S. (Kentucky, 1980)					
Adams v. Adams (Lousiana, 1978)					
Hall v. Hall (Michigan, 1980)					
N.K.M. v. L.E.M. (Missouri, 1980)					
L. v. D. (Missouri, 1982)			[shaded]	[shaded]	
In re Jane B. (New York, 1976)		[shaded]			
DiStefano v. DiStefano (New York, 1978)					
Roberts v. Roberts (No.Carolina, 1975)		[shaded]			
Jacobson v. Jacobson (No.Dakota, 1981)					
M.J.P. v. J.G.P. (Oklahoma, 1982)				[shaded]	
Ashling v. Ashling (Oregon, 1969)					
Dailey v. Dailey (Tennessee, 1981)	[shaded]			[shaded]	

3. *Negative Allegations About Mothers (Chart III)*

The information collected in this chart tells us something about lesbian personality and parenting behaviors as seen through the eyes of opponents and courts, and something about the effects of bad conduct allegations, of whatever kind and weight, on case outcomes.

a. *Lesbian Personality and Parenting Behaviors*

In roughly two thirds of these cases, the essence of the opponent's complaint, straightforwardly put or distilled in the course of trial, was that the mother was actively or passively a lesbian. Bad parent allegations other than allegations going simply to the fact of the mother's sexual orientation or relationship were made in ten cases,[81] but such allegations tracked classic stereotypes[82] in only five of these.[83] The specific behavior complained of in the remaining six cases was not particularly distinguishable from that which might be charged against a heterosexual mother or her new mate in a hotly contested custody struggle.[84] Given the adversarial nature of custody proceedings, and the importance of the interests at stake, it is reasonable to assume that opponents who had damning information, and particularly stereotype-specific information, would have offered it. The absence of such information from two thirds of the cases, and the non-stereotypical bad conduct charges made in most others, therefore tends to confirm the findings on lesbian personality and parenting behaviors discussed in Section I.

b. *The Effects of Bad Conduct Allegations on Case Outcomes*

Viewed statistically, there is a rather striking difference between the losing and the winning cases. Nonsexual bad conduct charges were noted in fifty percent of the decisions in which mothers lost custody but appeared in only fifteen percent of the cases where they won. Although there is a temptation, on that basis, to assume that lesbian mothers who are well adjusted, socially responsible, competent and caring parents have a good chance to prevail, such a conclusion is obviously over-optimistic. In eight cases where mothers lost, the proofs went to their sexual relationships without much more. In these cases, the negative outcomes were uniquely a function of decision makers' attitudes about homosexuality itself. This writer also suspects, but cannot document, that, in at least some of the cases

81. Chaffin v. Frye, 45 Cal. App. 3d 39, 119 Cal. Rptr. 22 (1975); Peterson v. Peterson, Civ. Act. No. D-66634 (Col. D. Ct. 1978); Bennett v. Clemens, 230 Ga. 317, 196 S.E. 2d 842 (1973); D.H. v. J.H., 418 N.E. 2d 286 (Ind. Ct. App. 1981); L. v. D., 630 S.W. 2d 240 (Mo. Ct. App. 1982); *In re* Jane B., 85 Misc. 2d 515, 380 N.Y.S. 2d 848 (1976); Roberts v. Roberts, 25 N.C. App. 198, 212 S.E. 2d 410 (1975); M.J.P. v. J.G.P., 640 P.2d 966 (Okla. 1982); Dailey v. Dailey, 635 S.W. 2d 391 (Tenn. Ct. App. 1982); Schuster v. Schuster, 90 Wash. 2d 626, 585 P.2d 130 (1978).

82. *See supra* note 22 and accompanying text.

83. Bennett v. Clemens, 230 Ga. 317, 196 S.E.2d 842 (1973) (where it was alleged that the mother had left her daughter, for an unspecified period of time, in the care of women friends who smoked pot, as did the mother, and who taught the child about "the gay life" and engaged in sexual acts with men and with each other in the child's presence); L. v. D., 630 S.W.2d 240 (Mo. Ct. App. 1982) (where the father's witness claimed that the mother had representations of nude women on a poster, a shower curtain and a toilet seat cover in her bathroom, that on one occasion she had taken her younger children to a gay church to their apparent discomfort, that she walked about her home partially clad, as did her lover, that she was sometimes visible to her children while lying in bed with her lover and that she defended her lifestyle to her children); M.J.P. v. J.G.P., 640 P.2d 966 (Okla. 1982) (where the charges were that the mother and her lover slept in the same room with the involved two and a half year old boy, separated from him only by a screen, and that the mother and her lover sometimes held hands, kissed and touched in the infant's presence); Schuster v. Schuster, 90 Wash. 2d 626, 585 P.2d 139 (1978) (in which the mother won but in which the dissent noted with disapproval that she had given interviews and lectures and made a film about her relationship and that she had involved her children in some of these activities).

84. In Peterson v. Peterson, Civ. Act. No. D-66634 (Col. D. Ct. 1978) where the mother won, the opponent's allegation was that, in the last stages of her marriage she had drunk to excess and on one occasion had attempted suicide by ingesting alcohol and valium. In Chaffin v. Frye, 45 Cal. App. 3d 39, 119 Cal. Rptr. 22 (1975), it was charged that the mother had a record of petty crimes, a spotty work history and a work related physical disability that caused her to black out periodically. It was also charged that her lover had several times slapped her younger child. In D.H. v. J.H., 418 N.E.2d 286 (Ind. Ct. App. 1981), the father alleged that the mother was a poor housekeeper and that during the marriage he had sometimes been required to see to his children's means and supervise their school work. In *In re* Jane B., 85 Misc. 2d 515, 380 N.Y.S.2d 848 (1976), the child's maternal grandmother signed a pretrial statement to the effect that her granddaughter was abused and neglected in the mother's custody. However, at trial, she recanted, saying that the statement had been given under stress, with the result that only the mother's sexuality, no doubt augmented by the aura of the grandmother's withdrawn averments, remained as a viable issue. In Roberts v. Roberts, 25 N.C. App. 198, 212 S.E.2d 410 (1975), the involved child's maternal aunt and uncle alleged that their niece, whose custody they sought, had been born out of wedlock and also that the child had a hearing defect for which she needed special attention that the mother was not providing. In Dailey v. Dailey, 635 S.W.2d 391 (Tenn. Ct. App. 1982), it was charged that the mother's lover was imposing unduly harsh discipline on the subject child.

where bad conduct proofs were noted, such charges were given inordinate weight in the presence of the sexuality issue with which decision makers were primarily preoccupied.

In sum, it is fair to conclude that a lesbian mother who is portrayed by her opponent, her sexuality aside, as a good parent probably has a somewhat better threshhold chance than does of mother who is portrayed as a bad parent, but it is also clear that her good parent status provides no guarantee that she will prevail. For decision makers who embrace traditional condemnations of homosexuality, a lesbian mother's custodian disentitlement may be, and may by law be permitted to be, a self proving proposition.

© JEB (Joan E. Biren) 1985

4. *Investigators and Experts (Chart IV)*

a. *Best Interests Investigators*

The function of court appointed investigators is to look into the home situations, family and social relationships, health, educational progress and custodial preferences of children who are the subjects of custody disputes and, when possible, to make dispositional recommendations to custody judges.[85] Because these investigators are lay people, it is expectable that their perceptions and judgments will reflect traditional attitudes about parents and parenting. It is interesting, therefore, that of the six decisions that alluded to best interests findings, only one reported an unequivocal recommendation against the mother.[86] In three cases where mothers won, best interests reports were favorable;[87] in one losing case the investigator reported favorably on the households of both the mother and her opponent;[88] and in one losing case the investigator declined to make a recommendation solely because of a lack of knowledge about homosexuality.[89] Here again, at least so far as these data reveal, the cases thus tend to confirm that the households and parenting patterns of lesbian mothers, as seen through the eyes of community representatives, are similar to the households and parenting patterns of divorced or divorcing heterosexual mothers.

b. *Experts*

Mental health professionals played a decisional role in eighteen of thirty cases, in seven of which their opinions about the effects on children of living with a lesbian mother were in conflict. The themes that ran through expert testimony offered against mothers were roughly these: living in a lesbian household is per se detrimental to children;[90] and/or, children who live with lesbian mothers may adopt their sexual behavior and are in danger of suffering gender confusion;[91] and/or, children who live with lesbian mothers may suffer damaging social stigma to which they should not, in fairness, be subjected.[92] Experts who testified for mothers universally rejected per se formulations, generally rejected the notion that homosexuality is learned by imitation and urged that the court judge children's well-being by current status rather than fears for the future.

It is apparent from these decisions that, despite the revised positions of their professional organizations, mental health professionals are less than fully unified about homosexuality and its consequences for children. Indeed, it is fair to say that their debate, as framed by their conflicting testimony, mirrors rather than illuminates the debate in which courts are presently engaged. For those who oppose the granting of custody to lesbian mothers, the formula seems to be that homosexuality is wrong, that homosexual outcomes are by definition undesirable and that children must be protected from parents who may "make" them homosexuals or cause them to be the objects of social disapproval inflicted by a society that knows (rightly) that homosexuality is wrong. Those who support the custodial rights of lesbian mothers question all of those premises. On balance, at least as revealed by these cases, experts who support the custodial rights of lesbian mothers appear to outnumber those who oppose such rights.

---

85. *See, e.g.*, N.J. Ct. R. 5:3-3.

86. *See* Chaffin v. Frye, 45 Cal. App. 3d 39, 119 Cal. Rptr. 22 (1975), (where the mother's opponents, maternal grandparents, had preexisting custody of the subject children).

87. Gay v. Gay, 149 Ga. App. 173, 253 S.E.2d 846 (1979); Unglaub v. Unglaub, No. 62380 (Minn. Dist. Ct. June 12, 1978); Belmont v. Belmont, No. M-16937-74 (N.J. Super., Ch. Div. July 22, 1989).

88. *In re* Jane B., 85 Misc. 2d 515, 380 N.Y.S. 2d 848 (Sup. Ct. 1976).

89. Jacobson v. Jacobson, 314 N.W. 2d 78 (N.D. 1981).

90. Peterson v. Peterson, No. D-66634 (Colo. Dist. Ct. Apr. 26, 1978); Doe v. Doe, 16 Mass. App. Ct. 499, 452 N.E. 2d 293 (1983); L. v. D., 630 S.W. 2d 240 (Mo. Ct. App. 1982).

91. S. v. S., 608 S.W.2d 64 (Ky. Ct. App. 1980), *cert. denied*, 451 U.S. 911 (1981); *In re* Jane B., 85 Misc. 2d 515, 380 N.Y.S. 8484 (Sup. Ct. 1976); Dailey v. Dailey, 635 S.W. 2d 391 (Tenn. Ct. App. 1981), *permission to appeal denied* 1982; Schneider v. Schneider, No. 77-1622 (Mich. Ct. App. Oct. 19, 1978).

92. S. v. S., 608 S.W.2d at 66; Dailey v. Dailey, 635 S.W.2d at 394.

IV: INVESTIGATORS AND EXPERTS

	EXPERT TESTIMONY NOTED		BEST INTERESTS REPORT NOTED	
WINNING CASES (13) (ARRANGED BY JURISDICTION)	For Mother	For Mother's Opponent	For Mother	For Mother's Opponent
Gerdes v. Gerdes (Arizona, 1979)				
Peterson v. Peterson (Colorado, 1978)	■	■		
Gay v. Gay (Georgia, 1979)			■	
Buck v. Buck (Georgia, 1977)				
Whitehead v. Black (Maine, 1976)	■			
Doe v. Doe (Massachusetts, 1983)	■	■		
Schneider v. Schneider (Michigan, 1978)	Unclear who expert was for			
Unglaub v. Unglaub (Minnesota, 1978)			■	
M.P. v. S.P. (New Jersey, 1979)	■			
Belmont v. Belmont (New Jersey, 1980)	■		■	
Spence v. Durham (No.Carolina, 1973)	■			
Schuster v. Schuster and Isaacson v. Isaacson (Washington, 1978)	■			
In re L.J.R. v. J.A.R. (Wisconsin, 1976)	Unclear who expert was for–probably opponent			
LOSING CASES (16) (ARRANGED BY JURISDICTION)				
Chaffin v. Frye (California, 1975)				■
Mueller v. Mueller (Colorado, 1981)	■	■		
Bennett v. Clemens (Georgia, 1973)	■			
D.H. v. J.H. (Indiana, 1981)				
S. v. S. (Kentucky, 1980)	■	■		
Adams v. Adams (Lousiana, 1978)				
Hall v. Hall (Michigan, 1980)				
N.K.M. v. L.E.M. (Missouri, 1980)	■			
L. v. D (Missouri, 1982)	■	■		
In re Jane B. (New York, 1976)	■	■	■	■
DiStefano v. DiStefano (New York, 1978)	Not clear who expert testified for			
Roberts v. Roberts (No.Carolina, 1975)				
Jacobson v. Jacobson (No.Dakota, 1981)			Investigator favored neither	
M.J.P. v. J.G.P. (Oklahoma, 1982)	■			
Ashling v. Ashling (Oregon, 1969)				
Dailey v. Dailey (Tennessee, 1981)	■	■		

V: THE LAW AND THE OUTCOMES

	Anti-Gay Law on state's books	Gays held per se unfit	Constitutionality Claims		Nexus standard use claims			Harm to children found by court			
WINNING CASES (13) (ARRANGED BY JURISDICTION)			Raised by mother	Honored by court	Present Harm	Poss. future harm	None stated	Social	Psych.	Moral	None stated
Gerdes v. Gerdes (Arizona, 1979)	■				■						■
Peterson v. Peterson (Colorado, 1978)					■						■
Gay v. Gay (Georgia, 1979)	■						■				■
Buck v. Buck (Georgia, 1977)	■						■				■
Whitehead v. Black (Maine, 1976)			■		■						■
Doe v. Doe (Massachusetts, 1983)	■				■						■
Schneider v. Schneider (Michigan, 1978)	■					■					■
Unglaub v. Unglaub (Minnesota, 1978)	■				■						■
M.P. v. S.P. (New Jersey, 1979)					■						■
Belmont v. Belmont (New Jersey, 1980)					■						■
Spence v. Durham (No.Carolina, 1973)	■						■				■
Schuster v. Schuster and Isaacson v. Isaacson (Washington, 1978)			■				■				■
In re L.J.R. v. J.A.R. (Wisconsin, 1976)	■				■						■
LOSING CASES (16) (ARRANGED BY JURISDICTION)											
Chaffin v. Frye (California, 1975)	■		■			■			■	■	
Mueller v. Mueller (Colorado, 1981)						■			■		
Bennett v. Clemens (Georgia, 1973)	■		■				■			■	
D.H. v. J.H. (Indiana, 1981)							■			■	
S. v. S. (Kentucky, 1980)	■					■		■	■		
Adams v. Adams (Lousiana, 1978)	■						■				■
Hall v. Hall (Michigan, 1980)	■						■				■
N.K.M. v. L.E.M. (Missouri, 1980)	■		■			■					■
L. v. D. (Missouri, 1982)	■		■			■			■	■	
In re Jane B. (New York, 1976)	■		■				■		■		
DiStefano v. DiStefano (New York, 1978)	■					■		■	■		
Roberts v. Roberts (No.Carolina, 1975)	■		■				■				■
Jacobson v. Jacobson (No.Dakota, 1981)						■		■			
M.J.P. v. J.G.P. (Oklahoma, 1982)	■						■		■	■	
Ashling v. Ashling (Oregon, 1969)			■				■				■
Dailey v. Dailey (Tennessee, 1981)	■		■			■		■	■		

5. *The Law and the Outcomes (Chart V)*

This chart shows the murkiness of the legal rules by which courts decide the custodial claims of lesbian mothers. Predictably, no court based its decision on constitutional claims made by mothers, nor did any court conclude that a mother was unfit because of her sexual preference or activity despite the fact that there were anti-gay proscriptions on the books of roughly two-thirds of the jurisdictions where these cases were decided. Indeed, in ten cases where mothers lost, courts stated explicitly or by necessary implication, that lesbianism did not amount, per se, to unfitness.[93]

Decisions favoring mothers were distinguishable from those that disfavored them, not by differing approaches to the formal legal status of homosexuals, but by differing levels of fear about homosexuality. In at least half of the cases where mothers lost, their lesbian sexuality was overtly viewed as presumptive proof that future harm, social, psychological or moral, would befall their children; and such presumed future harm, whether standing alone as in eight of the losing cases[94] or in concert with bad conduct allegations, was deemed sufficient to deprive them of custody. By contrast, in twelve of the thirteen cases where mothers won, courts, overtly or by necessary implication, required proof of specific, present harm, attributable to the mother's sexual conduct, to deprive her of custody. In important respects, these differing standards of proof parallel the differing in-court views of mental health professionals. Basic attitudes about the "wrongness" of homosexuality and the "rightness" of negative social attitudes toward homosexuals inform judicial responses in much the same fashion as they inform the in-court views of experts, albeit in differing ratios. The critical difference between the two groups lies, of course, in the power of courts to take children from mothers. Clearly those who justify such a taking solely on the basis of presumed future harm, but who at the same time assert that lesbian mothers are not per se unfit, engage in double speak.

6. *Summary of the Cases*

Lesbian mothers have only recently surfaced as custody contenders. Their emergence forces decision makers to grapple with ancient taboos. Case outcomes are a function of the degree to which these taboos have been internalized by decision makers. The legal framework allows for this. In the exercise of judicial discretion, judges may treat lesbian sexuality as an impenetrable barrier to custody. In courts that purport not to invoke an irrebuttable presumption against them, lesbian mothers may bear the impossible burden of establishing that their children will not be "exposed" to lesbians and gay males[95] or involved in the struggles of lesbians and gay males for legal rights;[96] will never witness displays of ordinary

93. Mueller v. Mueller, No. 79 DR 1246 (Colo. Dist. Ct. Mar. 27, 1980); D.H. v. J.H., 418 N.E.2d 286 (Ind. Ct. App. 1981); Hall v. Hall, 95 Mich. App. 614, 291 N.W.2d 143 (1980); L. v. D., 630 S.W.2d 240 (Mo. Ct. App. 1982); N.K.M. v. L.E.M., 606 S.W.2d 179 (Mo. Ct. App. 1980); DiStefano v. DiStefano, 60 A.D.2d 976, 401 N.Y.S.2d 636 (1978); *In re* Jane B., 85 Misc. 2d 515, 380 N.Y.S.2d 848 (Sup. Ct. 1976); Roberts v. Roberts 25 N.C. App. 198, 212 S.E.2d 410 (1975).

94. Mueller v. Mueller, No. 79 DR 1246 at 4-5; S. v. S., 608 S.W.2d 64 (Ky. Ct. App. 1980), *cert. denied*, 451 U.S. 911 (1981); Adams v. Adams, 357 So.2d 881 (La. Ct. App.), *cert. denied*, 359 So.2d 1309 (1978); Hall v. Hall, 95 Mich. App. at 144; N.K.M. v. L.E.M., 606 S.W.2d at 183; DiStefano v. DiStefano, 60 A.D.2d at 977; Jacobson v. Jacobson, 314 N.W.2d at 79; Ashling v. Ashling, 42 Or. App. 47, 599 P.2d 475 (1979).

95. *See, e.g.*, N.K.M. v. L.E.M., 606 S.W.2d 179 (Mo. Ct. App. 1980); *see also* Dailey v. Dailey, 635 S.W.2d 391 (Tenn. Ct. App. 1981); DiStefano v. DiStefano, 60 A.D.2d 976, 401 N.Y.S.2d 636 (1978) (mothers lost custody because of lesbian relationships and were also ordered not to allow their lovers to have contact with their children during visits). In *N.K.M.*, upon divorce, the mother received custody of the pre-adolescent daughter of the marriage upon the conditions (1) that she discontinue all association with a fellow-employee whom her husband characterized as her lesbian lover (an allegation that she denied); (2) that she never allow her daughter to be in her purported lover's presence; and (3) that she not live with any female persons other than members of her family. Subsequently, her ex-husband moved to take custody from her, asserting, *inter alia*, that the first two conditions had been violated. The daughter, then ten years old, expressed a strong preference to remain with the mother. A psychiatrist testified that the child was well adjusted, had appropriate friends, was doing well in school and should be permitted to remain in the mother's custody. Both the trial and appellate courts disagreed, the appellate court concluding that the second condition, that the purported lover be kept away from the child, was essential to the child's welfare.

96. *See, e.g.*, Schuster v. Schuster, 90 Wash. 2d 626, 585 P.2d 130 (1978), a case the mothers won, wherein the dissent lamented thus:

> In the instant case, Sandra Schuster and Madeleine Isaacson were and are living together in a homosexual relationship. The respondents are living together with their children as a family unit.
>
> The respondents have been engaged in publicizing the homosexual cause in general and their lesbian relationshp. They have given a series of lectures and granted interviews where they discussed their own homosexual lifestyle. The

affection between mothers and their lovers[97]; will be protected from stigmatization in a society that routinely ostracizes lesbians and gay males;[98] will grow up heterosexual;[99] and will adopt the time-encrusted belief that homosexuality is immoral.[100] When such assurances are required, the opportunity to defend is obviously a sham.

Even where negative decisions purport to rely upon proofs of specific, completed harm, the harm found may be rooted in homophobic perceptions. *Bennett v. Clemens* provides a classic example. There a majority of the appellate court, affirmed the trial court's transfer of custody from mother to paternal grandparents and concluded:

> The evidence is voluminous and, while conflicting in some aspects, authorized the findings that the mother, after divorcing her husband, moved to Atlanta and lived for several years in different locations with the child, worked for a so-called "underground" newspaper, that she had been arrested for possession of marijuana and on a plea of nolo contendere was given a two year sentence on probation, that in the fall of 1970 she left Atlanta to attend the funeral of her mother in Delaware, leaving the child with 4 female friends one of whom was a defendant, and thereafter went to San Francisco where she has since lived while pursuing her vocation

---

> children have accompanied respondents at some of these engagements, and the respondents and their children participated in making a movie which depicts the lifestyle of two families bound together by homosexual parents.
>
> They have advertised in a brochure entitled "The Gay Family A Valid Life-Style?" in which they offered interested persons a booklet, "Love is for All", and information about a film entitled "Sandy and Madeline's Family", and also offer to make personal appearances. An article in the San Francisco Chronicle with the headline "The Lesbian Love of Two Mothers" explained the appearance of the two women visiting in the Bay Area publicizing their film.
>
> From such publicizing it can be readily seen that they are not content to pursue their lifestyle but are also using their children for the purposes of advocating and proselytizing that style.

90 Wash.2d 626, 634, 585 P.2d 130, 134.

97. *See, e.g.*, Ashling v. Ashling, 42 Or. App. 47, 599 P.2d 475 (1979) where the appellate court, noted that "[t]he mother has . . . behaved affectionately toward . . . women in front of the children, although she insists such activity is merely friendly, not passionate or sexual" and affirmed the order of the trial court that during visitations of her children in her home, the mother may "not have with her, in her home, or around the children any lesbians." 42 Or. App. 57, 111, 599 P.2d 475, 476.

98. The opinions in Mueller v. Mueller, No. 79 DR 1246 (Colo. Ct. App. Nov. 19, 1981); S. v. S., 608 S.W.2d 64 (Ky. Ct. App. 1980); Jacobson v. Jacobson, 314 N.W.2d 78 (N.D. 1981); M.J.P. v. J.G.P., 640 P.2d 966 (Okla. 1982); Dailey v. Dailey, 635 S.W.2d 391 (Tenn. Ct. App. 1981), all reflect judicial preoccupation with social stigmatization as a predictable consequence of the mothers' lesbian relationships; *cf.* M.P. v. S.P., 169 N.J. Super. 425, 404 A.d 1256 (App. Div. 1979), where the mother won. In *M.P.*, reversing a trial level decision against the mother, the appellate court said:

> [W]ithin the context of a loving and supportive relationship, there is no reason to think that the girls will be unable to manage whatever anxieties may flow from the community's disapproval of their mother.
>
> . . .
>
> If, defendant retains custody, it may be that because the community is intolerant of her differences these girls may sometimes have to bear themselves with greater than ordinary fortitude. But this does not necessarily portend that their moral welfare or safety will be jeopardized. It is just as reasonable to expect that they will emerge better equipped to search out their own standards of right and wrong, better able to perceive that the majority is not always correct in its moral judgments, and better able to understand the importance of conforming their beliefs to the requirements of reason and tested knowledge, not the constraints of currently popular sentiment or prejudice.
>
> Taking the children from defendant can be done only at the cost of sacrificing those very qualities they will find most sustaining in meeting the challenges inevitably ahead. Instead of forbearance and feelings of protectiveness, it will foster in them a sense of shame for their mother. Instead of courage and the precept that people of integrity do not shrink from bigots, it counsels the easy option of shirking difficult problems and following the course of expedience. Lastly, it diminishes their regard for the rule of human behavior, everywhere accepted, that we do not forsake those to whom we are indebted for love and nurture merely because they are held in low esteem by others.
>
> We conclude that the children's best interest will be disserved by undermining in this way their growth as mature and principled adults. Extensive evidence in the record upon which we have not commented amply confirms the trial judge's finding that defendant is a worthy mother. Nothing suggests that her homosexual preference in itself presents any threat of harm to her daughters or that in the ordinary course of their development they will be unable to deal with whatever vexation may be caused to their spirits by the community.

169 N.J. Super. at 436-439, 404 A.2d at 1262-3.

99. Judicial concern that children might emulate parental sexual behavior and emerge, in adulthood, as homosexuals was expressed, explicitly or implicitly, in Muller v. Mueller, No. 79 DR 1246 (Colo. Ct. App. Nov. 19, 1981); S. v. S., 608 S.W.2d 64 (Ky. Ct. App. 1908); *In re* Jane B., 85 Misc. 2d 515, 380 N.Y.S. 2d 848 (Sup. Ct. 1976); and Dailey v. Dailey, 635 S.W.2d 391 (Tenn. Ct. App. 1981), in all of which the mother lost custody.

100. *See, e.g.*, Chaffin v. Frye, 45 Cal. App. 3d 39, 119 Cal. Rptr. 22 (1975); Hall v. Hall, 95 Mich. App. 614, 291 N.W.2d 143 (1980); M.J.P. v. J.G.P., 640 P.2d 966 (Okla. 1982). The rather clear implication of all of these decisions against mothers was that, because homosexuality is or is regarded as immoral, it is not in a child's best interests to live with a homosexual parent (Thereby internalizing the parent's immorality? The parent's view that homosexuality is not immoral?).

> as a writer and poet, that the female friends with whom she left the child smoked pot on occasions, engaged in sexual acts with men and with each other in the presence of the child and otherwise taught the child about "the gay life."[101]

The dissent, on the other hand, found that:

> .... the natural parents are both graduates of Harvard where they were National Merit Scholars. They were divorced several years ago and the mother was awarded permanent custody of the child.
>
> There is no evidence in the record that the child's physical needs were not being cared for. There is no evidence in the record that the child was neglected, abused, or mistreated in any way. There was evidence in the record that the child attended school and church regularly.
>
> Dr. Lila Bonner Miller, an Internist and Psychiatrist, after examining the child and those with whom she lived, testified: "... She is a great little girl and whatever happened to her is really pretty fine. I don't think it is one person. I think it is a lot of people loving her. This child has felt love, there is no doubt about that." Dr. Miller further testified that in her opinion the child had been living in a healthy environment.
>
> The natural mother testified that she wanted to retain custody of the child and was capable of properly caring for the child.
>
> The natural father testified that he wanted his former wife to have custody of the child, that if she could not have custody then he wanted custody of the child, but that he did not want his 65-year-old parents to have custody of the child.[102]

These judges reviewed the same trial record. What accounted for the extraordinarily sharp divergence of their views about its meaning? In all likelihood, the explanation lies in their differing assumptions about lesbian behavior. Judges, like others, see what they expect to see. If, at a visceral level, they embrace condemnatory images of homosexuals, allegations of debauched conduct become entirely credible. If they do not, such allegations are scrutinized critically.

Nor does an articulated proof-of-present-harm requirement offer reliable protection, for despite the apparent hospitality of that standard, judges remain free to find the requisite harm in children's exposure to a homosexual environment.[103] It is, of course, encouraging that, in almost half of these cases, sometimes with the support of best interests investigators or mental health professionals, judges declined to do so; but these outcomes are by no means a binding instruction to oust homophobia from the decision-making process. Ultimately, that instruction must come from the Constitution. The final section of this paper explores the route to constitutional rights.

IV. PROSPECTS

Although lesbian mothers prevailed in roughly forty-five percent of the cases examined in this article, they have yet to secure any rights of sexual choice that custody courts "are bound to respect."[104] Rather, success in these cases appears to be an unguided response to the current gay rights initiative of which the demand of lesbian mothers for custody of their children is a part. That initiative, undertaken in the late

101. 230 Ga. 317, 319, 196 S.E.2d 842, 843.

102. *Id.* at 320-21, 196 S.E.2d at 843-44.

103. *See, e.g.*, Mueller v. Mueller, No. 79 DR 1246 (Colo. Ct. App. Nov. 19, 1981) (mother loses custody of her nine and four year old children because her lover sometimes stays overnight in the maternal household); Adams v. Adams, 357 So.2d 881 (La. App. 1978) (evidence that, during marriage, mother had lesbian relationship is grounds, in divorce action, to award custody to father); Hall v. Hall, 95 Mich. App. 614, 291 N.W.2d 143 (1980) (mother's commitment to her lesbian relationship, together with mother's undetailed "lifestyle" held sufficient to deprive her of custody); Ashling v. Ashling, 42 Or. App. 47, 599 P.2d 475 (1979) (fact that mother had sexual relations with her lover in privacy of her own bedroom out of her children's presence but while they were at home, together with affectionate conduct toward other women in her home held sufficient to deprive her of custody); N.K.M. v. L.E.M., 606 S.W.2d 179 (Mo. Ct. App. 1980) (mother's violation of court order to keep lesbian lover from contact with children held sufficient to derpive her of custody); DiStefano v. DiStefano, 60 A.D.2d 976, 401 N.Y.S.2d 636 (1978) (mother's failure to keep lesbian lover out of presence of children held sufficient to deprive her of custody); S. v. S., 608 S.W.2d 64 (Ky. Ct. App. 1980); Jacobson v. Jacobson, 314 N.W.2d 78 (N.D. 1981); M.J.P. v. J.G.P., 640 P.2d 966 (Okla. 1982) (fact that mothers had live-in lesbian mates held sufficient to deprive them of custody of their children).

104. Dred Scott v. Sanford, 60 U.S. (19 How.) 393 (1857).

1960's at enormous risk to liberty, safety, employment, community reputation and personal relationships, was an open and organized challenge to legalized homophobia, and it has borne some fruit. Realities have begun to displace stereotypes as homosexual men and women surface in increasing numbers, and the perceptions of some judges have changed. In the courtrooms of those judges, lesbian mothers can sometimes win. But the law does not demand such hospitality, and episodic victories have little precedential value in other cases before other judges. Until sexuality *qua* sexuality is no longer a constitutionally permissible decisional criterion, the custodial rights of lesbian mothers will always depend on unfettered judicial reactions, hostile or benign, to their sexual alliances. Obviously, such scant protection is not enough.

The task of achieving rules under which sexual orientation no longer counts is, to say the least, formidable. A campaign to install such rules in discretion-laden custody courts has almost no chance of success unless it is linked to a larger campaign directed toward achieving recognition of a general right to private, adult, consensual sexual expression. That larger campaign should be a priority of the women's movement, not only because feminists need to speak out against discrete cruelties visited upon lesbians and gay males but because a right to sexual autonomy is essential to the achievement of full gender equality.

The history of women's subordination is a history of male co-option and regulation of female sexuality. Demure and closeted ideal nineteenth-century women were groomed to be delivered, virginity intact, from the houses of their father to the houses of their husband. There, under patriarchal protection, women were to be exalted servants, sexual consorts, and breeders of children but never decisionmakers in the larger community where the rules for their governance were devised.[105] Men might be forgiven, even admired by other men, for their sexual experiences outside of marriage, but women risked their futures when they engaged in non-marital sex.[106] Social conditioning taught women that the unwise use of their only bargaining chip would make them "damaged goods" and render them ineligible for wifehood,[107] the only respectable vocation to which they were given full access.[108] A complex mythology about women's weakness and need for male protection, their childlike vulnerability to temptation, their mercurial natures and their devine reproductive mission supported their social and sexual subordination to male desires and male systems of control until well into the twentieth century.[109] Indeed, the contemporary women's movement is even today struggling to reform the conditions of marriage, child bearing, child rearing and social participation imposed by the law and mythology of the past.[110] To advance that struggle, it behooves women to take deliberate aim at the sexual proscriptions through which their conduct continues to be managed by a male dominated legal system. Jacqueline Jarrett and legions of like-situated women have reason to know how those proscriptions can still be used to require women to conform to male norms of female behavior.

Earlier, I suggested that strictures against non-marital heterosexual intercourse are a barrier to the achievement of lesbian and gay rights. Here I suggest that the inverse is also true. Establishment of a right of sexual autonomy for heterosexuals would bring the Supreme Court face to face with the irrationality of anti-homosexual suppressions, a confrontation for which it is apparently not yet ready. The Court's toleration of proscriptions against all non-marital sex obviates

105. L. Kanowitz, Women and the Law, The Unfinished Revolution (1969) [hereinafter cited as Kanowitz].

106. McAllister v. McAllister, 71 Tex. 695, 10 S.W. 294 (1888) (upholding the different treatment of men and women who have been adulterous when divorce was sought).

107. *See* Kanowitz, *supra* note 106, at 96 (where the author notes: "At least 14 of the American states have at one time or another permitted a husband, but not a wife, to be granted a divorce upon proof of the other's unchaste condition at the time of marriage.") (citing U.S. Women's Bureau, Legal Status of Women: Analyses of Sex Distinctions 5 (mimeo Jan. 1, 1944); *see also* M. Mead, Male and Female 290-95 (1949) (where the author discusses twentieth century echoes of nineteenth century mores with respect to female sexuality).

108. Rothman, Woman's Proper Place (1978) [hereinafter cited as Rothman].

109. *See* Kanowitz, *supra* note 105, at 35-93 and Rothman, *supra* note 109.

110. The Quest Staff, Building Feminist Theory: Essays From Quest (1981); Chafe, *Looking Backward in Order to Look Forward*, in Women and the American Economy (J. Kreps ed. 1976).

the necessity to examine the legitimacy of legalized homophobia, but at continuing cost to women.

I argue here, not for the renunciation of consensual marital relationships, but for the premise that women are also entitled to choose non-marital modes of sexual expression without governmental interdictions that echo a past in which they were valued, judged, used and controlled as sexual and reproductive commodities. In the struggle for recognition of that right of choice, heterosexual and lesbian women are bound together by a common necessity, because the rights of each are pawn to the withheld rights of the other.

Can a right of sexual autonomy, such that women's sexuality and sexual orientation will no longer disable them in custody litigation, be achieved? History teaches that it can, but only through that arduous political process by which constitutional rights are always won by disfavored groups. *Palmore v. Sidoti*,[111] decided by the Supreme Court in the Spring 1984 term, is the paradigm. Here is the High Court's recitation of the state proceedings in that case:

> When petitioner Linda Sidoti Palmore and respondent Anthony J. Sidoti, both Caucasians, were divorced in May 1980 in Florida, the mother was awarded custody of their three-year-old daughter.
>
> In September 1981 the father sought custody of the child by filing a petition to modify the prior judgment because of changed conditions. The change was that the child's mother was then cohabiting with a Negro, Clarence Palmore, Jr., whom she married two months later. Additionally, the father made several allegations of instances in which the mother had not properly cared for the child.
>
> After hearing testimony from both parties and considering a court counselor's investigative report, the [Florida] court noted that the father had made allegations about the child's care, but the court made no findings with respect to these allegations. On the contrary, the court made a finding that "there is no issue as to either party's devotion to the child, adequacy of housing facilities, or respect[a]bility of the new spouse of either parent." App. to Pet. for Cert. 24.
>
> The court then addressed the recommendations of the court counselor, who had made an earlier report "in [another] case coming out of this circuit also involving the social consequences of an interracial marriage. *Niles v. Niles*, 299 S.2d 162." *Id.*, at 25. From this vague reference to that earlier case, the court turned to the present case and noted the counselor's recommendation for a change in custody because "[t]he wife [petitioner] has chosen for herself and for her child, a life-style unacceptable to her father and to society. . . . The child . . . is, or at school age will be, subject to environmental pressures not of choice," Record 84 (emphasis added).
>
> The [Florida] court then concluded that the best interests of the child would be served by awarding custody to the father. The court's rationale is contained in the following:
>
>> The father's evident resentment of the mother's choice of a black partner is not sufficient to wrest custody from the mother. It is of some significance, however, that the mother did see fit to bring a man into her home and carry on a sexual relationship with him without being married to him. Such action tended to place gratification of her own desires ahead of her concern for the child's future welfare *This Court feels that despite the strides that have been made in bettering relations between the races in this country, it is inevitable that Melanie will, if allowed to remain in her present situation and attains school age and thus more vulnerable to peer pressures, suffer from the social stigmatization that is sure to come.* App. to Pet. for Cert. 26-27 (emphasis added).
>
> The Second District Court of Appeals affirmed without opinion, 426 So. 2d 34, thus denying the Florida Supreme Court jurisdiction to review the case. *See* Fla. Const., Art. V, § 3(b)(3) (West Supp. 1983); *Jenkins v. State*, 385 So. 2d

111. 104 S. Ct. 1879 (1984).

> 1356 (Fla. 1980). We granted certiorari, — U.S. —, 104 S. Ct. 271, 78 L. Ed.2d 253 (1983), and we reverse.[112]

The reasoning of the Florida court about the need to take children from mothers to protect them from social disapproval of maternal sexual relationships has a familiar ring. It is the very reasoning invoked to preclude the custodial rights of lesbian mothers. In *Palmore*, the Supreme Court's unanimous response was to declare that reasoning violative of the Constitution. It said,

> The State, of course, has a duty of the highest order to protect the interests of minor children, particularly those of tender years. In common with most states, Florida law mandates that custody determinations be made in the best interests of the children involved. Fla. Stat. § 61.12(2)(b)(1) (1983). The goal of granting custody based on the best interests of the child is indisputably a substantial governmental interest for purposes of the Equal Protection Clause.
>
> It would ignore reality to suggest that racial and ethnic prejudices do not exist or that all manifestations of those prejudices have been eliminated. There is a risk that a child living with a stepparent of a different race may be subject to a variety of pressures and stresses not present if the child were living with parents of the same racial or ethnic origin.
>
> The question, however, is whether the reality of private biases and the possible injury they might inflict are permissible considerations for removal of an infant child from the custody of its natural mother. We have little difficulty in concluding that they are not. The Constitution cannot control such prejudices but neither can it tolerate them. Private biases may be outside the reach of the law, but the law cannot, directly or indirectly, give them effect. . . .[113]

Thirty years ago, a Supreme Court decision expelling racial criteria from custody courts would not have been attainable. It has finally come, not as the product of sudden enlightment, but as the result of a long and encompassing struggle, conducted in courts, in legislatures and in the streets of the nation, to dismantle an entire system of racial oppression.[114] An important component of that system was the regulation by white males of the sexual lives of black men and women.[115] *Palmore* is a formal announcement that white males are no longer in charge of black sexual choices.

Women's right of sexual choice can come by the same route, but only if heterosexual women join lesbians and gay men in their quest for full scale, across-the-board constitutional protections, as large numbers of whites ultimately joined the small and embattled ranks of black civil rights activists whose freedom campaign made *Palmore* a constitutional reality. For their own more complete liberation, it is in the long range interests of heterosexual women to do so.

112. *Id.* at 1880-81.
113. *Id.* at 1881.
114. *Id.* at 1880-81.
115. *Id.* at 1881.

WOMEN AS FATHERS: Motherhood and Child Care Under a Modified Patriarchy

BARBARA KATZ ROTHMAN
Baruch College and the
Graduate Center, City University of New York

Although the modern American kinship system is nominally a bilateral system, the very definition of kin ties is based on the principles of patriarchy. Women do not gain their rights to their children in this society as mothers, but as father-equivalents, sources of genetic material. In child rearing as in childbearing, women may take on the role of fathers to their children, substituting poorer women to do the traditional mothering work. The resultant recasting of the classic Oedipal drama in terms of class and race rather than gender is described. This article argues that we must move beyond patriarchal definitions of the relations between parents and children, and beyond modifications that simply extend to some women, most often at cost to other women, some of the privileges of patriarchy.

Patriarchal kinship is the core of patriarchy: Paternity is the central social relationship. A very clear statement of patriarchal kinship is found in the book of Genesis, in the "begats." Each man, from Adam onward, is described as having "begot a son in his likeness, after his image." After the birth of this first-born son, the men are described as having lived so many years and begot sons and daughters. The text

---

AUTHOR'S NOTE: *Portions of this article appear in* Recreating Motherhood: Ideology and Technology in a Patriarchal Society *(W. W. Norton, 1989). I want to thank Maren Lockwood Carden, Betty Leyerle, Eileen Moran, Wendy Simonds, and Mary White, for their helpful discussions of these issues, and Mary Cunnane and Judith Lorber, for both substantive and editorial assistance.*

REPRINT REQUESTS: *Barbara Katz Rothman, Department of Sociology, Baruch College, CUNY, 17 Lexington Ave., New York, NY 10010.*

GENDER & SOCIETY, Vol. 3 No. 1, March 1989 89-104
© 1989 Sociologists for Women in Society

then turns to that first-born son and, in turn, his first-born son. Women appear as the "daughters of men who bore them offspring." In a patriarchal kinship system, children are reckoned as being born to men, out of women. Women, in this system, bear the children of men.

The essential concept here is the *seed,* the part of men that grows into the children of their likeness within the bodies of women. Such a system is inevitably male dominated, but it is a particular kind of male domination. Men control women as daughters, much as they control their sons, but they also control women as the mothers of men's children. It is women's motherhood in particular that men have to control to maintain patriarchy. In a patriarchy, because what is valued is the relationship of a man to his sons, women are a vulnerability that men have: To beget these sons, men must pass their seed through the body of a woman.

In matrilineal societies, where a shared mother produces the lineage or family group, men may still rule the women and children who are related to them through their mother's line, but they do not rule as fathers. Women in such a system are not a vulnerability, but a source of connection. As anthropologist Glenn Petersen says, in a matrilineal system "women, rather than infiltrating and subverting patrilines, are acknowledged to produce and reproduce the body of society itself" (1982, p. 141).

The modern American kinship system is not classically patriarchal, though remnants of patriarchy can be seen in such things as the strong pressure to give children their father's surname. Modern American kinship is a bilateral system, in that individuals are considered to be equally related to their mother's and their father's kin. Yet the very definition of kin is based on the principles of patriarchy. A patriarchal kinship tie is, by definition, a connection by seed. Kinship through fathers is based on the act of impregnating and therefore glorifies genetic connections. In a mother-based system, a "blood" relative does not mean simply a genetic relative; the blood tie is the mingled blood of mothers and their children; children grow out of the blood of their mothers, of their bodies and being. When the shared bond of kinship comes through mothers, the tie is based on *the growing of children,* a social connection.

Each type of kinship system leads to different ideas about kin identity and what a person is. In a mother-based system, a person is what mothers grow—people are made of the care and nurturance that brings a baby forth into the world and turns the baby into a member of the society, intimately related to those who had the same care and nurturance. In a patriarchal system, a person is what grows out of men's seed. The belief that the essence of the person is there in the seed when it is planted in the mother still persists as the various forms of genetic determinism. And of course remnants of the classical patriarchal valuing of men's seed also persist. In 1987, the director of a California sperm bank distributed T-shirts with a drawing of sperm swimming on a blue background and the words "Future People." Using the principles of a patriarchal system, shared identity—or kinship—comes from shared genes.

THE SEEDS OF WOMEN

Modern procreative technology has been forced to go beyond the sperm as seed. Modern science has had to confront the *egg* as seed also. Modern scientific thinking cannot possibly hold onto old notions of women as nurturers of men's seeds. The doctor who has spent time "harvesting" eggs from women's bodies for *in vitro* fertilization fully understands the significance of women's seed.

The old patriarchal kinship system had a clear place for women: the nurturers of men's seeds, the soil in which seeds grow, the daughters of men who bear them offspring. When forced to acknowledge that women's genetic contribution is equal to men's genetic contribution, Western patriarchy could have foundered. *But the central concept of patriarchy, the importance of the seed, was kept by extending it to women.* Valuing the seed of women, the genetic material women also have, extends to women some of the privileges of patriarchy. That is, when the significance of woman's seed is acknowledged in her relationship with her children, women too have paternity rights in their children.

In this modified system based on the older ideology of patriarchy, women can be seen to own their children, just as men do. Unlike what happens in a mother-based system, this relationship between a woman and her children is not based on motherhood *per se*, not on the unique nurturance, the long months of pregnancy, the intimate connections with the baby as it grows and moves inside her body, passes through her genitals, and sucks at her breasts. Instead, women are said to own their babies, have "rights" to them, just as men do: based on their seed.

This modification does not end patriarchy, and it does not end the domination of the children of women by men. Instead, by maintaining the centrality of the seed, the ideology maintains the rights of men in their children, even as it recognizes something approaching equal rights of women in their children. Since men's control over women and the children of women is no longer based simply on their (no longer) unique seed, their economic superiority and the other privileges of a male-dominated social system become increasingly important. Children are, based on the seed, presumptively "half his, half hers"—and might as well have grown in the backyard. Women do not gain their rights to their children in this society as *mothers*, but as *father-equivalents*, sources of seed.

And if the nurturance of mothers is discounted, so too is the nurturance of fathers. There are men who mother, who intensively, actively, intimately nurture children, as prime parents in families in which the mother's role is reduced (Pruett 1987), or as single fathers (Risman 1987), or in families in which parenting is shared equally (Ehrensaft 1987). In our society their fatherhood has no more legal weight than that of men whose paternity is based entirely on a single sexual encounter. The principles of a patriarchal kinship system denigrate all nurturance, that of women and that of men, in favor of genetic ties.

THE GENETIC TIE

What precisely is the physical nature of this highly valued genetic connection between parent and child? The closest genetic connection a human being can have is an identical twin. Identical twins have the same chromosomes, the same genes. The next closest relations are those between parent and child and between full siblings. If an individual carries a certain gene, the chances that a sibling will carry the same gene are fifty-fifty, the same as the parent-child relationship. Genetically, "there is nothing special about the parent-offspring relationship except its close degree and a certain fundamental asymmetry. The full-sib relationship is just as close" (Hamilton 1978 p. 191).

Genetic connections exist as percentages: the 100 percent connection of identical twins, the 50 percent connection of siblings and of parents to children, the 25 percent connection of grandparents to grandchildren and of half-siblings to each other, and the 12.5 percent connection of cousins, and so on. In strictly genetic terms, your sister might as well be your mother. The genetic connection is the same.

But parenthood is not just a genetic connection, a genetic relationship. And it is more than the age difference that distinguishes the parent-child relationship: The social relationship of a 17-year-old sibling to a newborn baby is quite different from the relationship of a 17-year-old parent to a newborn baby. It may be six-of-one and a half-dozen of the other genetically, but socially these are not interchangeable relationships.

The parent-child relationship is invested with social and legal rights and claims that are not recognized, in patriarchal societies, in any other genetic relationship, because that genetic connection was the basis for men's control over the children of women. The contemporary modification of traditional patriarchy has been to recognize the genetic parenthood of women as the equivalent of the genetic parenthood of men. Genetic parenthood replaces paternity in determining who a child is, who it belongs to. I believe it is time to move beyond the patriarchal concern with genetic relationships.

We can recognize and appreciate the genetic tie without its being a determinative connection. We do so in most of our genetic relations. American society recognizes no special claims of a sibling on a sibling, or of aunts and uncles on nieces and nephews. We recognize the relationship, we allow people to make what they will of it, but it carries no legal weight. If my brother does not like the way I am raising his niece, so it goes. Even though he is closely related to her genetically, even if she were to look more like him than like me, it gives him no legal claim to her. He has no basis to challenge my claim as parent, short of showing me to be unfit—and even then, no assurance that the child would be turned over to him. And if my son doesn't like the way I am raising my daughter, even though he is just as closely related to her genetically as I am, there is no way he can challenge my custody, and that will still be true when he is over 21. But my husband, as her father, can challenge my custody, without showing me to be unfit, as I can challenge his custody. We are parents, and that gives us special legal rights over the child.

While not denying the reality of genetic relationships, the question I am asking is: Just how much weight do we need to give this one genetic tie? How much can it hold? Stripped of all the social supports, is that genetic tie sufficient to define a person, to establish a kinship? Am I my mother's daughter because of our chromosomes? If a woman donates an egg, or a man turns over to a lab technician a vial of semen, does that make the person a *parent* to the child created with those chromosomes?

PATRIARCHAL IDEOLOGY AND NEW PROCREATIVE TECHNOLOGY

The new procreative technology being developed is based on the patriarchal focus on the seed. The seed, the genetic material, is the one absolutely irreplaceable part of procreation, as science now approaches it. The procreative technology continues to substitute for one after another of the nurturing tasks, but makes no substitute for the seed. Breasts became unnecessary quite some time ago, as artificial formula substituted for human milk. The act of giving birth becomes increasingly unnecessary, as doctors work on surgical removal of babies to the point that now one of five American babies is born by cesarean section. The nurturance of late pregnancy becomes increasingly unnecessary as neonatal intensive care units develop the skill to maintain younger and younger, smaller and smaller premature babies or "extrauterine fetuses" in incubation. And the nurturing environment of the Fallopian tube becomes replaceable by the nurturance of the glass dish, the *in vitro* environment. None of these techniques of artificial nurturance works as well as the natural mothering experience, though as one or another becomes "faddish" there is a tendency for doctors to proclaim its superiority over the natural mother.

But once a substitute for a thing is possible, the thing itself loses its mystique. By their very existence, substitutes denigrate the original. What is the value of penmanship in a world with typewriters? Who needs to do long division in a world with calculators? What happened to oral history in a world that learned to write? Once a substitute, or just a "man-made" alternative is available, what's so "special" about the original? Think about how the mystical meaning of flight changed once airplanes streaked over eagles.

That is what is happening to mothering—as a physical and as a psychological experience—as we offer substitute after substitute, surrogate after surrogate. What's so *special* about motherhood?

I recently listened to a group of lawyers trying to figure out if there was anything special or unique about a mother's relationship to her fetus, compared with anyone else's relationship to that fetus. The context was the issue of prenatal torts—the ever-present legal question of who can sue whom for what. The question was: If a child can sue someone—for example, a manufacturer of a chemical—for harm that was done to that child when it was a fetus, then is there any reason that the child cannot sue the mother for harm she did to the child when it was a fetus? Is there anything that makes a mother's relationship to her fetus unique? They had a hard time thinking of anything.

We have focused on the seed, on the embryo, and then on the fetus, and reduced all of the nurturance, all of the intimacy, all of the *mothering*, to background, environmental factors. And so as we substitute for this and for that environmental factor, substitute for this or for that mothering experience, we wonder what is so special about the original, what is so unique about the mother's relationship.

The problem has been brought to a head with the new procreative technology, which forces us to confront questions about mothering and the mother-child relationship. But these are not new problems or new questions. People in our society have been substituting for aspects of mothering for a long time, always along the same patterns. Upper-class women have bought the services of lower-class women to provide one or another mothering service for their children. Or, it might more accurately be said in some circumstances, upper-class men have bought the services of lower-class women to supplement the services of their wives. Some societies have let men have mistresses while their wives mothered; some have let men hire servants to do the mothering so that their wives could be reserved for their sexual and social uses.

Wet nurses are the earliest example of such a biological substitution of one woman for another in mothering. Sometimes wet-nursing has been done in an exchange among women—two or more women occasionally nursing each other's babies, substituting for one another. "Commercializing" wet-nursing meant putting a price tag on that service. When the wet nurses were slave mammies, the price was factored into the cost of the slave. In the case of a hired wet nurse, the cost was per feeding, per hour, or per week. In these cash exchanges, the "breast-feeding relationship" was considered unimportant. Milk became a product to be bought and sold, not the basis of an intimate relationship. Once one buys the milk, the producer of the milk is reduced in status—her relationship is not a mothering relationship, not a relationship to the child, but a relationship to her product, her milk. Her value lies in the quality of her milk, not the quality of her relationship with the child. And so wet nurses were inspected, like animals, for the quality of their product.

In a patriarchal society, men use women to have their children. A man can use this woman or that woman to have *his* children. He can hire this woman or that woman to substitute for one or another aspect—biological, social, or psychological—of the mothering his child needs. From the view of the man, his seed is irreplaceable; the mothering, the nurturance, is substitutable.

And from the woman's point of view? We can use this man's sperm or that man's to have our children. With this or that man as father, our bellies will swell, life will stir, milk will flow. We may prefer one man's seed to another, just as a man may prefer one woman's nurturance to another for his child. For men, the nurturers are substitutable, interchangeable. For men, what makes the child *his* is his seed. For women, what makes the child ours is the nurturance, the work of our bodies. Wherever the sperm came from, it is in our bodies that our babies grow, and our physical presence and nurturance that makes our babies ours.

But is that inevitable? Did not some women substitute other women's bodies when they hired wet nurses? Don't some women substitute other women's arms, other women's touch, when they hire housekeepers and baby-sitters and day-care workers? And now the new procreative technology lets us cut our seeds loose from our bodies and plant them in other women's bodies. Now the seed, the egg, of one woman can be brought to term in the body of another.

Then who is the mother? The person who supplied the egg or the person in whose body the fetus grows and is nurtured? Our answer depends on where we stand when we ask the question.

When we accept the patriarchal valuing of the seed, there is no doubt—the real mother, like the real father, is the genetic parent. When we can contract for pregnancy at the now-going rate of $10,000, we can choose which women to substitute for us in the pregnancy.

For which women are these substitutes available? Who can afford to hire substitutes for the various parts of mothering? The situation today is exactly what it has been historically: women of privilege, wealthy or fairly wealthy women, hiring the services of poor, or fairly poor women. Upper-class women can have some of the privileges of patriarchy. Upper-class women can have, can buy, some of the privileges of their paternity, using the bodies of poorer women to "bear them offspring." And upper-class women can, as they so often have, be bought off with these privileges, and accept men's worldview as their own. And so we have women, right along with men, saying that what makes a child one's own is the seed, the genetic tie, the "blood." And the blood they mean is not the real blood of pregnancy and birth, not the blood of the pulsing cord, the bloody show, the blood of birth, but the metaphorical blood of the genetic tie.

This is the ultimate meaning of patriarchy for mothers: Seeds are precious. Mothers are fungible.

Today we are more familiar with the nonbiological services that we hire from mother-substitutes. We hire baby-sitters, day-care workers, nannies, housekeepers, to "watch" our children. The tasks are the traditional tasks of mothering—feeding, tending, caring, the whole bundle of social and psychological and physical tasks involved in the care of young children. When performed by mothers, we call this mothering. When performed by fathers, we have sometimes called it fathering, sometimes parenting, sometimes helping the mother. When performed by hired hands, we called it unskilled.

We devalue these nurturing tasks when we contract for them. When we do them ourselves because we want to do them, we see them as precious, as special, as treasured moments in life. That is the contradiction that allows us to value the children so highly, to value our special time with them, to speak lovingly of the child's trust, the joys of that small hand placed in ours—and hire someone to take that hand, at minimum wage.

In sum, the ideology of a patriarchal society goes much deeper than male dominance. It means far more than just having men in charge, or men making more decisions than women do. The ideology of patriarchy is a basic worldview, and in a patriarchal system that view permeates all of our thinking. In our society, even as a modified patriarchy, the ideology of patriarchy provides us with an understanding not only of the relations between men and women but also of the relations between mothers and their children.

MOTHERWORKS AND THE NEW OEDIPUS COMPLEX

There's a new kind of mother-manual out these days. The old ones told us how to mix formula, or how many minutes on a breast, how to be firm, be gentle, protect the baby, toughen the baby, let it cry, never let it cry—how to mother by the current rules of the game. The new how-to tells us all that, and also how to find, use, train, supervise, relate to, (not) compete with, our children's caregivers.

I find reading through such books, listening to such conversations, deeply troubling. It's not just the "good-help-is-so-hard-to-find" elitism of it all. What troubles me is listening to women talking about the mothering work of other women in a way that sounds to me like Victorian fathers must have talked about their wives. Maybe this is indeed the way rich people have always talked, and I am just new to the world of servants. But whether historically rooted or a new variant, this is not what I thought the women's movement was about.

First of all, as in every other discussion of child care, it is focused entirely on the mother. The slipping in and out of nonsexist language and assumptions is extraordinary. The *Big Apple Parent's Paper* ran a multipart series, "In Search of Mary Poppins." It was Part III that happened to come my way, "Training and Supervising Your Caregiver" (Morin and Davis 1987). The article began, "The time has come for you to go back to work—and to turn your baby over to someone else to be the mother all day." So much for nonsexist assumptions. The "parent" reading is a woman, and the caregiver is also a woman. The reader is quite specifically told, "First get your husband's agreement that only one person—namely you—will be supervising the caregiver. Any suggestions or complaints on his part about her performance should go through you." It is not just being patronizing, saying that caregivers "perform" and employers' suggestions or complaints need to come from one person only. It is specifically saying that one person, "namely you," is the mother. Parents genuinely sharing child care, as they are apparently sharing responsibility for earning money, are not addressed.

And how are we to treat Mary Poppins? "The general principle here is to relate to your caregiver in the way you want her to relate to your baby." That sentence is in italics. We are to relate to her as we do to the baby, but fortunately, it is a good era for babies: "Give her the feeling that you respect her and have confidence in her. Praise her for the things she does well. Don't attack or criticize her for mishaps or mistakes." Just what is meant by doing well, what are the standards for performance? Apparently, making sure that she is doing things "your way." The reader, presumably not an old hand at raising babies or she would not be looking for direction, will begin the relationship with the caregiver by taking three days off and "training her to take care of your baby by letting her see how YOU do it." The caregiver, never quite described here, is actually not Mary Poppins, or even a trained nanny, who would have flown off the job, but more likely a mother, maybe a grandmother, a woman who has herself raised children, her "own" or others on her previous jobs. But we will not learn from her, not even work with her, not even teach her. We will train her.

As I read this—and similar articles are appearing monthly in women's magazines and books for new parents—I am reminded of the centuries of child-rearing advice from male experts to mothers (Ehrenreich and English 1979). I think specifically of the control exercised by fathers, who without actually taking care of children, set the boundaries, the moral tone, the values, under which mothers could rear the children of men.

But unlike the Victorian father who was assured of his control, and did not want or expect intimacy or expressions of love from the child, women today want good child care, want deeply to have someone the child can feel secure with, feel love for—and yet feel threatened by that love. The mother needs to be reassured of her control, her very motherhood. Barbara Berg, for example, in *Crisis of the Working Mother,* tells the reader, "Knowing that you are in control—that your values and standards are being transmitted to the child even in your absence—will reinforce the feeling that YOU are the mother even if you are away at work" (1986, p. 74).

But something more disturbing than the "transmission of values and standards" is happening. The pull and tug for the child's affection, as it is described today, reads like the Oedipal drama, stripped bare of sex, and left with the naked power issue. The mother wants the child to identify with her, not with the caregiver. The child expresses its attachment to the caregiver. Berg describes her own feelings of distress when her child spoke of the babysitter as his other mother. She describes other women firing a caregiver precisely because the child grew to love her "too much." But as Freud showed us, if the child has access to power, it will turn to power, just as sons turned to fathers.

In the Oedipal situation as described by Freud, sons begin by loving their mothers, their caregivers, more than their fathers—the attachment, the identification, is with the nurturant, caring mother. The father, unlike the mother, is a distant and frightening figure who has power over the child and over the mother. Once the boy-child learns of the physical difference between the sexes, Freud hypothesized, he fears the powerful father will mutilate the son just as the boy believes the mother has been mutilated, by the loss of the treasured penis. Eventually the boy resolves his dilemma and fear by changing his identification to his father. The boy comes to want to grow up to be a man like his father, possessing for himself a woman like his mother.

In the modern American family drama I am describing, both mother and father are sources of power—but it may be that neither is the daily care giver. The child is asked to switch identification from the nurturant care giver to the powerful parent, mother or father. And so, in this situation, the source of power is not sex, but class.

Berg says she was comforted by someone assuring her that "just as you can't feel the same way about another person's child as you do about your own, your child can't feel the same way about a caretaker as he or she feels toward you." But what does "own" mean here? Berg is no biological determinist. Her child was adopted (Berg 1986). What makes this mother-child bond so sure, so irreplaceable? It is not biology. It is not sex. It is power. We won't be "vulnerable to our children's declarations that they love their baby-sitters more than they love us" (1986, p. 73), because we will remember who is in control: "Remember that they are really our assistants rather than our substitutes" (1986, p. 73). And so might Freud have reassured the Victorian father who left his son in the care of the mother all day: The young child may profess his love of his mother, but he will grow up to walk in his father's footsteps.

If we are the fathers here, setting the tone, controlling the mothers, what are we looking for in these mothers? Compliance, certainly, is always important if we are going to remain in control. But there are other qualities, too, that vary with the age of the child. Berg describes Sara's search for a suitable replacement caregiver for 3-year-old Gary. It is a discussion I've heard in playgrounds, school lobbies, kitchens, and offices for years. Lucy, the first caregiver, "had been wonderful with Gary when he was a baby. She was warm, nurturing, honest, neat and willing to follow directions. But she was not really interested in engaging Gary in intellectual or creative activities, and Sara felt these to be increasingly important as Gary grew older" (1986, p. 75).

These qualities of Lucy are distinctively, stereotypically feminine qualities. Sara, and the dozens of women like her I know, are doing what men have long done. Men left the babies and little children to the warm, neat, nurturant, compliant care of women. Men took their sons, as they outgrew the world of women, the private world, and sent them off to schools, apprenticeships, the world of men, the public world. The new mother, the managerial mother, is doing the same thing, only she wants her daughters too to leave the world of nurturant women and enter into the world of intellect and creativity.

Racism and classism enter into this equation, too. Poor women, women of color, are often valued for their nurturant qualities. I've been told it is "Island women," or "old Southern Black women," or "Mexican women" who are especially motherly. But by the age of 3 or 4—just the age the "Oedipal" drama begins—these aren't the qualities being sought, just as these feminine qualities were inappropriate for the Victorian boy. At this point, European, American, middle-class caregivers are wanted, class and ethnicity replacing sex in this revised Oedipal drama. Given the expense of a trained English-type nanny, or even a graduate student live-in, preschool and "enriched" day-care programs become the choice for entry into the world of intellect, what used to be the world of men, the public world.

What happens to the discarded outgrown caregiver? Some move from job to job, home to home, child to child. One friend told me about firing the baby-sitter her child had outgrown. A responsible employer, she found another job for the woman, in a household with a baby, and for more money. The child took it well, the mother told me—didn't seem more than passingly sad to hear the baby-sitter was leaving. But the baby-sitter—she cried. She said she'd miss the child. She'd cared for her for four years. She'd grown to love her.

In *The Managed Heart: Commercialization of Human Feeling,* Hochschild describes the work of flight attendants. In the work that they do, "processing people, the product is a state of mind" (1983, p. 6). They are paid for what she calls emotional labor:

> This labor requires one to induce or suppress feeling in order to sustain the outward countenance that produces the proper state of mind in others—in this case, the sense of being cared for in a convivial and safe place. This kind of labor calls for a coordination of mind and feeling, and it sometimes draws on a source of self that we honor as deep and integral to our individuality. (1983, p. 7)

The emotional work of mothers done without pay, or the emotional work of hired caregivers done for a salary, is a more intense version of the flight attendant's work, with the similar goal of creating a state of mind in others—just that sense of being safe and cared for. To do such work, Hochschild shows, calls for "deep acting," as Stanislavski described it: We must call forth the emotion from our very selves. To soothe a child, we must make it feel cared for, nurtured, loved. To make the child feel that way, the caregiver must act as if she indeed loves the child. To act so convincingly—and children are a tough audience—requires calling forth her own feelings of love, of tenderness.

What do caregivers *feel* when they rock an infant to sleep, kiss a boo-boo to make it better, play peek-a-boo, settle a playground spat? What did slave mammies feel as they nursed these white babies, lullabied them with songs of their own children's suffering, sale, even death? Perhaps it is not so surprising that child caregivers, from Roman slave nurses (Joshel 1986) to the American black slave and on to today's minimum wage caregivers are so often remembered with affection by their grown charges. At unknown costs to themselves, they were mothers.

CHILD CARE AND MIDREVOLUTIONARY MORES

Is the moral of the story that no one should ever hire anyone to do child care? Maybe, in a better world, it would be. Maybe, in a better world, communities of people, not just mothers, would be responsible for children. I haven't the imagination to envision how such a world would work. The highly imaginative author, Ursula LeGuin, says that she could only make that work in a novel by having the people live in small communities (1988). Small communities of like-minded people have no great appeal for me personally, a resident by choice in one of the largest and probably the most diverse city on the planet. So maybe, in a better world, there would be no need of paid child care—but I don't even know what such a world would, could, or even should look like.

In this world, something has to be done about child care, and paying for it is an inevitable part of what needs to be done. Some would like to see all child care institutionalized, with marvelous, state-subsidized or state-run day-care centers for children from birth on up. I'm not so sure. I worry about what values state-run, or worse yet, corporate-run day care, might teach children. I wonder about maintaining intimacy, warmth, individuality, nurturance in an institutional setting. I know it has been done in some fine centers, with a committed, excited staff developing wonderful programs. But on a large scale? For all children?

For now, we need to keep our options open. We need a variety of child-care arrangements: cooperatives or communes of mothers and fathers sharing child care, small in-home child-care centers; housekeepers; baby-sitters, parental leave time, day-care centers—all of that, and whatever else our imaginations can conjure up. And it all takes money.

Child care is either going to have to be provided in a fundamental rearrangement of the way we live—for starters, flexible time and households and shared responsibility for children beyond just one or even two harried parents—or with a rearrangement of our money. We cannot go on calling for "affordable child care" if what is meant is that one person's salary must be big enough to cover another's. That inevitably sets the salary to be made in child care as too low for that person in turn to be able to afford good child care. If children are to be something other than a burden, the costs must be a societally shared responsibility. Monetary subsidies for child care would help us break out of the class- and race-based pattern, in which the children of the rich are cared for by the poor, and the children of the poor too often go untended.

But it's not just money. *Someone* has to watch the children. *Someone* has to actually, truly, literally be there with them. Whoever that person is—mother, father, adoptive parent, sibling, housekeeper, teacher, baby-sitter, day-care worker, grandparent—*whoever* is with that child, that person must never be thought of as being there in place of someone else. The person who is there, is there. The person is not a substitute-someone-else. The person caring for a child is in a first-person, one-on-one, direct relationship to that child.

That relationship deserves respect, just as that work deserves to be valued. And respect, as we know from what has happened to mothers historically, is not adequately taken care of with politeness and a Mother's Day card. It must come with legally recognized rights. Someone who has been raising a child has moral rights invested in that child. At a minimum, we have to protect child-care workers from arbitrary firing, from loss of visitation rights to the children they raise, from having the relationship with the child used as a source of exploitation.

We must move beyond patriarchal definitions of the relations between parents and children, and beyond modifications that simply extend to some women, most often at cost to other women, some of the privileges of patriarchy. Otherwise we seem to be caught in a trap of endlessly remaking the world in the same image: some people in the public sphere, the world of power, of importance, and some people in the private sphere, rocking the cradle but never really ruling the world.

REFERENCES

Berg, Barbara. 1983. *Nothing to Cry About*. New York: Bantam.

———. 1986. *Crisis of the Working Mother: Resolving the Conflicts Between Family and Work*. New York: Summit Books.

Ehrenreich, Barbara and Deirdre English. 1979. *For Her Own Good: 150 Years of the Experts' Advice to Women*. Garden City, NY: Doubleday Anchor.

Ehrensaft, Diane. 1987. *Parenting Together: Men and Women Sharing the Care of Their Children*. New York: Free Press.

Hamilton, W. D. 1978. "The Genetic Evolution of Social Behavior." In *The Sociobiology Debate: Readings on Ethical and Scientific Issues*, edited by Arthur L. Caplan. New York: Harper & Row.

Hochschild, Arlie Russell. 1983. *The Managed Heart: Commercialization of Human Feeling*. Berkeley: University of California Press.

Joshel, Sandra. 1986. "Nurturing the Master's Child: Slavery and the Roman Child-Nurse." *Signs: Journal of Women in Culture and Society* 12:3-22.

LeGuin, Ursula K. 1988. Anthropology Department Colloquium, University of Rochester, January 12.

Morin, Eleanor and Harold Davis. 1987. "In Search of Mary Poppins: Part III. Training and Supervising Your Caregiver." *Big Apple Parent's Paper* 2(4):10-11.

Petersen, Glenn. 1982. "Ponepean Matriliny: Production, Exchange and the Ties That Bind." *American Ethnologist* 9:140-50.

Pruett, Kyle D. 1987. *The Nurturing Father: Journey Toward the Complete Man*. New York: Warner Books.

Risman, Barbara. 1987. "Intimate Relationships from a Microstructural Perspective: Men Who Mother." *Gender & Society* 1:6-32.

Barbara Katz Rothman is a Professor of Sociology at Baruch College and the Graduate Center of the City University of New York. In addition to Recreating Motherhood: Ideology and Technology in a Patriarchal Society *(Norton, 1989), she has written two other books on motherhood,* In Labor: Women and Power in the Birthplace *(Norton, 1982), also published as* Giving Birth *(Penguin, 1982), and* The Tentative Pregnancy: Prenatal Diagnosis and the Future of Motherhood *(Viking, 1986). She is actively seeking a totally new research topic.*

Webster v. Reproductive Health Services: Devaluing the Right to Choose

Gayle Binion

ABSTRACT. This article analyzes the constitutional significance of *Webster v. Reproductive Health Services* (1989). The decision of the U.S. Supreme Court has two prominent features. First, the Court upheld the restrictions on the right of women to abort by devaluing the provisions of the challenged Missouri law. It is suggested that these provisions are far more significant than the Court majority acknowledged, and that their significance became apparent soon after the Court's resolution of the case. Second, the Court's approach to resolving *Webster* suggests a reconceptualizing of fundamental rights questions wherein *state action* jurisprudence has been turned upside down and *compelling state interest* analysis has been eschewed. While *Roe v. Wade* (1973) was not overruled in *Webster*, it appears that the contemporary Court's approach to operationalizing the right of choice is unlikely to restrict the power of the states to regulate in this area.

Throughout the month of June, 1989, a vigil was maintained outside the U.S. Supreme Court in anticipation of the resolution of *Webster v. Reproductive Health Services* (1989). All eyes focused on the high Court and on the possibility that *Roe v. Wade* (1973) would be overruled and the constitutional right to abortion negated. The frenzy surrounding *Webster* soon subsided, giving way to reflection on its legal significance.

While research on the long term consequences of *Webster* (1989) must necessarily await another day, a near-consensus quickly emerged in the law review literature on the immediate legal significance of the decision. While Estrich and Sullivan (1989) stressed

---

Gayle Binion is affiliated with the University of California at Santa Barbara.

Women & Politics, Vol. 11(2) 1991
© 1991 by The Haworth Press, Inc. All rights reserved.

that the actual *holding* of *Webster* was minimal, and chose instead to continue a dialogue with Justice O'Connor, hoping to sensitize her to the cause of reproductive choice, the predominant view of legal scholars has been that *Webster* substantively changed the course of abortion law. Dellinger and Sperling (1989) and Johnsen (1989) both found the absence of *strict scrutiny* in *Webster* most disturbing. In the former case, this was accompanied by a condemnation of the Court's portrayal of abortion rights as uniquely or especially problematic to define and enforce (Dellinger and Sperling 1989, 85-88). In the latter case, special concern was raised that *Webster* has heightened the trend toward turning mother and fetus into legal adversaries (Johnsen 1989). A somewhat different emphasis is placed on *Webster* by Frances Olsen, who argues that the unraveling of the *Roe* (1973) compromise on abortion devalues women and their reproductive labor. She also fears that the lack of concern for women's rights in *Webster* will serve to undermine women's respect for the objectivity of the law (Olsen 1989, 111). In stark ideological contrast with Estrich and Sullivan, Dellinger and Sperling, Johnsen, and Olsen, are Bopp and Coleson (1989) of the "right to life" movement, who applaud the Court for returning decision making on abortion to the states' legislatures. While they share the view of their liberal counterparts that *Webster* substantially modified restrictions on state power to regulate abortion, they reject the contention that this undermines the constitutional rights and status of women.

My own view of the *Webster* decision is that it is very important in two respects. First, by upholding the challenged Missouri statute, the Court significantly undermined the right of women to *exercise* their freedom to choose to abort, wherever state legislatures choose to promote their "preference for childbirth." This was accomplished by the Court's systematic devaluing of the provisions of the Missouri anti-abortion law and their implications for women's rights. Second, *Webster* is significant as well for the way in which it reconceptualized how constitutional rights questions, or perhaps just *this* constitutional rights question, are to be resolved. Although *Roe* (1973) was not formally overruled, the majority of the Justices disposed of *Webster* as if a fundamental right to reproductive choice does not exist.

UNDERMINING THE EXERCISE OF FREEDOM OF CHOICE

Rehnquist's opinion for the Court's majority of five,[1] reversed all of the major findings of the Eighth Circuit Court of Appeals, and systematically minimized the significance of the challenged Missouri law for the right of women to exercise reproductive choice. In their zeal to persuade the reader of the relative unimportance of their decision, they underestimated and devalued the apparent and predictable *meaning* and *consequences* of the statute in question.

1. The Preamble

The Missouri law in question "finds" that life begins at conception and that except to the extent that the U.S. Constitution or the "decisional interpretations thereof by the U.S. Supreme Court," or "specific provisions . . . in the statutes and Constitution" of Missouri hold to the contrary, "the unborn child" is entitled to "all the rights, privileges and immunities" of all other persons. The Preamble further obligates the courts of the state to interpret all of the state's laws to protect these rights, including the rights to "life, health and well-being" (Mo.Rev.Stat. 1986, 1.205).

The Supreme Court in *Webster* concluded that this Preamble was of no *legal* consequence and constituted only a statement of a governmental "value judgment favoring childbirth over abortion" (*Webster* 1989, 3050, citing its approval of such in *Maher v. Roe* 1977). This is a curious conclusion for several reasons. First, nothing in the Preamble mentions a "value preference" for childbirth over abortion . . . neither abortion nor childbirth are mentioned and there is no weighing therein of the relative merits of either. The Preamble says nothing about using state machinery to encourage childbirth as a preferable outcome for a pregnancy.[2] Far from being a statement of *values*, the Preamble is a clear directive to all state agents in Missouri, presumably including all legislators, judges, administrators, etc. to interpret state law to *effect* the rights of personhood for "unborn children." The Court's conclusion that it does not of itself direct *action* is quite remarkable, as is its conclusion that the Preamble itself speaks to childbirth versus abortion. It is *action* it commands, and the *generic personhood of fetuses* that it defines, with implications for a wide variety of state policies: on abortion, birth control, alternative modes of reproduction, and otherwise.

It was no surprise to either critics or supporters of the Court's resolution of *Webster* that within months of the Court's decision, a case law on the Preamble was already developing, demonstrating its *consequences*. Two Missouri judges separately acquitted a total of twenty-five abortion clinic protesters charged with trespass. The judges ruled that the refusal of the U.S. Supreme Court to invalidate the Preamble supported the claim of the clinic protesters that they had a right to prevent women in Missouri from procuring abortions. The judges accepted a claim of *necessity*, to wit, that the barring of the clinic entrances was necessary to prevent the greater evil of abortion (Johnsen 1989, 184). Johnsen also notes the variety of other legal claims, perhaps predictably, being made under the Missouri Preamble. Pregnant women who are incarcerated are arguing that their fetuses, *people* from conception, are illegally being incarcerated with them. They seek release from jail (Johnsen 1989, 185). Minors charged with underage drinking have made the claim that because "life begins at conception" under Missouri law, they are actually nine months older than the state had previously recognized (Johnsen 1989, 185). Finally, state legislators in Missouri have had to turn to the state's Attorney General, William Webster, who argued the case for Missouri before the U.S. Supreme Court, to inquire what the reach is of the Preamble (Johnsen 1989, 186, n.27).

In the same vein, it should be noted that the assumption of Justice Rehnquist that the Preamble would not restrict the use of abortifacients, such as IUDs or low-estrogen birth control pills, belies the terms of the law itself. Because the Supreme Court has never specifically addressed or upheld the right to use such post-fertilization methods of "birth control," the Missouri statute in question does not exempt them from the edict that fertilized eggs have a right to life. Rehnquist rejects this concern with the observation that it is not before the Court until the law is actually applied to such practices (*Webster* 1989, 3050). Justice O'Connor, while echoing Rehnquist's view that the application of the Missouri law's Preamble to post-fertilization "contraception," or for that matter to alternative modes of reproduction such as *in vitro* fertilization, which "sacrifice" fertilized ova, is not properly before the Court, observes that *overbreadth* is the basis for a declaration of unconstitutionality in only the limited context of the First Amendment (*Webster* 1989, 3059-60). What O'Connor neglects to demonstrate is *why* facial overbreadth doctrine should be applicable to abridgements of the First Amendment and not to restrictions on the right to reproductive choice, which entail equally *chilling* ramifications and potentially far more serious and permanent personal consequences.

In sum, the Preamble, while directly threatening abortion rights, also extends well beyond the abortion question. The implications for the right of women, more generally, to control their fertility, apparent in the Preamble, were simply dismissed by the Court as either not consequential or not ripe.

But it is not only the Court's minimization of the scope and consequence of the Preamble that is troubling. It is also odd that the Court chose to ignore its implications for constitutional federalism. The terms of the Preamble are reminiscent of the Virginia and Kentucky Resolutions in which the states maintained that their authority was co-equal to that of the federal courts in interpreting the Constitution. The Missouri Preamble, it should be noted, allows for exceptions to the rights, privileges and immunities of fertilized ova when the U.S. *Supreme* Court has found that the Constitution requires otherwise. Noticeably absent from constituting restrictions on the state's autonomy are decisions of the federal *district* courts or courts of *appeals*. Has the U.S. Supreme Court in *Webster* unwittingly endorsed the principle that states are not bound by such decisions? Similarly, the restrictions in the Preamble on effectuating the rights of "unborn children" do not include federal statutory provisions that would run contrary to the recognition of such fetal rights. Only *state* statutory exceptions are recognized in the Missouri law. Has the U.S. Supreme Court unwittingly reversed the more than century-old principle of federal preemption? Would a federal *Freedom of Choice Act* be null in Missouri to the extent that the state's judges are directed to disregard it? In upholding the Missouri Preamble, the Supreme Court has understated or ignored the significance of its provisions not only for abortion but as well for federalism.

2. The Restriction on Public Monies for Abortion Counselling

In *Webster* the Supreme Court vacated the ruling of the Court of Appeals that declared unconstitutional the state's prohibition on spending public monies for abortion counselling unless the life of a pregnant woman is at stake.[3] The Court concluded that the *interpretation* of the state of Missouri that the restriction would apply to only those who dispense the public funds rendered moot the initial claim of the appellees that their rights were adversely affected by this provision of the law. The appellees, thus, had to acknowledge that under the state's definition of the target of the law they were not *directly* affected by the law. The Supreme Court, in deeming the issue moot, but taking the opportunity to vacate the decision below on this point,[4] did not address the *chilling effect* of such a law on the exercise of First Amendment rights by the employees whose *salaries* are paid by those charged with enforcing the ban on the use of public funds for abortion counselling. While the right of privately funded practitioners to discuss abortion, even in public facilities, and the right of *their* patients to hear this discussion are, presumably, protected by the rulings in the court of appeals that are *not* appealed by the state, the publicly employed health care providers and their patients are left without the same First Amendment rights.[5] As in the case of its interpretation of the Preamble, the Court's disposition of this counselling provision of the Missouri law underestimated the reach of the law and the consequences of its actions.[6]

3. Gestational Age Testing

Perhaps most unconvincing in the Court's systematic devaluing of the reach of the Missouri anti-choice statute was its interpretation of the requirement of *viability* testing.[7] Under the provisions of Missouri law any physician who "has reason to believe" that a woman is twenty or more weeks pregnant is clearly *required* to perform the (expensive, invasive) tests necessary to determine age, weight and lung capacity of the fetus. Both the Court plurality and Justice O'Connor tortured the statute in suggesting that the words "shall perform" (with respect to such tests) in the second sentence of the law are modified by the requirement that physicians use "care, skill, and proficiency" in the first sentence. The dissenters, as well as probably all other knowledgeable readers of the Rehnquist and O'Connor opinions, were equally surprised by the disingenuity apparent in their interpretation that "shall" in the law is merely a recommendation. No physician practicing obstetrics in Missouri would be legally advised that the choice is hers or his as to whether to perform the viability tests. The Justices surely knew this and, thus, knew, not only what the law says, but also how the law would operate to restrict medical practice. Again, the Court majority simply dismissed as of minimal consequence a potentially far-reaching provision of the Missouri law.[8]

RECONCEPTUALIZING CONSTITUTIONAL RIGHTS

In reviewing 188.210,215 of the Missouri statutes in *Webster*, the Court concluded that the ban on the use of public personnel and public facilities for the performance of an abortion (not "necessary" to save the life of a pregnant woman) was constitutional under its previous rulings in *Maher* (1977), *Poelker* (1977) and *Harris* (1980). This assessment of these provisions of the Missouri law not only minimized the significance of what Missouri had in fact done through this prohibition,[9] but signaled as well that the Court was redefining the manner of its resolution of civil liberties disputes. Whether its *new* way of approaching questions of individual rights was intended for only the resolution of reproductive rights cases is not entirely clear, but it is apparent that it has taken a novel approach to state action jurisprudence, and to fundamental rights analysis.

1. State Action Jurisprudence – Upside Down

State action jurisprudence has, for more than forty years, been the means by which private parties, infected with state power, were held accountable for the protection of constitutional rights (see, e.g., *Smith v. Allwright* 1944; *Burton v. Wilmington Parking Authority* 1961; and *Lombard v. Louisiana* 1963). Those who, *inter alia*, ran political parties, leased public land, or who may have taken advice from a mayor advising segregation, were all obligated to protect constitutional rights as if they were themselves governmental agents. State action theory would, thus, suggest that health providers associated with state government would be obligated to respect the constitutional rights of privacy and reproductive choice. In stark contrast with this constitutional mandate, the state of Missouri has obligated, for example, state lessees to *deny* this right. In upholding 188.210,215 the Supreme Court has turned *state action* jurisprudence upside down. The Court has not simply permitted the state to leave private actors free to decide whether to perform abortions, even though they are arguably clothed with *state action*, which should necessitate that they respect constitutional liberties; it has, rather, permitted the state to reach out a long arm definition of *state action* and, in so doing, obligate those with whom it deals to *deny* to others the rights protected by *Roe*.

The basic flaw in the Court's analysis with respect to its implications for *state action* theory is that it has confused *neutrality* with *hostility* and has not even developed a persuasive theory of the constitutional limitations on official *hostility*. Rehnquist *et al.* reasoned that the power of the state to refuse to spend money on abortions for the medically indigent (see, *Maher* 1977; *Poelker* 1977; *Harris* 1980) is constitutionally indistinguishable from the power of the state to enforce a ban on abortions wherever it can reach the behavior of health care providers who are *not* paid by the state. While I strongly disagree with the Court's decisions in *Maher et al.* (Binion 1989, 14-20; Binion 1982, 55-57) its own justification for them was that the state could express its preference for childbirth over abortion by funding the former and not the latter, *because* no one has a right to public funding for the exercise of a constitutionally protected right. What was presumably inviolate even in those anti-choice decisions was that the state could not similarly prevent a pregnant woman from procuring an abortion by interfering with the ability of her private physician to perform one. The general theme of the inviolate nature of medical judgment and practice with respect to abortion services with only minor exceptions[10] has run throughout the history of abortion rights litigation. Yet, the Court summarily concluded that the preference for childbirth over abortion that the states may articulate through funding may be manifested as well by a prohibition on physicians' practices within institutions administratively reachable by the state. In sum, a refusal to pay for most abortions, accompanied by a neutrality toward women procuring such in the free market, upheld in *Maher, Poelker* and *Harris*, has been converted into an open hostility toward such procurement by Missouri. By the sufferance of the Court, it has been permitted to see *state action* where it is only remote, such as in private institutions on land leased from the state, and to use it for the purpose of preventing, not furthering, the exercise of constitutionally protected rights.[11]

2. A New Approach to Fundamental Rights Analysis?

While a majority of the U.S. Supreme Court did not overturn *Roe v. Wade*[12] in *Webster*, it failed to apply ordinary principles of fundamental rights' analysis to the regulations on abortion that were involved in the challenged Missouri statute. By the Supreme Court's own tradition, when a fundamental right is at stake, states must demonstrate that a *compelling state interest* is *actually furthered* by abridgements of this right, *and* that there exists *no less restrictive alternative* that will substantially accomplish the state's *compelling* objective (see, for example, *Skinner* 1942; *Loving* 1967; and *Zablocki* 1978). While civil libertarians are often justifiably disappointed by the application of the test to actual cases in which the Court may too easily find a state goal *compelling* (for example, *Storer v. Brown* 1974) or does not sufficiently inquire as to whether the challenged regulation will bring about the compelling end (for example, the goal of requiring parental consent for abortions on minors is allegedly to "promote familial communications," see *Hodgson* 1990 and *Ohio* 1990), *compelling state interest jurisprudence* is probably the most sound and persuasive standard to apply to abridgements of constitutional rights. If properly operationalized, it recognizes the sanctity of fundamental rights, it requires the government to meet a very high burden of proof when abridging these guarantees, and it obligates the state to seek to use all possible alternatives to the abridgement of these rights as a way of meeting *other compelling interests*.[13]

Because *Roe* found that a constitutional right to privacy extended to abortion, and that the state's interest in potential human life was not *compelling* prior to viability, it restricted the state from prohibiting abortions prior to viability.[14] *Roe* held further that the life and health of pregnant women, even after the point of fetal viability, is superior to the state's compelling interest in fetal life. In sum, under *Roe*, the right to choose an abortion is fundamental; the power of the state to prevent it arises only after presumed viability, and is *never* superior, *never more compelling*, than a woman's claim to life and health. Despite the clarity of *Roe*, none of the majority writing in *Webster*: Rehnquist, Scalia and O'Connor, applied *compelling state interest* jurisprudence to the Missouri law. Scalia who is prepared to overturn *Roe*, and to subject abortion to state regulation as he would most public policies, had no call to do so. O'Connor, who believes that, awaiting a reconsideration of *Roe*, the right to abortion means only that the state not unnecessarily subject its procurement to "undue burden" (*Webster* 1989, 3063), has not fully defined what this means.[15] She also views the state's interest in "potential human life" as always compelling, therefore preempting the critical first question in strict scrutiny, to wit, has the state demonstrated the *compelling* quality of its interest.[16]

Most problematic was the rationale of the Rehnquist plurality as to why the regulations in the Missouri law were not to be subjected to *strict scrutiny*.[17] While tortured interpretations of the state's regulations allowed the plurality to minimize what they were in fact reviewing and deciding in *Webster*, they nevertheless seized the opportunity to suggest that they fundamentally disagreed with the jurisprudence underlying *Roe*. While claiming that their decisions in *Webster* did not necessitate overturning *Roe*,[18] they did aver that *Roe* was badly decided, especially with respect to the trimester framework, and believed that the framework should be jettisoned (*Webster* 1989, 3056).

In announcing their disapproval of *Roe*, the plurality suggested that constitutional decisions should be reconsidered when "unsound in principle and unworkable in practice" (*Webster* 1989, 3056). It is interesting that none of the three precedents they cited in support of this observation involved reversing the recognition of a constitutional right.[19] In fact, the only occasion on which the Supreme Court has reversed the granting of a constitutional right was in *Hudgens v. NLRB* (1976) wherein the Court overruled *AFEU v. Logan Valley* (1968) and held that shopping centers are not *public fora* under the First Amendment, hence their use for the purposes of communications were subject to the control of their owners.[20] Between *AFEU* (1968) and *Hudgens* (1976) the Supreme Court had decided *Lloyd v. Tanner* (1972) which significantly narrowed *AFEU* and functioned as both a harbinger of things to come and as a trial balloon for limiting First Amendment rights. One is prompted to ask whether *Webster* is the *Lloyd* of reproductive choice? Is the rejection of the trimester framework a prolegomenon to the effective end of the constitutional right recognized in *Roe*?

Because the plurality viewed the state's interest in potential human life as *compelling* throughout pregnancy, not just from viability (*Webster* 1989, 3057), and because it saw the right to procure an abortion, not as a fundamental right but rather as "a liberty interest" (3058), there is little question but that the right to choose to abort would regularly, predictably, and necessarily lose to state regulation under the Rehnquist formulation. This is structurally predetermined. Ordinarily, under *compelling state interest* jurisprudence the state alleges an interest *apart* from preventing the exercise of constitutionally protected behavior. Its claim for its authority to abridge constitutionally protected right(s) is that in the process of effectuating its *other* compelling goal it unavoidably has to interfere with a right. The Court, if it is persuaded that the interest is *compelling*, will then seek evidence that the means chosen will reach the identified *compelling goal*, and that these means are the least restrictive possible (see, Binion 1989, 16-24, and notes). In an approach to abortion rights in which the state is defined as having, from conception, a *compelling interest* in "potential human life," the *goal* of the state is *itself* the prevention of abortion, the prevention of the exercise of a constitutionally protected right. This *goal*, forbidden under *Roe*, has become for the Rehnquist plurality, *compelling*.

The irony, perhaps, of the Rehnquist approach is that he did not take *strict scrutiny* through its paces. He simply concluded that with the *Webster* decision the balance has been restored to a proper one with respect to returning substantial authority to the legislatures to regulate abortion (*Webster* 1989, 3058). He set no guidelines for, and offered no scrutiny of, whether the state will actually accomplish its allegedly *compelling* goal of preventing abortions through the restrictions in the Missouri statute, as opposed to just harassing women, and most importantly for civil liberties, he has paid no attention to the question of whether, if one accepts *arguendo*, that the states have a compelling interest in preventing abortion despite the inconsistency with a constitutional right to abort, the states ought to demonstrate that less restrictive alternatives, such as incentives to have children, should not have to be employed.[21]

The key to understanding the plurality's fundamental rejection of abortion rights is that their explanation for their condemnation of the trimester/viability framework of *Roe* rejects abortion rights *per se*. Such arguments as "trimesters and viability are not found in the text of the Constitution" (*Webster* 1989, 3057) lead one to note that the right to privacy on which *Roe* is based is also not in the text of the Constitution. By this argument they have rejected not trimesters and viability; they have rejected the right to privacy as it underpins, *inter alia*, abortion, contraception and the right to raise one's children.[22] As Dellinger and Sperling so aptly note, the "no textual reference" argument would leave no constitutional rights standing (1989, 90).[23]

Similarly, Rehnquist *et al.* asserted that *Roe* and its progeny have led to "a web of legal rules that have become increasingly intricate, resembling a code of regulations rather than a body of constitutional doctrine" (*Webster* 1989, 3057). What they failed to note is that to the extent that the law of abortion rights is "intricate" it is no more so than the law of the First Amendment, the law of defendants' rights or of any other controversial provisions of the U.S. Constitution. All levels of government test the limits of their authority when the U.S. Supreme Court issues general declaratory judgments in constitutional cases. Dellinger and Sperling are incredulous that the Rehnquist Court would counsel the abandonment of a constitutional principle because those dissatisfied with a Court decision pass laws "at the margin of its application," thus forcing the development of intricate rules (1989, 86). More to the point, the Burger and Rehnquist Courts, generally unsympathetic to claims for individual rights as against governmental authority, have invited legislatures, executive agents and, indeed, law enforcement officers to test the limits of constitutional rights by pushing those margins. The Court has been an active and willing participant in this process of creating intricacy. The law of defendants' rights is a particularly apt example. Although numerous precedents from the Warren Court era set *general principles* under the Fourth, Fifth, Sixth, and Eighth Amendments for federal, state and local law enforcement to follow, the Burger and Rehnquist Courts have developed a *creative* web of distinctions and exceptions to these rules, such as the limited "good faith" exception to the Fourth Amendment requirement that a search warrant be valid (*U.S. v. Leon* 1984) and have, as in the abortion cases prior to *Webster*, encouraged the testing of the limits of the general constitutional principles enunciated by the Warren Court. Their disposition of *Webster*, unless it is the harbinger of a complete rejection of a constitutionally based right to procure an abortion, leaving the state to regulate the subject free of constitutional restraint, will, like the Court's handling of the defendants' rights cases, serve further to complicate the relevant law.

CONCLUSION

Webster (1990) was not an inconsequential decision. While it did not overrule the doctrine of *Roe v. Wade* (1973), its devaluation of the significance of the restrictions on reproductive choice in the challenged Missouri statute has served to undermine the substantive protections of that right as against governmental obstacles and burdens. When this devaluation is coupled with the refusal of a majority of the Court to apply *compelling state interest* jurisprudence to such restrictions, it is apparent that for this Court, at this time, the litigative burden will remain with women to demonstrate that each state legislated "preference for childbirth," however operationalized, is not sustainable. In a one year period after the *Webster* decision, 350 bills were introduced in 41 states to express this preference.[24] And, while nearly all were defeated, this activity encouraged by the Court's resolution of *Webster* continues.

NOTES

1. The group of five Justices joined together on Parts I and II A, B, C of Rehnquist's opinion. While O'Connor and Scalia, respectively, disagreed with the position of Rehnquist *et al.* on the need to reconsider the principles enunciated by *Roe*, they all minimized the importance of the provisions of the challenged Missouri statute.

2. I am not suggesting that I would find such a law to be constitutional, (see Binion 1989, 16-20), but rather that a law expressing a preference for childbirth is quite different from one granting *personhood*, and enforceable rights, to fetuses. It is also instructive to note that *Maher* (1977), and its progeny, on which the Supreme Court depends for its assertion that states may express a "preference for childbirth over abortion" involved state policies that funded childbirth for the medically indigent but not abortion. That action constituted an official statement of "preference." The Missouri Preamble has no such statement therein.

3. The state of Missouri had chosen not to appeal the decision of the Court of Appeals invalidating the provisions of the law prohibiting public employees from discussing abortion within the course of their employment (Mo. Rev. Stat. 1986, 188.210) and prohibiting such counselling within a public facility (Mo. Rev. Stat. 1986, 188.215). It chose to appeal the invalidation of its ban on public expenditures for such counselling.

4. It should be noted that the high Court was unanimous on this procedural point.

5. The question of whether receipt of governmental health care funding can be conditioned on a prohibition on abortion counselling will be addressed by the U.S. Supreme Court during the 1990 term in *Rust v. Sullivan* and *New York v. Sullivan*.

6. Only Justice O'Connor acknowledged that such an interpretation of the reach of 188.205 is not impossible but holds, nevertheless, that when and if the Missouri Supreme Court applies it to muzzle state employed health care providers, it will be time enough to consider whether such an action violates the U.S. Constitution (*Webster* 1989, 3060). Olsen notes that, on the basis of the *Webster* decision, administrators at Southwest Missouri State University have ordered their employees not to discuss abortion with their clients, patients or others (Olsen 1989, 114, n.40, citing article in L.A. Times, September 28, 1989, at 24, col. 3).

7. Under Mo. Rev. Stats. 1986, 188.029, a physician with reason to believe that a fetus is twenty weeks (or more) may not perform an abortion without first determining if the fetus is viable by assessing its age, weight, and lung capacity.

8. That the devaluing of restrictions on reproductive choice in *Webster* signaled a new direction, rather than a temporary aberration, is apparent in the Court's resolution of the minors' abortion rights cases in 1990. In *Ohio v. Akron Center for Reproductive Health* (1990) the Court upheld a parental consent statute despite attacks on its numerous procedural requirements. Among the provisions challenged were those relating to the judicial bypass mechanism, open to minors who are unable or unwilling to have their parents notified about their pregnancies.

Attacked as burdensome to minors was, *inter alia*, the existence of three different judicial bypass forms: one for a minor claiming to be mature enough to make the abortion decision on her own, another for a minor seeking a judge's determination that an abortion would be in her best interests, and a third form for a minor who sought both. While the Court acknowledged that some minors might well find this scheme confusing, it minimized the significance of the obstacle with the observation that it trusted that the state would allow amended pleadings after the minor had had the advice of counsel (*Ohio* 1990, 2982). Similarly, it concluded that the possibility of up to a three week delay in the judicial procedure was not a serious problem because it trusted that the state would act expeditiously (*Ohio* 1990, 2981). Finally, in acknowledging that the requirement that a petitioning pregnant minor sign her name to her pleading form compromised the anonymity that the Court had required in several previous cases, the Court, nevertheless, ruled that the provision was not unconstitutional. It concluded that the distinction between anonymity and confidentiality is minimal and that the latter could be protected by state laws barring disclosure of these records by their employees (*Ohio* 1990, 2980).

9. The definition of *public facility* in the Missouri law broadly includes "any public institution, public facility, public equipment, or any physical asset owned, leased or controlled by this state or any agency or political subdivision thereof." (188.200(2)) This would appear to include the Truman Medical Center, in Kansas City, a private facility on land *leased* from the government. Ninety-seven percent of all hospital abortions at 16 weeks or later were performed there the year before Missouri passed the challenged law (Blackmun, dissenting in *Webster* 1989, 3068, n.1). Its application to Truman was clear at the time of *Webster* and reinforced shortly thereafter when a pregnant woman, who discovered at 22 weeks' gestation that her fetus was anencephalic, could not obtain an abortion anywhere in Missouri (L.A. Times, September 28, 1989, at 1, col. 1, cited in Olsen 1989, 114, n.45). In Justice O'Connor's concurring opinion she noted that the question of the constitutionality of bans on abortions at private facilities that use public sewers or lease state property "need not be decided here" (O'Connor, concurring in *Webster* 1989, 3059). Unlike the majority opinion, which she joins on the question of *public facilities*, she is simultaneously apparently willing to consider the reaches of the concept of "public" should an allegation of *overreach* come to the Court.

10. Under the relevant precedents, not addressed for overruling in *Webster*, except only indirectly by the Rehnquist plurality of three, the state may require certain intrusions into medical judgments *only* where necessary to protect the health of the woman or to preserve the life of a potentially viable fetus. In the latter case, extraordinarily rare in American society, a second doctor may be required during the abortion procedure, but only if this does not entail any additional danger for the woman, as might happen when a delay is thereby necessitated (*Thornburgh* 1986).

11. Interesting in Rehnquist's opinion for the Court is his observation that "a different analysis might apply" if either of two conditions existed in Missouri. He

reasoned that banning physicians, who performed abortions *elsewhere*, from using public facilities, or the existence of socialized medical practice might require a different decision by the Court (*Webster* 1989, 3052, n.8). To reach this conclusion is to assume that women in Missouri actually retained a choice and that the ban on the use of such facilities would not prevent them from obtaining abortions, otherwise the hypothetical of socialized medicine is misplaced. His analysis paid no attention to the information brought to the Court by various *amici* that the state's ban on the use of such facilities and personnel will have the effect of denying choice. Distinctly naive was his assumption that the Missouri law as written leaves women no worse off than if there were no public hospitals and facilities, therefore cases such as *Maher* (1977) are controlling. His point is not persuasive because in fact hospital construction and operation does have a *socialized* quality in that minimum census is needed for a hospital to operate. The existence of public hospitals, rather than encouraging competition, makes the building and operation of additional hospitals less likely. He also ignores the common interweaving of public and private in the building, funding, and operation of hospitals in the United States, particularly voluntary or private, non-profit facilities. The Court's decision, in sum, is not that Missouri may keep its own hands off abortions, but rather that states may limit, perhaps prevent, the existence of abortion options by their involvement in the *private* practice of medicine. This goes well beyond the logic of *Maher* and its progeny and has little to do with the premise that one does not have a right to governmental aid in the procurement of an abortion.

12. Only Justice Scalia *overtly* advised this conclusion in his concurring opinion in *Webster* (3064-67).

13. I use the term "other" compelling interests because I believe that a fundamental flaw in the way in which some jurists have applied the *compelling state interest* test is that they have assumed that the *protection of civil liberties* is itself *not* a compelling interest of the state and that *this* goal is necessarily adversarial to the state's interests. This suggests a dangerous assumption of a schism between the interests of people and the interests of the governments under which they live.

14. The states were limited to regulations to protect only the *actual* health needs of pregnant women, needs that the Court until *Webster* left largely to the medical profession to define. See, especially, *Thornburgh v. American College of Obstetricians and Gynecologists* (1986).

15. But see Estrich and Sullivan who argue that for O'Connor "undue burden" is a finding of quantity, not justification, and (contrary to her statement in *Webster* that an "undue burden" is unconstitutional) its finding only triggers *strict scrutiny* (1989, 135). In her dissent in *Thornburgh*, O'Connor stated her position as that described by Estrich and Sullivan. Because she believes that the state has a *compelling interest* in potential human life throughout pregnancy, restrictions on abortion need meet only a *rationality* test; "undue burdens" are for her, apparently, subject to *strict scrutiny* (1986, 2214). Her votes and her opinions in the abortion cases during her nine years on the Court suggest that few regulations by the state will rise to "undue burden" proportions. While in her *Thorn-*

burgh dissent she expressed some concern about the informed consent provisions of the Pennsylvania law that she voted to uphold, it is easily argued that her concern was more about First Amendment rights and the impropriety of state propaganda than it was about reproductive choice (1986, 2215). O'Connor's concurring opinion in *Hodgson v. Minnesota, inter alia*, suggested that a two-parent notification requirement for minors seeking abortions is an undue burden that does not further a "compelling state interest." Despite that observation she, nevertheless, concluded that the existence of a judicial bypass mechanism, for those teens unwilling or unable to tell their *second parent*, negated the "undue burden" and rendered the two-parent requirement constitutional (1990, 2950-51). While Estrich and Sullivan, in their very provocative analysis of O'Connor's abortion jurisprudence and her concept of "undue burden," make a valiant attempt to demonstrate to the Justice that she ought to see the "undue burdens" she seems reluctant to recognize, in my view they are unduly optimistic that the seeds of such exist in her decisions to date.

16. The Court's eschewing of strict scrutiny in *Webster*, like the devaluing of the restrictions in the challenged statutes, manifested a new direction in the abortion law. Its subsequent decisions in 1990 (*Hodgson and Ohio*) assessed a minor's right to secure an abortion under a standard of *reasonableness*. While minors' constitutional rights are often deemed by the Court to be less than those of adults, the usual standard applied in such instances is *middle level* scrutiny (*Bellotti* 1976 and *Carey* 1977).

17. It is not that these justices have consistently eschewed *compelling state interest* jurisprudence for abridgements of fundamental rights. In fact, during the October, 1988, term, they strictly applied the standard to an affirmative action program in the city of Richmond, Virginia. In *City of Richmond v. J.A. Croson Co.* (1989) the Court rigorously reviewed the actions and justifications of the city of Richmond and determined that the city could not demonstrate that it needed to use racial set-asides in granting contracts. This differential, perhaps ideological use of *strict scrutiny* by the Rehnquist Court, has been noted by legal scholars analyzing *Webster* (Estrich and Sullivan 1989, 144; Dellinger and Sperling 1989, 104). It appears that vis-à-vis *other* constitutional rights, more favored by the contemporary majority of the high Court, the right of reproductive choice is devalued.

18. They did acknowledge that revisions in *Colautti v. Franklin* (1979) and *Akron v. Akron Center for Reproductive Health* (1983) were necessary to support their decision on viability testing in *Webster*. In the two previous decisions the Court had frowned on regulations on the medical practice of abortion, especially when the cost of obtaining an abortion could thereby be increased. It had also held that even during the second trimester of pregnancy, regulation of such would be upheld only if shown to be necessary for the preservation of maternal health, not for the preservation of fetal life. In upholding Missouri's regulations on viability testing, the Rehnquist plurality acknowledged that they had, *necessarily*, rejected the authority of the two precedents on these points (a point disputed by O'Connor, *Webster* 1989, 3062-3064).

19. They cited *Garcia v. SAMTA* (1985), involving the applicability of federal employment laws to state employees; *Solorio v. U.S.* (1987), changing the jurisdiction of military tribunals; and *Erie v. Tompkins* (1938), holding that the law to be applied in federal *diversity* cases is the state law, in support of the principle that constitutional decisions should be open to reconsideration by the courts.

20. It might be suggested that the qualified liberty of contract as against governmental labor regulation, the contours of which were shaped by *Holden* (1898), *Lochner* (1905), *Muller* (1908), *Bunting* (1917), and *Adkins* (1923) was effectively ended in *West Coast Hotel v. Parrish* (1937). I am indebted to Professor Leslie Friedman Goldstein for this observation.

21. These might include such policies as free pre-natal and post-natal care for women and their children, free childcare as well as clothing, food and shelter for children. The proposition here is that while the plurality's logic is fundamentally flawed in recognizing a *compelling state interest* in preventing the exercise of a constitutionally protected right, fundamental right, or liberty interest (*Webster* 1989, 3058), it is nevertheless obligated to determine if the state has actually furthered the goal with its policies and perhaps more critically whether it could substantially reach the same level of success in reaching its goal by means less restrictive of choice.

22. The plurality added to the argument that "trimesters and viability are not found in the text of the Constitution" "or in any place else one would expect to find a constitutional principle" (*Webster* 1989, 3057). This reader has no idea where "else" Rehnquist *et al.* deem it appropriate to look for constitutional principles.

23. For example, one of the bedrocks of the First Amendment, that there is a freedom of association, is not to be found within the text of the Constitution.

24. "The Fundamental Questions Is Still: Who Decides?" *New York Times*. July 1990, 18.

REFERENCES

Adkins v. Children's Hospital. 1923. 261 U.S. 525.

Akron v. Akron Center for Reproductive Health. 1983. 462 U.S. 416.

Amalgamated Food Employees Union v. Logan Valley. 1968. 391 U.S. 308.

Bellotti v. Baird. 1976. 443 U.S. 662.

Binion, Gayle. 1982. "The Disadvantaged Before the Burger Court: The Newest Unequal Protection." *Law & Policy Quarterly* 4:37-69.

Binion, Gayle. 1989. "Reproductive Freedom and the Constitution: The Limits on Choice." *Berkeley Women's Law Journal* 4:12-41.

Bopp, James, and Richard Celeson. 1989. "What Does *Webster* Mean?" *University of Pennsylvania Law Review* 138:157-177.

Bunting v. Oregon. 1917. 243 U.S. 426.

Burton v. Wilmington Parking Authority. 1961. 365 U.S. 715.

Carey v. Population Services International. 1977. 431 U.S. 678.

City of Richmond v. J.A. Croson Co. 1989. 109 S.Ct. 706.

Colautti v. Franklin. 1979. 439 U.S. 379.

Dellinger, Walter, and Gene Sperling. 1989. "Abortion and the Supreme Court: The Retreat from *Roe v. Wade*." *University of Pennsylvania Law Review* 138:83-118.

Erie v. Tompkins. 1938. 304 U.S. 64.

Estrich, Susan, and Kathleen Sullivan. 1989. "Abortion Politics: Writing for an Audience of One." *University of Pennsylvania Law Review* 138:119-155.

Garcia v. San Antonio Metropolitan Transit Authority. 1985. 469 U.S. 528.

Harris v. McRae. 1980. 448 U.S. 297.

Hodgson v. Minnesota. 1990. 110 S.Ct. 2926.

Holden v. Hardy. 1898. 169 U.S. 366.

Hudgens v. National Labor Relations Board. 1976. 424 U.S. 507.

Johnsen, Dawn. 1989. "From Driving to Drugs: Governmental Regulation of Pregnant Women's Lives After *Webster*." *University of Pennsylvania Law Review* 138:179-215.

Lloyd v. Tanner. 1972. 407 U.S. 551.

Lochner v. New York. 1905. 198 U.S. 45.

Lombard v. Louisiana. 1963. 373 U.S. 267.

Loving v. Virginia. 1967. 388 U.S. 1.

Maher v. Roe. 1977. 432 U.S.464.

Missouri Rev. Stats. 1986. 1.205, 188.029, 200, 210, 215.

Muller v. Oregon. 1908. 208 U.S. 412.

New York v. Sullivan. No. 89-1392, *cert. granted*, U.S. Supreme Court, May 29, 1990.

Ohio v. Akron Center for Reproductive Health. 1990. 110 S.Ct. 2972.

Olsen, Frances. 1989. "Comment: Unraveling Compromise." *Harvard Law Review* 103:105-135.

Poelker v. Doe. 1977. 432 U.S.519.

Roe v. Wade. 1973. 410 U.S. 113.

Rust v. Sullivan. No. 89-1391, *cert. granted*, U.S. Supreme Court, May 29, 1990.

Skinner v. Oklahoma. 1942. 316 U.S. 535.

Smith v. Allwright. 1944, 321 U.S. 649.

Solorio v. U.S. 1987. 483 U.S. 435.

Storer v. Brown. 1974. 415 U.S. 724.

Thornburgh v. American College of Obstetricians and Gynecologists, 1986. 476 U.S. 747.

Turnock v. Ragsdale. #88-790, motion to defer further proceedings granted, U.S. Supreme Court, December 1, 1989.

U.S. v. Leon. 1984. 468 U.S. 897.

Webster v. Reproductive Health Services. 1989. 109 S.Ct. 3040.

West Coast Hotel v. Parrish. 1937. 300 U.S. 379.

Zablocki v. Redhail. 1978. 434 U.S. 374.

IMAGES OF MOTHERS IN POVERTY DISCOURSES

MARTHA L. FINEMAN*

I. INTRODUCTION

This Essay focuses on the construction of the concept of "Mother" in poverty discourses. It addresses the role of patriarchical ideology in the process whereby a characteristic typical of a group of welfare recipients has been selected and identified as constituting the cause as well as the effect of poverty. I am particularly interested in those political and professional discourses in which single Mother status is defined as one of the primary predictors of poverty. This association of characteristic with cause has fostered suggestions that an appropriate and fundamental goal of any proposed poverty program should be the eradication of the status and practice of single motherhood. This goal is to be accomplished through appropriate coupling of the single mother with the child's father—who would thereby assume his "rightful" place in the family and fulfill his financial obligations. By his so doing, the paramount welfare reform objective—letting the state off the economic hook—will have been achieved.

Although many commentators prefer sanctified mother/father relationships, the coupling of the single mother and the financially-endowed male anticipated by such reforms need not be accomplished through the formation of a formal marital bond. The objective of such proposals is the creation of a legal tie between the male who is presumed to be economically-viable, and the dependent single mother and child. Through this legal tie, child support obligations can be established and enforced. Neither the mother's nor the father's wishes regarding the establishment of a legal tie are considered relevant.

The paternity proceeding is the legal process through which this tie is accomplished. Increased reliance on this process has been a mainstay of recent welfare reforms. It is an essential step in assuring that the family, and not the state, takes the responsibility for children. Generally,

---

Copyright © Martha L. Fineman 1991.

* Professor of Law, Columbia University School of Law. I would like to thank Christine Harrington and Henry P. Monaghan for their comments on an earlier draft of this Essay, and Mindy Dutton, my research assistant, for her valuable support and good humor.

This Article is part of a larger project to be published as a book entitled *The Neutered Mother, The Sexual Family and Other Twentieth Century Tragedies.*

recent welfare reforms assume as a normative matter that children should be firmly anchored (financially, morally, and legally) within the nuclear family. The socially and economically based deprivations that poor children and their mothers suffer are thereby transformed into deprivations attributable to and based upon their deviant family form. Economic and social salvation will therefore come about through the restructuring of these families to conform as closely as possible to the "natural" (nuclear) family, thereby perpetuating traditional roles. The reforms are driven by the perceived compelling need to reestablish patriarchy, redefined for the contemporary context in which not only is divorce common, but increasing numbers of never-married women choose to become mothers.[1]

In earlier work, I have explored the transformation of the concept of "Mother" that father-centered reforms have wrought in the divorce context.[2] When I began to read poverty discourses, I was struck by both the similarity in the articulation of the "problem" (absent father) and in the creation of an ideal "solution" (bring him [back] into the family in some form). The ideological spaces occupied by the discourses of poverty and divorce, with their two unique categories of single mothers, were different enough so that the existence of a common articulation of problems and solutions suggested to me the need for further thought.

Perhaps the similarity in the creation and definition of problems and solutions should not have been surprising because many single-mother families are created by divorce and many of these mothers will eventually need state assistance. In poverty discourses, however, the single-mother family under consideration is not typically presented as the once-married, formerly middle-class housewife and mom, who, with her children, now finds herself upon hard times as the result of divorce. The single mother constructed and located within poverty discourses is differentiated by race and by class from her divorced sister. Yet, despite these

1. In an article summarizing the latest statistics, Myron E. Wegman noted:

> During the past decade, an increasing proportion of all births has been to unmarried women; in 1987 the birth rate to this group increased on the order of 5%. In fact, the whole increase in births and birth rates for the year is approximately equivalent to the increase to unmarried women. By contrast, both number of births and birth rates to married women were less than 1986 levels. The increase in births and birth rates among unmarried women has been relatively continuous since 1980 and the proportion has increased from less than one fifth in 1980 to nearly one quarter in 1987.
>
> Difference in races continued to be great; almost one fifth of all white births were to unmarried women, but more than three fifths of black newborns had unmarried mothers. On the other hand, since 1980, there has been a large, 40% increase in the birth per 1000 unmarried white women age 15 to 44 years, to a level of 24.6 whereas the increase among the black population was only 2%, to 84.7 in 1987, as against 82.9 in 1980.

Wegman, *Annual Summary of Vital Statistics - 1988,* 84 PEDIATRICS 943, 944-45 (1989).

2. *See* M. FINEMAN, THE ILLUSION OF EQUALITY: THE RHETORIC AND REALITY OF DIVORCE REFORM (1991).

differences, the core and common problem facing mothers within each group is identified as the missing male. It follows, therefore, that the solution to the problem for both categories of single mothers lies in the legally-coerced (re)establishment of a paternal presence, physically outside of, but metaphysically completing, the family structure.

As a result of this observation, I became interested in what I now call the phenomenon of "cross-over discourses." I define this phenomenon as the propensity for rhetorical images associated with being female in our culture, which are generated and perpetuated in one context, to spill over and define our understanding of women in other contexts. Any process in which "Mother" is explicitly the focus of attention generates images that ultimately are significant in shaping societal attitudes toward regulation of motherhood through the creation of rules governing such things as reproduction, child custody, and other areas of the law in which the institution of "Mother" is implicated. In other words, "Mother," as a socially defined and symbolic institution, has trans-substantive implications. The concept of "Mother" conceived of as "true" in forging images in poverty discourse will inevitably be a definitive presence in our understanding and grand construction of the institution of "Mother."

I have concluded that the tendency of discourses in this area to share images and ideas is directly related to the continued vitality of patriarchical ideology. Although our social circumstances have substantially altered during the past several decades, patriarchical concepts remain at the center of how we define and understand families in our culture. As developed in the last section of this Essay, the ideology of patriarchy is the most instrumental force in the creation and acceptance of discourses about Mothers in our society.

In focusing on the institution of "Mother," I do not mean to suggest that it represents a necessary or the only appropriate role for individual women. I realize that not all women are mothers. However, as an institution with significant and powerful symbolic content in our culture, motherhood has an impact on all women—independent of the individual choice about whether to become a mother. It comes from the durability and tenacity of the assumptions made about any individual woman that are forged in the context of the cultural and social forces that define the "essential" or idealized woman. For this reason, all women should care about the social and cultural presentation of the concepts of motherhood that are part of the process that constructs and perpetuates a unitary, essentialist social understanding of women. "Mother" is so interwoven with the notion of what it means to be a woman in our culture that it will continue to have an impact on individual women's lives, even where that impact is manifested primarily in the extent of resistance to the dominant cultural ideology.

II. The New Poverty Discourses

A. *Poverty Reform Discourse*

The Family Support Act of 1988 was the first major piece of legislation addressing poverty to pass Congress in several decades.[3] It reflects the belief that welfare dependency is a significant societal problem that requires a dramatic reorientation of welfare policy. The Act's primary objective is to link poverty with the lack of a work ethic, thereby attaching welfare recipients to a new workfare scheme. This is accomplished by mandating that the single mother work (or train for work) and/or by substituting state support with support from fathers, transforming a child's primary source of support from public to private hands. The legislation's focus on reinforcing the work ethic and dominant individualistic norms of self-sufficiency through the imposition of "workfare" provisions for mothers of young children has been the major emphasis of most commentators.[4] The work requirement sections provide greater

3. Family Support Act of 1988, Pub. L. No. 100-485, 102 Stat. 2343 (codified in scattered sections of 42 U.S.C.).

4. The Job Opportunities and Basic Skills Training Program component in Title II, 42 U.S.C. § 602 [hereinafter JOBS program], of the reform legislation represents the latest instance in the long history of welfare to promote work and discourage welfare dependency among the poor. The term "workfare" is generally reserved for the requirement that recipients work for their benefits, usually by accepting some form of community work assignment. *See* Mead, *The Logic of Workfare: The Underclass and Work Policy*, 501 ANNALS AM. ACAD. POL. & SOC. SCI. 156 (1989) [hereinafter *The Logic of Workfare*]. The JOBS program has been called the "new workfare" because although it requires work, it also offers opportunities for education, job training, skill development, job counseling, and placement in the private sector, along with other supportive services, such as extended child care and health insurance. The JOBS program is thus workfare plus support services. The JOBS program replaces the Work Incentive Program (WIN), Title IV, Parts A & C of Social Security Act of 1967, 42 U.S.C. § 602 (1967), that was initiated in 1967.

Prior to the passage of the Family Support Act, policy discussions emphasized research that highlighted the effectiveness of "new workfare" programs in moving people from welfare to work. *See* J. GUERON, REFORMING WELFARE WITH WORK (1987) (occasional Paper Number Two, Ford Foundation Project on Social Welfare and the American Future); L. MEAD, BEYOND ENTITLEMENT: THE SOCIAL OBLIGATIONS OF CITIZENSHIP (1986); Mead, *The Potential for Work Enforcement: A Study of WIN*, 7 J. POL. ANALYSIS & MGMT. 264 (1988); Wiseman, *Workfare and Welfare Policy*, 9 FOCUS 3 (1986). Mead, in particular, has stressed that low-cost programs that emphasize work requirements are generally cost-efficient and likely to provide considerable movement from welfare to work. *The Logic of Workfare*, *supra*, at 164. For a negative assessment, see Jencks & Edin, *The Real Welfare Problem*, 1 AM. PROSPECT 31 (1990). Jencks and Edin indicate that the increase in income produced by employment workfare programs are hardly a solution to the poverty problems of poor, female-headed families if we consider the earnings needed to enable these families to get out of poverty. They suggest that mothers of these families would, on average, require jobs that pay approximately two to three times the minimum wage before they could reasonably be expected to leave welfare and cover their expenses, including child and health care. They argue:

supportive services for recipients than previous schemes; liberals could therefore conclude that it offers more than merely a state policy that discourages welfare.[5]

Other provisions of the Family Support Act mandate stricter enforcement of child support orders, including wage-withholding.[6] Begin-

---

> The essence of the so-called "welfare trap" is not that welfare warps women's personalities or makes them pathologically dependent, though that may occasionally happen. The essence of the "trap" is that while welfare pays badly, low-wage jobs pay even worse. Most welfare mothers are quite willing to work if they end up with significantly more disposable income as a result. But they are not willing to work if working will leave them as poor as they were when they stayed home. . . . All these calculations lead to one inexorable conclusion. An unskilled single mother cannot expect to support herself and her children in today's labor market either by working or by collecting welfare. If she wants to make ends meet, she must either get help from someone else (usually an absent father, parent, or boyfriend) or she must combine work and welfare. At present, the only way she can combine work and welfare is to collect AFDC and then work without telling the welfare department.

Id. at 43-45.

5. *See, e.g., Welfare, Reform or Replacement? (Child Support Enforcement—II): Hearing Before the Subcomm. on Social Security and Family Policy of the Senate Comm. on Finance,* 100th Cong., 1st Sess. (1987) (statement of Irwin Garfinkel, Professor at the University of Wisconsin). Some feminists criticized the earlier workfare provisions, even with training, as unwarranted and inappropriate. *See, e.g.,* Abramovitz, *Why Welfare Reform is a Sham,* THE NATION, Sept. 26, 1988, at 24 (Officials running the workfare program in California (the GAIN program) calculated that a minimum of $11.00 per hour was needed for graduates to stay off the welfare rolls—yet graduates averaged only $6.50 an hour.); Walsh, *Take This Job Or Shove It,* MOTHER JONES, Sept. 1988, at 30, 32 (noting that a 1986 study found that low wages pushed 43% of educational training graduates back onto welfare, so program officials imposed a wage floor requiring that training contractors place women in jobs paying at least $6.00 an hour.).

6. Family Support Act of 1988, Title I - Child Support and Establishment of Paternity, Pub. L. No. 100-485, 102 Stat. 2343, 2344. Subtitle A, entitled *Child Support,* provides in section 101 that the wages of an absent parent are subject to withholding in enforcing payment of child support orders and that the Secretary of Health and Human Services will conduct a study to determine the feasibility of requiring immediate income withholding of all child-support awards in a state. Exceptions are allowed where both parties agree to other arrangements or where one party shows good cause why income should not be withheld. *Id.* at 2344-46. Section 102 provides that the first $50.00 of each child support payment to a family receiving public assistance will not be counted against their entitlement, even if that payment is made more than once in a single month, e.g., payments made for prior months. *Id.* at 2346. Section 103 requires that the state establish support payment guidelines and provides for a judicial "rebuttable presumption" that such guidelines are correct. This presumption may be overcome by showing that enforcement of such standards would be "unjust or inappropriate in a particular case." *Id.*

Subtitle C, entitled *Improved Procedures for Child Support Enforcement and Establishment of Paternity,* provides various standards for state support programs, *id.* at 2351, including that states establish time limits for response to requests for investigations and for when payments of support money collected by the state must be made. The Act also requires that states set up an automated data processing and information retrieval system, or an equivalent alternative system. Section 125 states that, "each State shall require each parent to furnish to such state . . . the social security account number . . . issued to the parent unless the State (in accordance with regulations prescribed by the Secretary) finds good cause for not requiring the furnishing of such number." *Id.* at 2353-54. Section 126 establishes a Commission on Interstate Child Support to make recommendations to Congress to improve interstate child support and to hold at least one conference on the issue to assist in formulating these recommendations. *Id.* at 2354-55. Section 128 requires that the Secretary of Health and Human Services:

ning in 1992, states are required to meet federal standards to establish paternity of children born out of wedlock as a method of obtaining child support from absent fathers.[7] The Family Support At mandates that as of October 1990, all states adopt the AFDC-UP Program, which provides welfare benefits to impoverished two-parent families in which the principal wage earner is unemployed.[8]

The significance of the paternity action and child support provisions of the Act in reinforcing traditional norms of the male-headed family and its responsibility for children is evident in the political rhetoric surrounding the reforms. In addressing the Family Support Act in the Senate, for example, Senator Moynihan began his address by commending President Bush for his remarks at the United Nations World Summit for Children. He stated:

> One sentence [of President Bush's remarks is] especially notable. "We want to see the day when every American child is part of a strong and stable family." The importance of this statement is elemental. Unlike the problems of children in much of the world; age-old problems of disease, new problems of ecological disaster, the problems of children in the United States are overwhelmingly associated with the strength and stability of their families. Our problems do not reside in nature, nor yet are they fundamentally economic. Our problems derive from behavior.[9]

After establishing his basic premise (that behavior, not nature or economics) accounted for the plight of America's poor children, Moynihan continued with a Karl Zinmeister quote:

> There is a mountain of scientific evidence showing that when families disintegrate, children often end up with intellectual, physical, and emotional scars that persist for life. . . . We talk about the drug crisis, the education crisis, and the problems of teen pregnancy and juvenile

---

> conduct a study of the patterns of expenditures on children in 2-parent families, in single-parent families following divorce or separation, and in single-parent families in which the parents were never married, giving particular attention to relative standards of living in households in which both parents and all of the children do not live together.

Id. at 2356. The Secretary also is required to submit policy recommendations based on this study.

7. *Id.* Subtitle B - Establishment of Paternity. Section 111 imposes standards of state compliance, requiring that as of 1991 each state's "paternity establishment percentage for such fiscal year equals or exceeds" a requisite amount. *Id.* at 2348. The "paternity establishment percentage" is the ratio of the total number of children born out of wedlock who receive some form of public assistance and for whom paternity has been established, to the total number of children born out of wedlock who receive some sort of public assistance. This section also requires that the child and all other parties in contested paternity cases submit to genetic tests, except where good cause has been shown that this is not in the best interest of the child. Those who do not receive AFDC may be charged for such tests. The statute encourages the states to adopt simple civil procedures for voluntary acknowledgement of paternity and for establishing paternity in civil cases.

8. Family Support Act, Part A - Aid to Families with Dependent Children, 42 U.S.C. § 602 (1988).

9. 136 CONG. REC. S14416, S14417 (daily ed. Oct. 3, 1990) (statement of Sen. Moynihan).

crime. But all these ills trace back predominantly to one source: broken families.[10]

Moynihan's rhetoric attributes the problems of the poor to their own behavior—thereby furthering the simplistic tendency in poverty discourses to categorize negatively mothers according to their relationship with men. Single mothers are generally deemed to present potential social problems for which they alone are largely responsible.[11] Women who have become single mothers due to the death of their spouses, however, are excused from such condemnation—considered worthy as sympathetic widows.[12] This view is expressed by Stuart Butler and Anna Kondratas:

---

10. *Id.* at 14418 (quoting Zinmeister, *Raising Hiroko*, AM. ENTERPRISE, Mar.-Apr. 1990, at 52, 53). For similar analyses, see *Senator Moynihan Discusses Children in Poverty*, 134 CONG. REC. S16919 (daily ed. Oct. 19, 1988) (statement of Sen. Glenn, introducing to the record an article by Sen. Moynihan); *The Family Security Act*, 134 CONG. REC. S7730 (daily ed. June 14, 1988) (statement of Sen. Specter); *The Need for Cooperation on Welfare Reform*, 134 CONG. REC. S4712 (daily ed. Apr. 26, 1988) (statement of Sen. Cochran); *Welfare Reform—The Need to Attack Family Dependency*, 134 CONG. REC. S3069 (daily ed. Mar. 25, 1988) (statement of Sen. Cochran).

11. For some politicians, this conclusion is evidenced by the assumptions made regarding the desire of welfare mothers to avoid work. For example, in opposition to the proposed Family Support Act, Representative Marge Roukema stated:

> [T]his bill, if enacted will trigger the law of perverse effects. There exists the potential that this bill will make it more attractive to go on welfare and stay on welfare. . . . H.R. 1720 provides that a welfare mother does not have to have training or take on a job if she has a child three years or younger.

133 CONG. REC. H11515 (daily ed. Dec. 16, 1987) (statement of Rep. Roukema).

After citing statistics that 50% of working women have children under one year of age and announcing that the proposed bill was inconsistent with present economic and social realities, she continued:

> Imagine this: A welfare mother could actually continue to have a child every 2 years and never have to work at all. That's wrong. And not only will that mother not have to work, but her children, born into impoverishment, will have little hope for their own futures. And the cycle of poverty and dependency continues. I ask: How much longer do you think the two-worker couple will tolerate the welfare state and its cost to them in taxes to support that welfare mother? . . . The answer is that they should not have to.

Id. at 11516. This is certainly the image of the "bad" single mother, not a victim, but a calculating individual who deliberately lives a parasitic existence, stealing from the "good" married citizens for her survival.

12. *See, e.g.*, McLanahan & Garfinkel, *Single Mothers, the Underclass, and Social Policy*, 501 ANNALS 92, 94 (1989). In defining the socio-economic classification of the "underclass," the authors assert that "weak attachment to the labor force" is one characteristic possessed by this group. The authors ultimately conclude that a small but significant percentage of single mothers (less than 5%) fall within their definition of underclass (a pejorative term), but only after excluding those who are single mothers as the result of the death of a spouse:

> Disabled workers, widows, and married homemakers may be indirectly attached to the labor force either through their personal work history or through the current or past employment history of their spouse. In the case of disabled workers and widows, the primary source of household income comes from social insurance, which is linked to the past work history of the individual and the individual's spouse, respectively. In the case of married homemakers, the primary source of income is the partner's current earnings.

Id. at 94.

> The typical AFDC parent today is not the "worthy widow" envisaged in the original legislation but a divorced, deserted, or never-married woman. Regardless of extenuating circumstances behind any particular out-of-wedlock birth or the justification for any particular divorce, the fact remains that illegitimacy and divorce have an element of personal choice and responsibility that widowhood does not. . . . Regardless of how difficult it is for individuals to pay for making irresponsible or unfortunate choices, that does not absolve those individuals of dealing with the consequences as best they can, before society is asked to step in to support them and their children. . . . Social assistance has always been based on social norms and expectations. One of the assumptions underlying AFDC was the idea that a mother has an important role to play in the upbringing and socialization of her children. It was the humane intention of the program in 1935 [designed for widows] to enable a mother to take care of her children—in other words, to encourage what was left of a family to stay together. It was, in today's parlance, a "pro-family" measure. But now the program finances a subculture whose citizens argue . . . that they want children but not marriage, because "you don't want the commitments" and "male figures are not substantially important in the family."[13]

In some discourses, however, poor women who become single mothers because of divorce are increasingly included within the category of deserving mothers.[14] This classification is permitted by the recent changes in the way we consider marital dissolution, epitomized by the advent of no-fault divorce, that was built upon the popular perception that the failure of a marriage is a joint responsibility. Although the insistence that there be a designated villain and victim between the spouses may have declined, children, however, continue to be portrayed as innocent victims, usually of their parents' selfishness and hostility during a divorce.[15]

The imagery of welfare discourse, however, remains laden with moral and normative judgments that are founded on stereotypical assumptions about single mothers in the poverty context. The decisions to become or remain a single mother, particularly when undertaken by a woman who has never been married, is the decisive issue for whether one

13. S. Butler & A. Kondratas, Out of the Poverty Trap: A Conservative Strategy for Welfare Reform 138-39 (1987).

14. This category of single mothers has received a great deal of sympathetic attention. *See, e.g.*, L. Weitzman, The Divorce Revolution, The Unexpected Social and Economic Consequences for Women and Children in America 184-87, 262-78 (1985) (discussing alimony and child support payments for women with children).

15. The villain and victim imagery continues to exist in divorce discourses, but it is not directed at the decision to divorce. For example, the contemporary divorce discourse of the helping professionals often casts mothers with custody as villains who wrongfully interfere with their ex-husband's relationship with his child (both of whom are cast as victims). For a fuller exposition of this rhetoric, see Fineman, *Dominant Discourse, Professional Language, and Legal Change in Child Custody Decisionmaking,* 101 Harv. L. Rev. 727 (1988).

is to be considered a "good" mother. If they are not single as a result of death (or, perhaps, divorce), poor, single mothers are deemed "bad" mothers. Their conduct is considered to have "earned" them this designation, and stereotypes about their motivations and behavior abound. Michael Katz reports the comment of "an otherwise sympathetic radio talk show host," who told him: " 'I don't mind paying to help people in need, but I don't want my tax dollars to pay for the sexual pleasure of adolescents who won't use birth control.' "[16] Katz concluded that the host's "outrage summed up popular stereotypes about the relation among adolescent pregnancy, welfare, and the underclass . . . ."[17]

The characterization of some single mothers as "bad" corresponds to the larger popular and political currents that classify the poor as either "deserving" or "undeserving." This classification is often based on the public's perception of whether the impoverished individual is poor because of her personal choices and actions, or as a result of forces beyond her control.[18] Furthermore, this distinction between deserving and undeserving has concrete and material implications attached to its ideological categories; the way that one becomes a single mother often dictates the source of public assistance to which one may turn. Widows are typically entitled to generous social security benefits; whereas mothers who are divorced or who never marry are left to the variability of the child support system or AFDC.[19] Such distinctions, therefore, go beyond social

16. M. Katz, The Undeserving Poor: From the War on Poverty to the War on Welfare 215 (1989).

17. *Id.*

18. In his analysis of the concept of the "underclass" that developed in the mid-1980s and the imagery that this new category evoked about the poor, Katz states:

> [T]he very poor evoked two different images among affluent Americans. When they appeared pathetic, they were the homeless; when they seemed menacing, they became the underclass. Although membership among the homeless and underclass overlapped, public discourse implicitly divided them by degree of personal responsibility for their situation. As long as they remained supplicants rather than militants, objects of charity rather than subjects of protest, the homeless became the new deserving poor.

Id. at 185-86.

19. In arguing for the Family Security Act, Senator Moynihan noted the inequities in financial support services for children. He pointed out that although Survivors Insurance benefits have increased by 53%, AFDC benefits were down an average of 13%. The result is that children receiving social security benefits get over twice the benefits of those relying on AFDC. The Senator questioned the equity of this situation, noting that "these are identically situated children. They are children living in a single-parent, female-headed household. . . . There is no way you can distinguish these children excepting one brutal [sic] reality: The majority of the children receiving Survivors Insurance are white, and the majority of the children receiving AFDC are nonwhite." 134 Cong. Rec. S7631, S7653 (daily ed. June 13, 1988) (statement of Sen. Moynihan). He questioned why this situation developed and answered: "Very simply it came about because the welfare program has become stigmatized as a program for people who did not work in a world where others did." *Id.* Although he recognizes this as a stigma, he concluded that "[it] also reflects a certain reality," and argued that to change the public's unwillingness to adequately fund these programs would require the adoption

stigma (which, in and of itself, is significant); they have a real impact on the material future of a large number of single mothers and their children.

Yet the stigma that society attaches to certain single mothers provides the conceptual basis for the disparate economic outcomes that such groups experience. The parsimonious treatment of poor women who chose to become mothers without marrying, or those whose choices are limited because of circumstances that grow out of poverty, is justified because they can be considered undeserving poor.[20] Recently, in poverty discourses emanating from a broad spectrum of groups, these single mothers have been lumped together with drug addicts, criminals, and other socially-defined "degenerates" in the newly-coined category of "underclass." The undeserving status of single-mother families in this context is established partly by their lack of relationship to the work force (either through their own jobs or through their attachment to a male breadwinner) and partly by their asserted role as mothers in the perpetuation of poverty. Consider, for example, this set of assumptions and conclusions:

> The link between female headship and welfare dependency in the urban underclass is also well established, leading to legitimate concerns about the intergenerational transfer of poverty. At the root of this concern is the paucity of employment among welfare mothers and how this affects attitudes of their children toward work.[21]

After citing statistics on this point, the author continued:

---

of a program that is primarily a jobs system with supplemental income assistance. *Id.* at S7653, S7663.

20. This distinction between categories of poor seems inevitable. As Paula Roberts and Rhonda Schulzinger noted, the principles established through the legacy of the English poor laws affected American thinking:

> First, the poor could be divided into two categories. The "worthy poor" were impoverished through no fault of their own (e.g., the sick and aged) or because they could not find work. "Paupers," on the other hand, were those who refused to work. Paupers represented a moral pestilence and had to be controlled, put in the poor house, and made to work.

Roberts & Schulzinger, *Toward Reform of the Welfare System: Is Consensus Emerging?*, 21 CLEARINGHOUSE REV. 3 (1987).

21. Kasarda, *Urban Industrial Transition and the Underclass*, 501 ANNALS 26, 44 (1989). The author's assumptions as to the composition of "work" is limited to that which is compensated by the market. Mothering, although a socially productive endeavor, is excluded. In noting the worsening of the "economic and social plight" of urban blacks, Kasarda also states that "persistently poor ghetto dwellers [are] characterized by substandard education and high rates of joblessness, mother-only households, welfare dependency, out-of-wedlock births, and crime." *Id.* at 27. The author asserts that one reason that poor urban blacks do not succeed in starting their own businesses and other ethnic groups do is because blacks lack the strong extended families of other ethnic groups. He concluded that the black "fragmented family" cannot pool their resources for labor, childcare, and funds as can other groups with extended families. *Id.* at 42-44.

> One does not require a deep sociological imagination to sense the attitudinal and behavioral consequences of growing up in an impoverished household where there is no activity associated with the world of work and a household that, in turn, is spatially embedded in a commercially abandoned locality where pimps, drug pushers, and unemployed street people have replaced working fathers as predominant socializing agents.[22]

The category of underclass is defined not only by the perception that its designated members are the chronically poor, but also by the belief that their poverty results from their own failings. Single motherhood is considered a sign of degeneration on the same level with crime and other social pathology. Single mothers are implicated by their asserted role in the intergenerational transmission of poverty through their complicity within a "culture of poverty."[23]

Many commentators assume that individual behavior not only causes poverty but also that choosing to make personal changes in their lives can solve the problems of this category of the poor. Consider, for example, the analysis of the problems of the underclass and the recommendations emanating from the American Enterprise Institute:

> Today, for example, significant numbers of American adults are not demonstrating the behaviors expected of free and responsible citi-

22. *Id.* at 45.

23. In a recent issue of *Signs*, Maxine Baca Zinn asserts that there are two models that have been used to describe the urban underclass: cultural and structural. Zinn, *Family, Race, and Poverty in the Eighties*, 14 SIGNS: J. WOMEN, CULTURE & SOC'Y 856 (1989). The cultural model first gained recognition with the "culture of poverty" theory popularized by Daniel Patrick Moynihan in the late 1960s. Moynihan, *The Negro Family: The Case for National Action*, in MOYNIHAN REPORT AND THE POLITICS OF CONTROVERSY 39-132 (L. Rainwater & W. Yancy eds. 1967). Zinn concludes that the cultural model blamed welfare for the disintegration of the family, for providing disincentives to work, and encouraging a lifestyle in which women have children out of wedlock and depend on the system to support the children, thus allowing men to avoid financial responsibility for children.

Loäc Wacquant and William Julius Wilson distinguish their articulation of the development of the underclass from the culture of poverty reasoning. In recounting the statistical data in support of their assertion that "[s]ocial conditions in the ghettos . . . are scaling new heights in deprivation, oppression, and hardship," they note that black communities in Chicago's South Side and West Side have experienced an increase in the number and percentage of poor families, an increase in the out-migration of working and middle-class households, stagnation of income, unemployment, and that "over two-thirds of all families living in these areas were headed by women." Wacquart & Wilson, *The Cost of Racial and Class Exclusion in the Inner City*, 501 ANNALS 8, 11 (1989). The thrust of their article is that the underclass is a product of the spacial concentration and isolation of the most socially excluded and economically marginal members of the dominated racial and economic group, not a new breed of individuals molded by all-powerful culture of poverty. *Id.* at 25. The authors asserted that a variety of factors—class, gender, race—contribute to the marginalization of groups who are concentrated into neighborhoods. Social, political, and economic structures are built up around these marginalized groups in such a way that they perpetuate their oppression. Female-headed households, however, although not blamed for the decline, are used as evidence that a decline has occurred.

> zens. Linked to poverty among an important fraction of the poor is a high incidence of dropping out from school, of failure to prepare themselves for future employment, of begetting children out of wedlock, of crime, of drug use, and of other visible disorders. Such persons—whose numbers appear to be growing—are the behavioral dependent, since their need for help from others springs in significant measure from their own behaviors.
>
> . . . .
>
> It is not entirely a mystery how many climb from poverty. Some specific behaviors empower them. The probabilities of remaining involuntarily in poverty are remarkably low for those who
>
> - complete high school
> - once an adult, get married and stay married (even if not on the first try)
> - stay employed, even if at a wage and under conditions below their ultimate aims.[24]

Predictably, marital status is central to the self-help regimes proposed for the poor. Liberal commentators often share this concern with the marital choices made by the poor and consider marital status an appropriate objective to be fostered by public policy. William Julius Williams writes:

> [P]erhaps the most important factor in the rise of black female-headed families [is] the extraordinary rise in black male joblessness. . . . [T]he decline in the incidence of intact marriages among blacks is associated with the declining economic status of black men. . . . [B]lack women nationally, especially young black women, are facing a shrinking pool of "marriageable" (i.e., employed) black men. This finding supports the hypothesis that the sharp rise of black female-headed families is directly related to increasing black male joblessness.[25]

B. *Singular Pathology*

In poverty discourses, such as those set out above, the label "Mother" is modified by her legal relationship (or lack thereof) to a male. Mothers are classified by whether or not they are single, a fact that is positioned as significant and even central to the discourses. The absence of the formal legal tie to a male is far more than just a descriptive term or classifying category. It has powerful ideological implications. In addition to providing a basis to determine who is undeserving in our culture, the rhetoric constructs single motherhood as dangerous and even

24. WORKING SEMINAR ON FAMILY AND AMERICAN WELFARE POLICY, THE NEW CONSENSUS ON FAMILY AND WELFARE: A COMMUNITY OF SELF-RELIANCE 4, 5 (1987) (The American Enterprise Institute for Public Policy Research).

25. W. WILSON, THE TRULY DISADVANTAGED: THE INNER CITY, THE UNDERCLASS, AND PUBLIC POLICY 104-05 (1987).

deadly, not only to the single mothers and their children, but to society as a whole.

Two recent news articles illustrate the typical ways that popular culture treats the threat of single motherhood. In the first, the Chancellor of the University of Wisconsin at Madison, Donna Shalala, was reported as stating that unwed teen pregnancy is "an issue that threatens the future of Milwaukee and is as dangerous and crippling to young lives as polio."[26] Describing the enormity of the problem as "frightening," Shalala warns that "the stakes are very high. The stakes are our future, not only in Milwaukee, but our future as a state."[27]

Joanne Jacobs, a columnist for the San Jose, California *Mercury News,* wrote an article entitled "Illegitimacy Biggest Killer Of Our Babies." She asserted that the leading killer of young children in America is not drugs, cancer, or accidents, but illegitimacy.[28] This article is offensive beyond the author's use of the antiquated term "illegitimate," a term that suggests that the marital status of the child's mother has implications for the quality of the child. Jacobs reported the results of a study by Nicholas Eberstadt of Harvard's Center of Population Studies and the American Enterprise Institute. In so doing, she popularized and publicized the rather amazing belief that the high death rate for "illegitimate" children was not related to the mother's age, race, education, or income, but was due to "the lack of care that leads to illegitimacy [and that] also leads to poor care for the kids."[29]

Jacobs dutifully lists, although ultimately disagrees with, Eberstadt's conclusions. She also rejects the policy recommendations of Nicholas Davidson that this lack of care warrants such drastic action as the denial of welfare benefits to mothers who "deliberately give birth to and raise a fatherless child."[30] In addition, she rejects Davidson's suggestion to alter the law "to grant divorce only for cause [such as] adultery, abuse or desertion—and to deny women custody of their children," thereby removing the "incentive" to divorce provided by easy no-fault divorce, maternal custody, and mandatory child-support payments.[31] Jacobs nonetheless concludes the column with her own normative observation that "we've become way too accepting of an 'alternative lifestyle' in which fathers are dispensable. It's bad for kids."[32]

26. *Teen Pregnancy Called Issue that Threatens State,* Milwaukee Sentinel, May 1, 1990, at 1, col. 6.

27. *Id.*

28. Jacobs, *Illegitimacy Biggest Killer of Our Babies,* WIS. ST. J., Feb. 9, 1990, at 11A, col 6.

29. *Id.*

30. *Id.*

31. *Id.*

32. *Id.*

These popular characterizations of single motherhood are crude echoes of the images of the single mother prevalent in political, legal, and professional discourses that speculate about the impact of single motherhood on the institution of the family, and ultimately on the fate of our society. We speak of the "broken" family, the "disintegration" of the family, the "crisis" in the family, the "unstable" family, the "decline" of the family, and, perhaps inevitably from some perspectives, the "death" of the family. Underlying such labels is the specter of single motherhood—statistically on the upswing—pathological and disease-like, contaminating society, contributing to its destruction and degeneration.[33] With this perspective so pervasive, it is little wonder that many reformers urge efforts to curb the practice of single motherhood. Even liberal commentators such as Irwin Garfinkel of the Institute for Poverty Research are mindful of the role of deterrents and incentives in welfare reform. Garfinkel states that "[t]he problem with providing more aid to single-parent families is that doing so creates incentives for the formation and preservation of single-parent families."[34]

Unlike some conservative commentators, Garfinkel recognizes that single-parent families are not necessarily bad:

> Of course, it is possible that society is better off—or at least not worse off—as a result of whatever additional single-parent families are created by more favorable treatment of those groups. Not all marriages are made in heaven. Some men beat their wives and children. In some of these cases, all the parties may be better off separate rather than together."[35]

Nonetheless, he concludes that "[d]espite the fact that increases in single parenthood may not be socially pernicious, prudence would suggest that in the face of ignorance we should seek to minimize incentives for single parenthood."[36]

Single motherhood has been designated as the source of other social phenomena such as crime and poverty. Indeed, in the public's mind, and despite overwhelming evidence to the contrary,[37] the face of poverty has

33. *See supra* text accompanying notes 9-10 (statement of Sen. Moynihan).

34. I. GARFINKEL, THE ROLE OF CHILD SUPPORT IN ANTIPOVERTY POLICY 12 (1982) (Institute for Research on Poverty, Discussion Paper, #713-82) (on file with author).

35. *Id.*

36. *Id.*

37. *See* M. KATZ, *supra* note 16, at 216-17. The statistics are summarized by Katz:

> [M]ost poor people in America do not live in families headed by adolescent mothers or even by women. In 1980, 37% of poor people lived in female-headed families. Poverty is not synonymous with single parents. Nor does adolescent pregnancy consume a large share of the social welfare budget or gross national product. In 1980, the money spent on AFDC represented only about 4% of all the costs of major public assistance and social insurance programs for the elderly, totally disabled, and all others, and only a fraction of AFDC payments go to adolescent mothers. In 1984, all means-tested cash transfer pay-

increasingly become that of a single mother, particularly the African-American single mother.

Single motherhood represents both the cause and the result of the disintegration of the family and society; it is a demographic category filled with political and moral significance, and as such is viewed as having both explanatory and predictive powers. This is true not only in popular discourse[38] but in more "reflective" areas of discourse such as the social sciences, policy, and law.[39]

> ments by federal, state, and local governments used 0.8% of GNP or two percent of the share of GNP spent by governments.
>
> Consider, next, the question of birth rates. Among blacks, adolescent birth rates have fallen; among whites they have increased. Between 1970 and 1980, the birth rate of unmarried black women dropped 13%, in contrast to a 27% increase among unmarried white women. Nonetheless, black marital fertility fell even faster (38%), which means that the fraction of births occurring to unmarried women increased. Among just 15- to 19-year-old black women, the nonmarital fertility rate (births per 1,000 women) rose from 76.5 in 1960 to a peak of 90.8 in 1970 and has declined since then; by 1980 it had dropped to 83.0. Among whites, the rate, always lower than blacks', has risen steadily from 6.6 in 1960 to 10.9 in 1970 and 16.0 in 1980. In other words, between 1970 and 1980, the fertility of unmarried black 15- to 19-year-old women dropped almost 10 percent, while the rate for unmarried white women of the same age increased by 48 percent. Although the black rate remained more than five times as great as the white, adolescent pregnancy is not an issue just for blacks.

Id. For a debunking of a variety of "welfare myths," see M. EDELMAN, FAMILIES IN PERIL: AN AGENDA FOR SOCIAL CHANGE 68-74 (1987).

38. *See, e.g.,* Passell, *Economic Watch: Forces in Society, and Reaganism, Helped Dig Deeper Hole for the Poor,* N.Y. Times, July 16, 1989, § 1, at 1, col. 4. The article reports the release of a document from the House Ways and Means Committee that shows a decrease in income for the poorest one-fifth of Americans and an increase in income for the most wealthy one-fifth between 1979 and 1987, and asks: "Why is the gap widening?" The author responds:

> Frank Levy, a professor at the University of Maryland, cites the growth of households headed by single parents who lack the skills and motivation to earn a decent living. The figures bear him out: From 1979 to 1987 the number of single parent families living below the poverty level rose by 46 percent.

Id. at 20. The article proceeds to cite several other factors, such as the increasing number of women competing for jobs in the work force, foreign competition, and changes in tax laws. *Id.* However, it is significant that the immediate response was to look to the increase in single motherhood as responsible for the increase in poverty.

39. The preeminence of the demographic category over material and structural circumstances, even when both factors are considered significant, is evident in the work of Renee Monson and Sara McLanahan. The authors state:

> [M]ost analyst agree that the proximate causes of poverty in mother-only families are: (1) the low earning capacity of single mothers; (2) the lack of child support from non-resident fathers; and (3) the paucity of public benefits.

R. Monson & S. McLanahan, A Father for Every Child: Dilemmas of Creating Gender Equality in a Stratified Society (unpublished paper) (available from author). Nonetheless, as they uncritically note: "The Family Support Act of 1988 attempts to redress [the first two] of these cases." *Id.* at 1. The authors identify as "[a] major weak link in the current child support system [the] failure to establish paternity for children born to never-married mothers." *Id.* The Act attempts to remedy this by "requiring states to (1) increase the proportion of AFDC cases with child support awards; (2) obtain social security numbers from both parents in conjunction with the issuance of birth certificates; and (3) require all parties in contested paternity cases to take a genetic test." *Id.* The Act also urges states to simplify civil paternity proceedings and gives incentives for states to set up programs to track down out-of-state fathers. *Id.*

Why has this casting of single motherhood as pathological, as a social disease, and as one of the core explanations of poverty been so readily accepted and perpetuated in such a variety of contexts, and by people who occupy a wide range of political viewpoints? It must seem evident to liberals that this stereotyping may be viewed as misogynist and racist. In attempting to answer this question, I have reached a number of conclusions about the role of ideology in constructing and perpetuating the images of single mothers in poverty discourses. The representation of single motherhood as pathological is inextricably linked to patriarchical ideology—a constellation of symbols and beliefs about the "natural" or "normal" family that is widely shared in our culture and through which all motherhood discourses are processed.

III. Ideology and Motherhood

I think of ideology as a rationalizing set of principles and concepts that link discourses to power. A close examination of discourse will thus reveal implicit aspects of an underlying ideology. Discourses are linguistic framings or stylized appeals to ideology; ideology, therefore, is what defines and structures the contours of discourse. A critical examination of poverty discourse discloses that patriarchy is the dominant family ideology that fixes the core concepts and images about motherhood. Patriarchy also facilitates, and even mandates, the cross-over effect of such discourses. In this regard, the discourses that concern poverty and single motherhood are consistent with other stereotypes and myths fashioned under the influence of patriarchical ideology in which the dominant family form is a male-dominated reproductive unit with defined gender roles. Although this traditional image has undergone some revisions in light of the modern concerns with gender equity and equality, the nuclear family form, with a sexual affiliation between man and woman as the paradigmatic intimate associational bond, is still the focus in policy discourses. Patriarchical ideology may have adapted to twentieth-century shifts in expressions of sexuality and redefinitions of gender roles, but the fundamental composition and nature of the core images remain constant. Mother and child are defined by their relationship to the patriarch.

When one considers the relationship between motherhood and patriarchy, it seems that motherhood has always been, and continues to be, a colonized concept—an event physically practiced and experienced by women, but occupied and defined, given content and value, by the core

Feminist critiques in this area are often poorly received. Monson and McLanahan complain, for example, that instead of supporting new child support legislation as "creating a partnership between the family and the state that would help ensure the adequate care of children," feminists have objected to the new law as an intrusion "on women's privacy and sexual freedom." *Id.* at 7.

concepts of patriarchical ideology. Single motherhood is viewed as a practice of resistance to patriarchal ideology, particularly because it represents a "deliberate choice" in a world with birth control and abortion. As such, the existence of single motherhood as an ever-expanding practice threatens the hold of the dominant ideology.

A. *Patriarchy as Ideology*

I view patriarchy as an ideology constituted by a set of concepts and symbols that are more complex and convoluted than the simplistic notion that men are the formal holders of power within the family and society (a parody of patriarchy). Gerda Lerner defines patriarchy as the "manifestation and institutionalization of male dominance over women and children in the family and the extension of male dominance over women in society in general."[40] I question the emphasis that this definition places on "manifestation" and "institutionalization" of dominance. This focus presupposes and mandates that societal structures such as the family become the focal point for critical assessment. Such an approach may foster the belief that structural changes alone can transform society. Yet, it seems essential that we go beyond a structural critique and consider the power of dominant ideology, a more elusive and resistive social and cultural product than any mere structure. It is the power of ideology that explains why an individual can resist or reject structure by, for example, refusing to participate in a nuclear family, but still find herself defined and ultimately controlled by the ideology underlying the structure.

My preoccupation with ideology leads me to be more pessimistic about the possibility for social transformation than social critics who focus on structure. Dominant ideologies are subtly and conclusively expressed and repressed in the very creation and recreation of social norms and conventions. They define the contours of culture and of society and its institutions. A dominant ideology is transmitted through everyday language, symbols, and images, as well as through the operations of formal institutions and structures of power.

Patriarchy is such a dominant ideology. It has as one of its organizing premises the belief that the primary affiliation in society is the sexual bond. In our legal and political culture, consistent with the demands of the ideology, it is this affiliation that is privileged as the paradigmatic manifestation of connection. In popular culture, sexual expression (particularly heterosexual expression traditionally realized through marriage) is portrayed as the quintessential indication of maturity, completeness,

40. G. Lerner, The Creation of Patriarchy 239 (1986).

success, and power. Historically, deviance from the heterosexual paradigm brought with it social and even legal sanctions,[41] as well as condemnation in the discourses of psychology, social work, and medicine. In recent years, the legal and heterosexual aspects of the paradigm have been challenged by demands to consider other forms of sexual intimacy as equally privileged as formal marriage. Alternative family legislation has been enacted that recognizes sexual affiliations other than heterosexual formal marriage.[42] I consider these reforms to be complicit with patriarchical ideology, however, because by duplicating the privileging based on heterosexual affiliation, these relationships merely affirm the centrality of sexuality to the fundamental ordering of society and nature of intimacy. The nexus or affiliating circumstances of these "alternatives," although not the traditional "conjugal family," is still the sexual connection.

Single motherhood, therefore, could be considered deviant and threatening simply because it is a rejection of the primacy of the sexual connection as the core organizing familial concept and the privacy basis for social organization. The very label "single mother" separates some practices of motherhood from the institution of "Mother" by reference to the mother's marital status. Mother, as constructed and defined in this discourse, is modified by her relationship (formal and legal) to the father. By contrast, the institution when practiced in its "normal" form is not so modified. No one speaks of a "married mother"; the primary connection of husband and wife is assumed in the unadorned designation of "mother." It is only the deviant form of motherhood that needs qualification and, by implication, justification. In this process of distinguishing the deviant variation of motherhood from the norm, a complementary cluster of stereotypical designated family roles are implied to form the contemporary images of the "ideal" family.

B. *The Role and Function of Ideology*

Thus far I have asserted a relationship between ideology and discourses about motherhood. As a methodological point, I have assumed that the study of rhetoric about motherhood reveals something about the existence and content of the dominant ideology. This in turn reveals

41. Legal sanctions are still possible for "deviant" sexual behavior. *See, e.g.*, Bowers v. Hardwick, 478 U.S. 186 (1986) (upholding constitutionality of state statute criminalizing consensual sodomy).

42. *See, e.g.*, MINNEAPOLIS MINN. CODE tit. 7, ch. 142 (1991) (recognizing the relationship of two non-married but committed adult partners); WEST HOLLYWOOD CA. MUN. CODE §§ 4220-4228 (1988 & Supp. 1991) (adopting procedures whereby adult non-married partners can declare themselves "domestic partners"); *see also* MADISON WIS. GEN. ORDINANCES § 3.38(25) (1988) (providing for the registration of "alternative families").

something about the location of power within society. In defining these relationships, I pursue the suggestion made by Terry Eagleton that "[d]iscourses . . . produce effects, shape forms of consciousness and unconsciousness, which are closely related to the maintenance or transformation of our existing systems of power."[43] Eagleton went so far to assert that ideology can be taken to indicate no more than the connective link between discourse and power.[44] I am also influenced by Christine Harrington and Sally Merry who cast legal ideology in terms of the mobilization of "symbolic resources by groups promoting different projects."[45]

My interpretation and application of these conceptions of ideology begins with an understanding of ideology as a system constituted by a complementary collection of symbols, beliefs, and assumptions that in combination rationalize and give meaning to discourses in the context of power. Ideology in this regard can be considered functional—a selection and sorting mechanism in that it provides coherence, structure, and form to social and political discourses. This understanding of ideology as having systemic implications, however, has theoretical as well as practical implications. On a practical level, it means that ideology is more likely to operate as a conservative force. It will serve to tame or domesticate discourses by exerting a confining pressure on their initial development, ultimately channeling even the most radical ideas into categories approved by the existing conceptual system. Within this perception of ideology, meaningful conflict and competition are confined to the level of discourse. To be labeled "dominant"—as providing the recognized links between power and discourse—an ideology may not be too sharply contested. A dominant ideology, therefore, is relatively stable, although it may be altered over time, adapting in response to the tensions generated by violent or prolonged demands from discourses.

In addition, if we posit ideology as a system of symbols and beliefs, it must be experienced as both complex and dynamic, which means that it will seldom be totally revealed in the context of a specific political debate or contest of discourses. It also will not be easily expressed abstractly, outside of the context of a specific set of assumptions. As a dynamic system, ideology represents a process that is not facilely reduced to a finite set of clearly stated principles that are operational outside of any specific application. A cluster of social and cultural symbols and beliefs, ideology is referenced in seemingly disassociated discourses; it

43. T. EAGLETON, LITERARY THEORY: AN INTRODUCTION 210 (1983).

44. *Id.* at 73.

45. Harrington & Merry, *Ideological Production: The Making of Community Mediation*, 22 L. & SOC'Y REV. 709, 714 (1988).

can be imagined as a series of synapses that together form possible permutations for various appeals (discourses) to the institutions of formal power in our society. The definitive shape and content of an ideology is only partially revealed in its symbols: fragmented and shadowed traces discernable within discourses from their initial formulation to their ultimate translation and transformation within institutions of power.[46]

My assertions have implications for the methodology that I have adopted in examining discourses about motherhood. The task is to uncover and understand the appeal to ideological components—the link to power. Ideology is a shared cultural construct. Its core images define the discourses produced by both the proponent of the status quo and its critic. For example, I have found in examining different discourses about motherhood that the underlying symbols and values are more uniformly shared than the differences in discourses would superficially indicate. Conflicts that are developed and located in these discourses concern the political implications or meanings of the symbols and beliefs that comprise the ideology, but do not dispute their existence or potency. Oppositional discourses reference the shared aspects of the ideology in what are attempts to compel and convince.

IV. Conclusion

An examination of motherhood in terms of its changing manifestations reveals that patriarchy as an ideology has adapted to ostensibly challenging discourses and absorbed seemingly significant challenges while ensuring the relatively undisturbed continuation of power distributions within the family structure. This adaptation or modification of patriarchy was necessitated by the "crisis" of the rising divorce rate in the

46. My use of the term "power" in this context is a narrow and explicitly political one—something manifested in and expressed by a dominant social group within the confines of institutions. In this regard, law and legal institutions are concrete manifestations or locations of power, tethered to ideology.

In the context of my examination of poverty discourses, this methodology has demanded an inquiry into the similarities between liberal and conservative political discourses about welfare and poverty. Both discourses attempt to forge links to the core concepts of an underlying ideology, and an examination of seemingly dichotomous rhetorical visions on an issue begins to reveal these key images and concepts. In addition, the threads of discourse by political actors are also reflected in popular discourses, mirrored in professional discourses such as those of the social sciences and psychology as well as other analyses of poverty, including those of the poor themselves. These discourses are all confined to the dominant core concepts of the ideology—a failure to reference those concepts will mean the discourse will lack the coherence forcefully to critique power, to make changes, to punish or to reward.

In addition, on a broader social and legal level, this referencing of core concepts is evident in other areas including discussions about child custody or child abuse and neglect. These discourses also relate to the overarching rationalizing ideology when the subject of discussion is motherhood.

twentieth century.[47] The traditional family formation was disrupted, but the acceptance of joint custody and the idea of shared parenting has ensured continued male control over children and, through them, over their mothers, even as divorce has become available virtually on demand.[48]

Mandatory paternity laws were fashioned in response to the increasing number of women who chose to become single mothers (this has also been designated as a crisis). These laws were first enacted in the context of AFDC single mothers, but are now expanding into other areas of the law. In Wisconsin, for example, under a rhetorical bludgeon that dictates that every child has a "legal right" to a father (or at least to a father's name on a birth certificate), legislation has been passed that requires all single mothers to participate in paternity proceedings.[49] These proceedings are mandated even if the state is not likely to ever be asked to provide economic support for the child. The core tenets of the ideology of patriarchy—that "natural" families have two parents and that the male position is one of financial primacy and family control—are thus ensured. The major architect of the legislation stated:

> We have now in Wisconsin a law that I authored that will become the model for the nation regarding paternity. It has as its foundation that every child born in Wisconsin has a legal right to a father. Children without legal fathers have started down a slippery slope that leads to poverty. Our new paternity law is a radical departure in that the interests of the child will become equal if not paramount to the interests of the natural parents. The law seeks to ensure that at the time of birth the state and the mother will pursue, for non-marital births, the establishment of paternity and the subsequent collection of support.
>
> But the philosophy of the law is not punitive. Rather, it assumes that families are natural and that it is not appropriate to have laws that have as their result that in one out of every two births outside of marriage the child will have no legal father—paternity will never be established. . . .
>
> Under our new law the birth certificate will be the vehicle to establish paternity early on. A presumption of paternity will be created with the filing of a statement. It will go with the birth certificate in most cases when the baby leaves the hospital. It will provide the basis

47. Even within the confines of the ideal family, the roles that wives and mothers are now expected to play have expanded. Modern female components of the ideal family are now likely to carry briefcases in addition to those cultural artifacts considered essential to present themselves as competent and caring (and therefore, of necessity, married) mothers and alluring wives. The goal of independence and autonomy for the individual has thus been tamed in its feminine presentation; such autonomy contained within the ideal family and confined by other roles is itself no more than a reflection of the resilience of the institution and the attractiveness of its head.

48. *See* M. FINEMAN, *supra* note 2, at 90-94; *see also id.* at 109-43 (discussing the use of social science information and the role of helping professionals in the resolution of custody disputes).

49. WIS. STAT. § 767.45 (1988).

> for courts to order child support. But most important, it will give that child a legal father.
>
> Both our child support law and our new paternity law are designed to ensure parental responsibility and to help families form and most importantly give new rights to children.[50]

The Legislator does not mention that the mother's cooperation in this paternity program will require her to reveal details of her sexual and personal life, or risk the potential sanction of incarceration through a court's contempt power. Nor does he note that the legal father's paternal involvement with the child will not be mandated by the court beyond (perhaps) payment of child support. Also absent is the fact that a legal father can exercise legal rights and become an unwanted or even abusive presence in the mother's and child's life merely because he fathered the child.

Changes in behavior precipitated by alterations in divorce laws and women's expanding economic security (which allow them to choose to become single mothers) have required the remodeling of patriarchy. The remodeling, however, has not substantially altered, let alone led to the rejection of, the core images of the ideology. The foundation remains stable. Society may now be grudgingly forced to accept single-mother households as an unfortunate byproduct of the social and economic dislocations that characterize the latter part of this century, but they are seldom treated as an acceptable, let alone a desirable, family form. In expressions of popular culture, as well as in public policy, single motherhood may be perceived at best as requiring an uneasy accommodation with the dominant ideology. The societal aspiration, however, remains to complete the "family" by the addition of a man.

In the context of the treatment of single mothers in poverty discourses, the focus on family form obscures the economic deprivations that make it difficult for single women to raise their children. Rather than addressing their needs as single mothers, reforms tend to try to push them into a model of family life that is increasingly discredited, even in the middle class from which it arose. The dominance of family imagery contained in the ideology of patriarchy has required the rejection of economic subsidies that would truly support single mother families. Instead, their deviation must be punished and deterred, and their practice of resistance curtailed.

50. Speaker Tom Loftus of the Wisconsin State Legislature, Remarks to the National Child Support Enforcement Association, New Orleans, La. (Aug. 23, 1988) (on file with author).

Reproductive Freedoms and African American Women

Charlotte Rutherford, Esq.†

I. Introduction

As an African American, a civil rights lawyer, a mother, and a feminist, I view reproductive freedoms for African American women from both a personal and professional perspective. Reproductive freedoms are life and death issues for many African American women and, as such, deserve as much recognition as any other freedom. Despite the importance of reproductive freedom in the lives of African American women, traditional civil rights and women's organizations have ignored these interests.

This paper grows out of a speech I delivered at the Third Annual Conference of Women of Color and the Law. I was asked to speak about "Women of Color[1] and Reproductive Rights." I struggled with using that title for this paper because the term "reproductive rights" is a factually and conceptually misleading term. Too often, reproductive rights are defined solely as a woman's right to an abortion. Poor women, however, do not have such a "right" in most states. Even though abortion is legal, poor women probably cannot afford one; in practice, therefore, no such "right" exists.

African American women seek "reproductive choice." We want the power to make genuine choices about our reproductive health. Such choices must be informed and supported by access to quality health care. Unfortunately, the term "reproductive choice" is used synonymously with reproductive rights, and the meaning of choice again is narrowed to focus on the right to abortion. Instead of the terms "reproductive rights" or "reproductive choice," I have selected "reproductive freedom" because of the term's powerful symbolic and historic image.[2] In addition, it accurately describes the desires and needs of

---

† Associate Counsel and Director, Black Women's Employment Program, NAACP Legal Defense and Educational Fund, Inc.; B.S. Portland State University, 1976; J.D. Howard University School of Law, 1983; LL.M. Georgetown University Law Center, 1985. The viewpoints expressed in this article are solely those of the author and do not necessarily reflect the views of the Legal Defense Fund. The author extends many thanks to Kim Barker, New York University Law School student, for her research and editing assistance as well as support in completing this paper.

1. The term "women of color" has been limited to African American women in the title of this paper because most collected and compiled data focuses only on African American and white women or white and non-white women. Data for all women of color groups in all of the subject areas discussed in this paper is not available. To the extent that it is available, however, data for other women of color groups is included.

2. Since the first days of slavery, the struggle for freedom has been the province of African Americans in the United States. That struggle is one of our most significant contributions to American history and

Copyright © 1992 by the Yale Journal of Law and Feminism

African American women.

This article challenges the reader to accept a definition of reproductive freedom which is broader than the right to abortion. African American women want and need reproductive freedoms that range from terminating unplanned and unwanted pregnancies to delivering healthy babies under healthy circumstances. This article also urges civil rights and women's groups to address in their respective agendas the reproductive health needs of African American women. Neither advocacy group has fought to recognize or secure reproductive freedoms for African American women as *we* define those freedoms. Without the active support of advocacy organizations, African American women find ourselves without a voice and often overlooked in discussions on reproductive issues.

When amicus curiae briefs for *Webster v. Reproductive Health Services*[3] were organized, African American women's groups asked the NAACP Legal Defense and Educational Fund, Inc. (LDF) to write a factual argument on behalf of African American women who would be adversely affected by abortion restrictions. After much internal debate, leaders of the organization realized that the staff and supporters of LDF, like much of the African American community, were not unified on the issue. Furthermore, LDF had not concerned itself with or defined its position on the issue of abortion. Ultimately, LDF did not enter the case.[4]

After LDF decided not to address the issues raised in *Webster*, a group of nationally recognized African American women active in public policy issues, including abortion rights, met with the LDF staff. They informed us that "[c]hoice is the essence of freedom"[5] and that African American organizations historically have struggled to achieve freedom of choice in all areas of life. They contended that reproductive freedoms for African American women are as important as the freedom to choose a seat on a public bus, to attend a public

---

culture. African American women have participated fully in the fight for freedom. Harriet Tubman, Sojourner Truth, Ida B. Wells, Mary McLeod Bethune, Rosa Parks, Daisy Bates, Ella Baker, Fannie Lou Hamer, Pauli Murray, Shirley Chisholm, Angela Davis, and Mary Frances Berry, to name a few of our best known champions, form a long and continuous line of freedom fighters. My use of the term freedom in the context of reproductive issues and African American women adds the unique historical and cultural dimension of Africans in America to the discussion of reproductive health issues.

3. 492 U.S. 490 (1989) (*see infra* Part VIII discussing restrictions in *Webster*).

4. Women of color issues were represented among the briefs that were filed in *Webster*, but those issues were not presented by the organization that historically has championed the rights of people of color, particularly African Americans. The Center for Law & Social Justice, the National Conference of Black Lawyers, the National Lawyers Guild, and the Center for Constitutional Rights (counsel of record) represented the following amici curiae: National Council of Negro Women, Inc.; National Urban League, Inc.; The American Indian Health Care Association; The Asian American Legal Defense Fund; Committee for Hispanic Children and Families; The Mexican American Legal Defense and Education Fund; The National Black Women's Health Project; National Institute for Women of Color; National Women's Health Network; Organizacion Nacional de la Salud de la Mujer Latina; Organization of Asian Women; Puerto Rican Legal Defense and Education Fund; Women of Color Partnership Program of the Religious Coalition for Abortion Rights; Women of all Red Nations, North Dakota; YWCA of the U.S.A.; and other organizations.

5. *See* AFRICAN AMERICAN WOMEN ARE FOR REPRODUCTIVE FREEDOM: WE REMEMBER 1 (on file with Nat'l Council of Negro Women, Inc., Wash., D.C.).

school, or to live or work without restriction. They maintained that reproductive freedoms are civil rights issues for African American women and that LDF's voice needed to be heard.[6]

This group of African American women acknowledged that the African American community, especially male African American civil rights and religious leaders, did not visibly support choice. They reminded us that African American women, particularly teens, use abortion services at disproportionately high rates, which demonstrates that access to abortion services is vital to African American women. Furthermore, as one participant cogently noted, "we [women] are not making these babies alone . . . yet."[7] Abortion is a family issue; many men support the choice to terminate unplanned and unwanted pregnancies.

Although not all African American and other women of color are poor, a disproportionate number of us are poor[8] or near poor.[9] Even many of us with upper incomes are closer to our poorer sisters than it might appear—missing two paychecks can make the difference between solvency and the brink of homelessness.[10] The amount of income and type of health

---

6. African Americans have struggled to obtain full citizenship status and rights since the days of slavery. One of the most horrendous aspects of slavery, particularly for female slaves, was the master's complete control of the slave's body, including her sexual and reproductive life—an individual's most personal domain. Without the legal sanction of slavery, women who had been slaves were no longer required to bear children in order to enhance the slave master's property holdings, and they were allowed to keep and raise their own children. Bodily integrity and control over her reproductive capacity were as essential for the newly freed African American woman as the ability to leave the plantation and to work for pay. Many African American women today view reproductive choice issues as civil rights issues in much the same way as slave women viewed the continuum of freedoms they sought. As the oldest legal organization advocating on behalf of African Americans, LDF is an essential player in the struggle to secure reproductive choices for African American women.

7. Author's recollection of statement by one of the participants.

8. In 1990, 29.3% of all African American families had incomes below the poverty level, compared to 8.1% of white families and 10.7% of families of all races. BLACK AMERICANS: A STATISTICAL SOURCEBOOK 285 (Alfred N. Garwood ed., 1992).

Fully 37% of all African American families had incomes below $15,000 in 1990, compared to 14% of white families and 17% of families of all races. At the same time, only 14.5% of all African American families had incomes above $50,000, while 32.5% of white families had higher incomes. *Id.* at 269.

Among all women of reproductive age (15-44), one in six had family incomes below the federal poverty limit in 1987. JACQUELINE D. FORREST ET AL., THE NEED, AVAILABILITY AND FINANCING OF REPRODUCTIVE HEALTH SERVICES 3 (Alan Guttmacher Institute, 1989).

9. The "near poor" are defined as people with incomes below 150% of the poverty level. In 1987, one in four women between the ages of 15 and 44 was below 150% of poverty, and one in three had an income less than 200% of the poverty level.

About one-quarter of all women of reproductive age are African American, Hispanic or non-white. The proportion who are African American is three times higher among low-income women than among those at over 150% of the poverty level. The proportion who are Hispanic is slightly more than twice as high among lower income women. FORREST ET AL., *supra* note 8, at 3.

10. In 1988, 44.5% of all African American households were able to acquire interest-earning assets at financial institutions, while 76.6% of all white households held such assets. Another 1.9% of African American households held other interest-earning assets while 10.5% of white households held such assets. BLACK AMERICANS: A STATISTICAL SOURCEBOOK, *supra* note 8, at 293.

Examining the mean value and net worth of those assets shows even greater disparities between African American and white households, demonstrating the economic vulnerability of relatively affluent African American families. The mean value of interest-earning assets at financial institutions for African American households in 1988 was $4,350, compared to $18,890 for white households. The mean net worth of African American households owning assets was $23,651, compared to $101,329 for whites. *Id.* at 294. Relatively

insurance coverage one has determine one's reproductive choices.[11] Irrespective of what the law might state one's reproductive rights to be, if money is needed to exercise those rights, most African American women and other poor women of color will not have many. The reproductive rights and choices of poor women of color are fairly limited and sometimes non-existent.

Moreover, the women also informed LDF that the question of race "colors" the definition of reproductive issues for African American women. The question of choice for African American women is not limited to the issue of abortion. African American women truly want choices—the choice of terminating an unplanned pregnancy as well as the choice of having a healthy baby. But as a result of our limited financial resources, many poor African American and other women of color find that reproductive health care that ensures healthy mothers and babies is unavailable. This leaves many African American women without meaningful choices.

Finally, the women informed us that white women's organizations fail to address their broader definition of choice (discussed fully below). Discussions and policy decisions in the area of reproductive rights generally focus solely on abortion. Important reproductive freedoms for African American women and other poor women of color are not identified, fought for, or secured by either traditional civil rights or white women's groups.

As a result of the dialogue with the persuasive sisters, LDF agreed to recognize reproductive health as an important civil right for African American families and to represent those interests. In order to determine the specific focus of LDF's reproductive health docket, we met with direct service providers, public health policy experts, consultants, and African American women involved in reproductive health issues. This paper represents the opinion of those advocates and experts with whom we met and of the African American sisters who persuaded us to address the issue. They all stated emphatically that reproductive choice for African American women must encompass a broad definition of reproductive health issues, rather than focus only on the issue of access to abortion services.

At a minimum, reproductive freedoms for poor women should include:

1) access to reproductive health care;
2) access to early diagnosis and proper treatment for AIDS, sexually transmitted diseases, and various cancers;

---

affluent African American households have not acquired significant financial cushions with which they can remain solvent during hard economic times caused by illness and loss of work.

11. In 1985, 73% of all women aged 15 to 44 had some form of private insurance, 10% were covered by public sources, primarily Medicaid, and 17% were uninsured. Almost all private insurance covers surgery, hospital care, and diagnosis and treatment of illness, but over 80% of policies do not include office visits or services for preventive care, such as contraception. "Adding an estimated 59 percent of all women who are not covered for preventive care (80 percent of those covered by private insurance) to the 17 percent who have no health insurance at all indicates that at least 76 percent of all women of reproductive age must pay themselves for preventive, nonsurgical reproductive health care." FORREST ET AL., *supra* note 8, at 18.

3) access to prenatal care, including drug treatment programs for pregnant and parenting drug abusers;
4) access to appropriate contraceptives;
5) access to infertility services;
6) freedom from coerced or ill-informed consent to sterilization;
7) economic security, which could prevent possible exploitation of the poor with surrogacy contracts;
8) freedom from toxins in the workplace;
9) healthy nutrition and living space; and
10) the right to safe, legal, and affordable abortion services.

I will not address all of the issues identified above in this paper. I have selected those issues that are either of concern to the greatest number of women or rarely ever discussed as they apply to African American women. I discuss why, at a minimum, a broad definition of reproductive freedoms must include access to adequate prenatal care; access to sex education and appropriate contraceptives; access to infertility services, including early diagnosis and proper treatment for preventable causes of infertility; concern about surrogacy and the potential exploitation of poor women of color; freedom from coerced or ill-informed consent to sterilization; freedom from reproductive hazards in the workplace; and, last but not least, the option of abortion. This paper will also provide some information on current legislative efforts to improve the delivery of services in some of the critical areas mentioned above.

II. PRENATAL CARE

As a result of inadequate prenatal care, African American mothers and babies unnecessarily die at disproportionate rates compared to white mothers and babies. Lack of prenatal care can result in low birth weight babies, infant mortality, and maternal mortality.[12] "Babies born to mothers receiving no prenatal care are three times more likely to die in infancy than those whose mothers do receive such care."[13] This problem is particularly relevant to African Americans, whose infant mortality is twice that of whites.[14] The fact

12. Low birth weight, defined as less than 5.5 pounds, is a leading cause of infant death. "Small babies who survive face an increased risk of being impaired for life by autism, retardation, cerebral palsy, epilepsy, learning disabilities, and vision or hearing loss." CHILDREN'S DEFENSE FUND, A VISION FOR AMERICA'S FUTURE: AN AGENDA FOR THE 1990s 38 (1989).

The maternal mortality rate (the number of deaths of mothers per 100,000 live births) for all African American women in 1986 was 19.3 compared to 4.7 for white mothers. U.S. DEP'T OF HEALTH AND HUMAN SERVICES, HEALTH UNITED STATES, 1988, at 71 tbl. 29 (1989).

13. Angela Davis, *Sick and Tired of Being Sick and Tired: The Politics of Black Women's Health*, *in* THE BLACK WOMEN'S HEALTH BOOK 21 (Evelyn White ed., 1990).

14. *Id.* at 21.

> In 1987, more than 38,000 infant deaths were recorded in the United States. Infant mortality is highest among Blacks—17.9 per 1,000 live births. For Whites, the 1987 infant mortality rate was 8.6; for all minority races as a group, it was 15.4. The overall infant mortality figure for American Indians and Alaska Natives in Reservation States is 9.7, based on 1984-86 data, but

that African American babies are twice as likely as white babies to be born to mothers who received late or no prenatal care contributes significantly to the disproportionately high infant mortality rate of African American babies.[15]

In 1986, African American women were 3.8 times more likely than white women to die from pregnancy-related causes. All women of color were 3.3 times more likely than white women to die from pregnancy-related causes. The leading causes for these maternal deaths are considered "preventable or probably preventable through routine medical care before pregnancy, early and continuous prenatal care, risk appropriate delivery procedures, and routine care after birth."[16]

As a result of inadequate prenatal care, African American mothers and babies are unnecessarily dying. Supporters of reproductive choice should support a woman's choice to bear a child and should advocate for better and more available prenatal care.

III. Contraceptives and Sex Education

Adequate education about the proper use of and access to contraceptives is essential to prevent unwanted pregnancies for sexually active heterosexual women. Each year, more than half of the over six million American women who become pregnant do so unintentionally.[17] Of the approximately 3.4 million women with unintended pregnancies each year, forty-three percent, or approximately 1.5 million, practiced some form of contraception.[18]

A society as divided as our own on the issue of pregnancy termination should encourage pregnancy prevention through education and easy access to contraceptives. Sex education and information about various forms of contraceptives ought to be widely available to younger as well as older women.

> when measured separately, the rate for Alaska Natives is unusually high. Available records indicate that infant mortality among U.S. Hispanics, at 7.9, compares favorably with rates among non-Hispanic Whites. Some Asian-American groups, including Chinese, Japanese, and Filipino, have unusually low infant mortality rates, ranging from 4.0 to 4.7 (1986 data).

U.S. DEP'T OF HEALTH AND HUMAN SERVICES, HEALTH STATUS OF MINORITIES AND LOW-INCOME GROUPS 87 (3d ed. 1991).

15. CHILDREN'S DEFENSE FUND, BLACK AND WHITE CHILDREN IN AMERICA: KEY FACTS 76 (1985). Nearly one African American baby out of ten is born to a mother who received late or no prenatal care. Among African American teenage mothers under age 15, the proportion increases to two in ten. *Id.* at 76.

> Blacks were more than twice as likely as Whites to have late (third trimester) or no prenatal care (11.1 percent vs. 5.0 percent), and the frequency of late or no care among American Indians was at least as high as that for Blacks—11.7 percent. Only 5.8 percent of Asian and Pacific Islander mothers delayed or did not obtain prenatal care. . . . Overall, 12.7 percent of Hispanic mothers had late or no care, while only 4.1 percent of non-Hispanic White mothers did. Non-Hispanic Black mothers had late or no care in 11.6 percent of cases. Within the Hispanic population, the percentages range widely, from 3.9 percent of Cuban mothers to 17.1 percent of Puerto Rican mothers.

U.S. DEP'T OF HEALTH AND HUMAN SERVICES, *supra* note 14, at 96.

16. CHILDREN'S DEFENSE FUND, THE HEALTH OF AMERICA'S CHILDREN: MATERNAL AND CHILD HEALTH DATA BOOK 10 (1989).

17. RACHEL B. GOLD, ABORTION AND WOMEN'S HEALTH 11 (Alan Guttmacher Institute, 1990).

18. *Id.* at 12, 13.

School-based clinics (SBC's) have proven effective in providing basic health care and contraceptive counseling to medically underserved teenagers, but unfortunately, the clinics are not widely available.[19]

So little money has been invested in research about contraceptive methods that American women today have fewer contraceptive options than European women.[20] Since the advent of the Pill in 1960, only one truly new contraceptive method—Norplant—has been approved by the Food and Drug Administration (FDA) for American markets. As discussed below, Norplant has not been welcomed unanimously and has the potential for misuse among poor African American women.

Approved by the FDA in December 1990, Norplant consists of six matchstick-sized rubber tubes filled with a steroid commonly used in contraceptive pills. The tubes are imbedded under a woman's skin, usually inside her upper arm. The steroid is released slowly for up to five years, preventing pregnancy by suppressing ovulation. It has a 99.8 percent success rate[21] and must be surgically removed.

Norplant operates like temporary sterilization and has been imposed as a criminal penalty for mothers who are convicted of child abuse[22] and women

---

19. CENTER FOR POPULATION OPTIONS, SCHOOL-BASED CLINICS ENTER THE '90S: UPDATE, EVALUATION AND FUTURE CHALLENGES 1 (1989). There are only 150 school-based clinics that operate in middle, junior, and senior high schools. Most of these provide:

> primary health care, physical examinations, laboratory tests, diagnosis and treatment of illness and minor injuries, immunizations, gynecological exams, pregnancy testing and counseling, referral for prenatal care, birth control information and referral, nutrition education, weight reduction programs and counseling for substance abuse. Some offer prenatal care on site; a few dispense contraceptives and provide day care for children of students.

Id.

On average, about half of the student body in schools with SBC's enrolls in the SBC and eight of ten of students enrolled use the clinic's services. "For about half of the enrolled students, school-based clinics are their sole or primary source of health care." *Id.*

20. Nine new contraceptive methods were made available during the period 1980-1990. Most of them are only available abroad.

> Injectable steroids, for example, are now available in 90 countries. France and China offer RU-486, an abortion pill, with Sweden, the Netherlands and Britain perhaps soon to follow. In the United States, only a low-dose oral contraceptive, an IUD called the copper-T830A, the contraceptive sponge—and last week Norplant—reached consumers. Some former options, including five types of IUD's, simply vanished in the litigious 1980s, although most were never shown to be unsafe, leaving women with fewer birth-control choices than they had five years ago. "The United States is the only country other than Iran in which the birth-control clock has been set back during the past 10 years," says Carl Djerassi, who helped synthesize the Pill in 1951.

Doug Podolsky & Marjory Roberts, *Sorry, Not Sold in the U.S.*, U.S. NEWS & WORLD REP., Dec. 24, 1990, at 65.

21. Matthew Rees, *Shot in the Arm: The Use and Abuse of Norplant*, NEW REPUBLIC, Dec. 9, 1991, at 16.

22. Darlene Johnson, a 27-year-old pregnant African American mother of four, was convicted of child abuse after beating two of her children with a belt and an electrical extension cord because she found them poking a wire in an electric socket. California Judge Howard Broadman gave Johnson the option of spending one year in prison and three years on probation, or four months in prison, three years on probation, and three years on Norplant. Johnson, unrepresented at sentencing, chose the Norplant sentence, but when she later changed her mind, the judge refused to change the order. The sentence has been appealed on the grounds that it violates her procreative rights, invades her privacy, and entails a judge practicing medicine without a license. Billy Allstetter, *Compulsory Conception: Does the Punishment Fit*

of childbearing age who have been convicted of drug possession.[23] Furthermore, legislation that provides financial incentives to women receiving welfare in order to encourage their "voluntary" use of Norplant has been introduced in Louisiana and Kansas and is under consideration in several other states.[24] While access to abortion services is denied public funding in most states, Norplant is funded by Medicaid in every state except California and Massachusetts.[25]

While providing poor women with access to Norplant (which currently costs about $350 for the device, $90 for insertion, and $84 for removal) is laudable, offering cash incentives to poor women who are on welfare to encourage their use of Norplant should not be allowed. From the woman's perspective, the primary issue is whether she has chosen voluntarily to restrict her procreative ability or has been unduly enticed to do so because of her limited economic situation. Another question is whether she had adequate information, including knowledge of other contraceptive options, with which to make an informed decision to use the device.

My primary concern, however, is that using cash incentives to encourage Norplant use among women receiving welfare is the wrong emphasis and sets the wrong tone. Limiting the procreative rights of poor African American women will not eliminate the root causes of poverty, which are racism, sexism, and classism. Fewer children will not ameliorate the lack of jobs and the poverty among African Americans, and we should not accept a policy position which implies that our disproportionate poverty rates can be solved if African Americans merely have fewer children. A policy targeting welfare women (mistakenly believed by most Americans to be primarily African Americans) is grounded in the same racism and classism that causes poverty and should not be permitted. At the same time, however, poor African American women should have the same procreative options which are available to affluent women, including Norplant if that is their choice.

Women must be able to control our reproductive lives; control must remain in our hands, not the hands of the doctor or the state. No coerced form of contraception should be approved or legislated. Even in rare cases where a woman has been convicted of child abuse or neglect, society should not permit courts to mandate her reproductive choices instead of ensuring through treatment that such behavior does not happen again. Women who want children

---

the Crime?, AMERICAN HEALTH: FITNESS OF BODY AND MIND, May 1991, at 32; Rees, *supra* note 21, at 16; Mark Stein, *Judge Stirs Debate With Ordering of Birth Control*, L.A. TIMES, Jan. 10, 1991, at A3; Mark Stein, *Judge to Let Birth Control Order Stand*, L.A. TIMES, Jan. 11, 1991, at A3.

23. Legislation introduced in Kansas would require Norplant use for women of childbearing age convicted of felony possession or distribution of cocaine, crack, or heroin. Rees, *supra* note 21, at 16.

24. *Id.* at 16.

25. *Id.* California Governor Pete Wilson allocated an extra $5 million to the Office of Family Planning to make Norplant available to poor women who are not eligible for Medicaid or Medi-Cal. *Id.* at 16. *See also* Editorial, *Welcome Contrast on Birth Control*, L.A. TIMES, May 20, 1991, at B4; Daniel Weintraub & George Skelton, *Wilson Favors Use of Birth Control Implant*, L.A. TIMES, May 17, 1991, at A1.

must be able to decide the number and timing of their children, and they must make those decisions fully informed of both their reproductive choices and the consequences of those choices.

Access to reproductive services and information, including information about the option of abortion, is being restricted for poor women, particularly women of color, who disproportionately rely on federally funded family planning clinics for contraceptives and general reproductive health services. The Supreme Court's 1991 decision in *Rust v. Sullivan*[26] forces family planning clinics to choose between receiving federal funding or providing unrestricted medical information to their pregnant patients; many clinics will reduce services rather than cease abortion counseling.[27]

Rust v. Sullivan upheld regulations issued by the Department of Health and Human Services under Title X of the Public Health Service Act.[28] The regulations restricted the speech of doctors and health professionals by prohibiting them from advising their pregnant patients about the abortion option, even if their patients expressly ask for such information.[29] Nevertheless, the Supreme Court found them constitutional. Moreover, the Court upheld the "Gag Rule" requirement that a federally funded Title X program must be separated from a clinic's range of other services, including

---

26. 111 S.Ct. 1759 (1991).

27. An informal telephone survey conducted by an LDF intern in October 1991 revealed that the "Gag Rule" provisions upheld in *Rust* had not yet been implemented and therefore clinics had not yet been affected. Most Planned Parenthood affiliates (representing 17% of all Title X clinics) reported that they would forego federal funds rather than compromise staff's ability to provide abortion counseling and referrals. Other clinics were concerned that the Gag Rule would create conflicts with state informed consent laws governing doctors. These providers also indicated that they would forego federal funds.

The Alan Guttmacher Institute released a memorandum on the "Effects of Implementing the Gag Rule" on July 15, 1991, which stated:

> Nationally, Title X contributed one-third of all public dollars spent for family planning in 1987 and well over half of all funds in some states. It is a major source of stable support for most clinics. (Medicaid, the Maternal and Child Health and Social Services block grants as well as state and local revenues contribute the balance.) Loss of Title X funding would so undermine the financial viability of many facilities that it would precipitate their closure.

Memorandum from the Alan Guttmacher Institute to Interested Parties 2 (July 15, 1991) (on file with author) (footnote omitted).

28. 42 U.S.C. §§ 300-300a-41 (1989).

29. 42 C.F.R §§ 59.7-.10 (1991). Section 59.8(a)(1) states that "a title X project may not provide counseling concerning the use of abortion as a method of family planning or provide referral for abortion as a method of family planning." Section 59.8(a)(2) states that Title X projects must refer every pregnant client "for appropriate prenatal and/or social services by furnishing a list of available providers that promote the welfare of mother and unborn child." A Title X project is expressly prohibited from referring a pregnant woman to an abortion provider, even upon specific request. When asked for a referral to an abortion provider, the Title X doctor or health professional must tell the pregnant woman that "the project does not consider abortion an appropriate method for family planning and therefore does not counsel or refer for abortion." 42 C.F.R. § 59.8(b)(5). The regulations broadly prohibit a Title X project from engaging in activities that "encourage, promote or advocate abortion as a method of family planning." 42 C.F.R. § 59.10(a). Section 59.9 requires that Title X projects be organized so that they are "physically and financially separate" from prohibited abortion activities and that "[m]ere bookkeeping separation of title X funds from other monies is not sufficient."

Challenges to the regulations contended that they were contrary to the underlying statute and that they violated First Amendment rights of doctors, staff, and patients and Fifth Amendment rights of pregnant women to make informed decisions about whether or not to exercise their constitutional right to abortion. Rust v. Sullivan, 111 S.Ct. at 1771-72, 1777.

publicly sponsored prenatal care programs and privately funded abortion services.[30]

For most clinics, the expense of establishing a separate facility with separate staff will be financially impossible. As a result of these regulations, most clinics that receive Title X funds will either eliminate abortion counseling or maintain their abortion counseling and services and lose federal funding altogether, jeopardizing overall financial stability and continued operation. Some programs will cease to operate entirely. Each alternative will have devastating consequences for poor women who rely on family planning clinics for contraceptives and general reproductive health services.

In an amici curiae brief on behalf of interested organizations, including women of color groups,[31] LDF argued that the regulations violated Congress' intent to provide comprehensive reproductive health care services for poor women, particularly women of color,[32] and that, as such, the regulations were invalid.[33] Not only did the brief cite the legislative history, which was replete with references to the "poor women" and "minority women" who were to be served by the bill,[34] but it also presented evidence that women of color are more likely to be served by these clinics.[35]

African American teenagers, in particular, rely heavily on clinic

30. 111 S.Ct. at 1769-1771.

31. Amici included NAACP Legal Defense and Educational Fund, Inc.; Mexican American Legal Defense and Educational Fund; Native American Community Board; National Urban League, Inc.; Black, Asian Pacific, Hispanic, and Native American Caucuses of the National Women's Political Caucus; Women of Color Partnership Program of the Religious Coalition for Abortion Rights; American Indian Health Care Association; Office of Church and Society of the United Church of Christ; National Institute for Women of Color; National Black Women's Health Project; New York City Commission on Human Rights; and National Medical Association. Amicus Brief for the Legal Defense Fund, Rust v. Sullivan, 111 S.Ct. 1759 (1991) (Nos. 89-1391 and 89-1392).

32. Congress indicated that Title X was specifically aimed at eradicating a dual system of health care for poor and wealthy women. *See* PPFA v. Bowen, 680 F. Supp. 1465, 1469 (D. Colo. 1988) (Title X designed to eliminate two-tiered system of family planning services).

Poverty among women heading families alone is staggering. In 1991, almost five million working mothers maintained their families alone and 22.3% of them lived in poverty. NAT'L COMM'N ON WORKING WOMEN OF WIDER OPPORTUNITIES FOR WOMEN, WOMEN, WORK AND FAMILY: WORKING MOTHERS—OVERVIEW (on file with author). In 1988, of all poor African American families, 75.6% were maintained by African American women alone, compared to 44% of poor white families and 47.8% of poor Hispanic families. WIDER OPPORTUNITIES FOR WOMEN, RISKS & CHALLENGES: WOMEN, WORK, & THE FUTURE 12 (1990).

33. Amicus Brief for the Legal Defense Fund, *supra* note 31, at 1, 5. Under the Administrative Procedures Act, 5 U.S.C. § 706 (1977), a reviewing Court may invalidate agency regulations that conflict with the underlying law or are beyond the scope of statutory jurisdiction.

34. Amicus Brief for the Legal Defense Fund, *supra* note 31, at 5-14. In targeting the poor with this legislation, Congress recognized that Title X necessarily and specifically targeted minority women. *See, e.g.*, 116 CONG. REC. 37,374 (1970) (statement of Rep. Schmitz) ("The people at whom this bill is specifically aimed are the poor . . . and . . . minorities").

35. Amicus Brief for the Legal Defense Fund, *supra* note 31, at 29-34. In 1988, African American women represented 13% of 58 million U.S. women of reproductive age (15-44 years). THE ALAN GUTTMACHER INSTITUTE, FACTS IN BRIEF: CONTRACEPTIVE SERVICES (Sept. 25, 1991). In 1988, an estimated 3.74 million women used a Title X clinic in their last family planning visit during the previous 12 months; of these, 28.1% were African American and 3.2% were other women of color. WILLIAM D. MOSHER, NATIONAL CENTER FOR HEALTH STATISTICS, USE OF FAMILY PLANNING SERVICES IN THE UNITED STATES: 1982 AND 1988 at 4 (1990).

services.[36] An essential prevention strategy for those teens who are sexually active includes access to contraceptive services and counseling.[37] Inadequate access to contraception services and counseling leads to large numbers of unintended pregnancies, disproportionately among young women of color.[38] In 1988, there was an increase in both the number of births to young women under age twenty and the percentage of that population to give birth. African American and Latina teens accounted for virtually all of the recent increase in birth rates among teens.[39]

Recognizing that poor women were most likely to be affected by the restrictions on the information a federally funded clinic may provide to its patients, the Court in *Rust* shifted the blame for having to receive restricted medical information onto the shoulders of the poor woman. It stated that "'[t]he financial constraints that restrict an indigent woman's ability to enjoy the full range of constitutionally protected freedom of choice are the product not of governmental restrictions on access to abortion, but rather of her indigency.'"[40]

The Court not only blamed poor women for being poor, but also told them that they have no right to expect the same information that a woman who pays for private medical care receives. To justify this conclusion, the Court "engage[d] in uninformed fantasy,"[41] holding that "the Title X program regulations do not significantly impinge upon the doctor-patient relationship . . . Nor is the doctor-patient relationship established by the Title X program sufficiently all-encompassing so as to justify an expectation on the part of the patient of comprehensive medical advice."[42] While acknowledging that "[i]t would undoubtedly be easier for a woman seeking an abortion if she could receive information about abortion from a Title X project . . .", the Court nevertheless holds that "the Constitution does not require that the Government . . . provide that information."[43] Finally, the Court rationalizes:

---

36. In 1988, 41% of all African American teenagers aged 15 to 19 had visited a family planning clinic within the last 12 months. MOSHER, *supra* note 35, at 2.

37. Relatively few pregnant teens practice regular contraception:

> Older abortion patients are more likely to have practiced some form of contraception than are younger patients who have been sexually active for a shorter time. Thirty-nine percent of abortion patients 17 years or younger said they had used some form of contraception in the month they had become pregnant, compared with 52 percent of patients 20-29 years old. Twenty-seven percent of patients 17 or younger said they had never practiced contraception, compared with only six percent of women in their twenties.

GOLD, *supra* note 17, at 12-14.

38. The birth rate among teens aged 15-17 rose 10% between 1986 and 1988, after having remained fairly stable for a decade. KRISTEN A. MOORE, FACTS AT A GLANCE 1 (Child Trends Inc., 1990). "Calculations by the Alan Guttmacher Institute indicate that in 1987, only 18% of the pregnancies to women under age 20 resulted in births that were intended, while 40% resulted in births that were not intended births, and 42% ended in abortions." *Id.*

39. *Id.*

40. Rust v. Sullivan, 111 S.Ct. 1759, 1778 (quoting Harris v. McRae, 448 U.S. 297, 316 (1980)).

41. *Id.* at 1782 n.3 (Blackmun, J., dissenting).

42. *Id.* at 1776.

43. *Id.* at 1777.

> Congress' refusal to fund abortion counseling and advocacy leaves a pregnant woman with the same choices as if the government had chosen not to fund family-planning services at all. The difficulty that a woman encounters when a Title X project does not provide abortion counseling or referral leaves her in no different position than she would have been if the government had not enacted Title X.[44]

In other words, poor women who use federally-funded family planning clinics for health services have no right to expect to be fully informed about all medical options which might apply, including the option of abortion. Amazingly, it is the poor woman's "indigency" which has restricted her right to be fully informed, not the federal government's Gag Rule. Even though her health may be affected by her lack of information, her "indigency" is to blame, not the government's restrictions on what her doctor can tell her.

The Title X regulations exacerbate poor women's already serious problem of inadequate access to contraceptives and health services, including abortion counseling. Furthermore, pregnancy at an early age for young women of color will most likely increase the probability of living in poverty.[45] Teenage mothers and their children are significantly more likely to live in poverty than older mothers and their offspring.[46]

The prevention of unwanted and unintended births to both teenagers and other poor women will increase their families' chances of escaping the effects of poverty.[47] At the same time, any restriction on health services and access to contraceptive and abortion information will have serious consequences for poor women who are already experiencing difficulty in locating affordable reproductive health care.[48]

---

44. *Id.*

45. Among households headed by individuals between 15 and 24 years of age, the poverty rate is staggering: 65.3% for young African American families and 28.5% for young white families. BLACK AMERICANS: A STATISTICAL SOURCEBOOK, *supra* note 8, at 285.

46. Of all women who received AFDC payments in 1988, 59% were 19 or younger when their first child was born. Nearly half of all children aged five or younger who lived in poverty in 1988 had mothers who began having children as teenagers. MOORE, *supra* note 38, at 2.

47. Poverty rates increase as the size of a family increases. In 1989, the poverty rate among all two-person African American families was 22.7% (compared to a poverty rate of 6.6% for a white two-person family). The poverty rate among six-person African American families was 44.3% (15.6% for comparable white families), and 51.2% among African American families with seven persons or more (compared to 25.5% for white families with seven persons or more). BLACK AMERICANS: A STATISTICAL SOURCEBOOK, *supra* note 8, at 285.

48. The tragic effects of what is truly a health care crisis for poor women are well known and widely documented. *See, e.g.*, U.S. DEP'T OF HEALTH AND HUMAN SERVICES, HEALTH STATUS OF MINORITIES AND LOW INCOME GROUPS, *supra* note 14, at 99; Irma J. Bland, *Racial and Ethnic Influences: The Black Woman and Abortion, in* PSYCHIATRIC ASPECTS OF ABORTION 171 (Nada L. Stotland ed., 1991); American Medical Association, Council on Ethical and Judicial Affairs, *Black-White Disparities in Health Care*, 263 JAMA 2344 (1990) ("[u]nderlying the racial disparities in the quality of health among Americans are differences in both need and access. Blacks are more likely to require health care but are less likely to receive health care services."); *see also* Harris v. McRae, 448 U.S. 297, 339 (1977) (Marshall, J., dissenting); Beal v. Doe, 432 U.S. 438, 455, 459 (1977) (Marshall, J., dissenting) (taking note of paucity

Advocacy groups must work to ensure that poor women have access to all medically necessary information with which to make informed decisions about their reproductive health.[49] Moreover, access to affordable and available reproductive health care services and pregnancy prevention programs must also be a high priority. The availability of such services is essential if African American and other poor women are to have meaningful choice in their reproductive lives.

IV. INFERTILITY SERVICES

The risk of infertility is one and a half times greater for African Americans than for whites,[50] yet discussions of reproductive issues concerning African American women seldom include the need for infertility services. Currently, the risk factors contributing to infertility among African American women include sickle cell anemia and other genetic disorders, alcohol and drug abuse, nutritional deficiencies, untreated infectious diseases such as gonorrhea and pelvic inflammatory disease, and infection after childbirth or after a poorly performed abortion.[51] According to Laurie Nsiah-Jefferson, other reasons for high infertility rates among poor women and women of color include "sterilization abuse, hysterectomies, IUD and birth control usage, lack of access to medical treatment, deleterious environmental and working conditions, and unnecessary surgery and medical experimentation."[52]

Basic information and adequate reproductive health care could eliminate many of these causes of infertility. For example, screening for sexually transmitted diseases, especially chlamydia, as part of routine gynecological

of abortion providers available to poor women and lack of "meaningful opportunity" to obtain abortion).

49. Congressional efforts are currently underway to remove the regulations' restrictive language on abortion counseling and to reinstitute the prior regulations' interpretation of the statute, which prohibited the performance of abortions with federal funds. On March 21, 1992, the *New York Times* reported that the Bush Administration announced that doctors in federally funded family planning clinics may give limited advice on abortions, but counselors and nurses may not. Detailed enforcement guidelines have have not been issued. In explaining the new stance, administration officials stated doctors could discuss abortion but would not be free to refer women to abortion clinics. Choice advocates called the stance a "transparent political ploy because very few of the health care workers who counsel low-income women in these clinics are doctors. The people who counsel women are still gagged." Philip J. Hilts, *White House Allows Some Advice At Public Clinics About Abortion*, N.Y. TIMES, Mar. 21, 1992, at A3.

50. U.S. DEP'T OF HEALTH AND HUMAN SERVICES, PUBLIC HEALTH SERVICES, HEALTH STATUS OF MINORITIES AND LOW-INCOME GROUPS 58 (1985) (23% of Black couples found infertile compared to 15% of white couples).

51. *Id.* Untreated pelvic inflammatory disease (PID) accounts for 15-30% of infertility cases (100,000-150,000 cases per year). PID is caused chiefly by gonorrhea and chlamydia; IUD infections also contribute a small number of cases. Untreated PID destroys fallopian tubes, causes ectopic pregnancies, and can lead to the need for hysterectomy. Philip J. Hilts, *Growing Concern Over Pelvic Infection in Women*, N.Y. TIMES, Oct. 11, 1990, at B12.

52. Laurie Nsiah-Jefferson, *Reproductive Laws, Women of Color, and Low-Income Women*, *in* REPRODUCTIVE LAWS FOR THE 1990S 17, 49 (Nadine Taub & Sherrill Cohen eds., 1988). *See also* BLACK WOMEN'S COMMUNITY DEV. FOUND., MINICONSULTATION ON THE MENTAL AND PHYSICAL HEALTH PROBLEMS OF BLACK WOMEN 19-26, 80-90 (1974).

exams could reduce the number of new infertility cases dramatically.[53] Women who have been infected with chlamydia are twice as likely to experience an ectopic pregnancy, which can result in infertility;[54] untreated chlamydia may account for as many as twenty thousand ectopic pregnancies each year.[55]

In general, adequate reproductive health care for poor African American women could prevent many avoidable cases of infertility. No woman should be allowed to become infertile against her wishes when adequate information and medical care can prevent infertility. Furthermore, low-income women who have difficulty conceiving should not be precluded from infertility services because of their poverty. The ability to have a child is as important as the ability to prevent having children. Advocacy groups should recognize that infertility treatment is an important element of meaningful reproductive choices and should ensure that adequate funds for infertility services are provided.

Whites and those with higher incomes are more likely to pursue infertility treatment than are African Americans and the poor.[56] Given that the average fee for each infertility treatment is between $2,055 and $10,000,[57] it is no wonder that poorer couples, a disproportionate number of whom are African Americans, do not pursue infertility treatment. Little information is available regarding the percentage of private insurance plans that cover infertility.[58] Medicaid and Title X regulations include infertility services among family planning services;[59] therefore, public funding for such services should be available. Unfortunately, "no precise data exist on the amount of public funds spent on infertility services,"[60] and relatively few private physicians accept Medicaid for infertility services.[61] Advocacy groups should also be mindful that some of the new reproductive technologies used to aid infertile couples, particularly surrogacy, may foster the exploitation of poor women.

V. SURROGACY

In the typical surrogacy arrangement, a married couple with a wife unable to bear children uses the husband's sperm to inseminate a fertile woman, who

---

53. *See* CONGRESSIONAL CAUCUS FOR WOMEN'S ISSUES, THE WOMEN'S HEALTH EQUITY ACT OF 1990 at 18 (1990).

54. *Id.* at 17-18.

55. *Id.*

56. About 75% of low-income women in need of infertility services have not received any services. "Among all higher income women, 47% have received no infertility services. . . . Women under 150% of poverty account for 22% of women in need of infertility services who have not received services, compared to 15% of all those in need." FORREST ET AL., *supra* note 8, at 71.

57. *Id.*

58. *Id.*

59. 42 C.F.R. § 440.40(c) (1991).

60. FORREST ET AL., *supra* note 8, at 71; *see also* Nsiah-Jefferson, *supra* note 52, at 50.

61. "While 45 percent of the physicians in the specialties surveyed provide infertility services, only 21 percent (47 percent of providers) accept Medicaid patients for infertility care and six percent (13 percent of providers) will lower their fees for low-income patients." FORREST ET AL., *supra* note 8, at 70.

becomes the "surrogate mother."[62] The surrogate mother carries to term the fetus conceived from her egg and the husband's sperm. Although the legal and constitutional status of surrogacy agreements is unresolved, most arrangements are governed by a contract whereby the surrogate mother is paid a sum of money and agrees to relinquish all parental rights to the child.[63] The "rate" is normally between $10,000 and $15,000, and the surrogate is paired with the infertile couple either through a fertility agency or privately. Because of the high cost of the procedure, affluent couples (who will most likely be white) are probably the only couples who will benefit from surrogacy arrangements. Further, poor women may be attracted to such arrangements because of the relatively substantial fees offered.

In this type of surrogacy, an African American surrogate mother will produce an African American child for a white couple because the egg is that of the surrogate mother. It is therefore likely that in most instances African American women will not be chosen as surrogates to be used with the egg donor type technology. Other women of color who genetically have less "color" may be exploited, however, particularly women from less industrialized countries where wages are low and the "labor supply more submissive."[64]

The *Baby M* case indicated to surrogacy brokers and couples seeking surrogates that healthy white women who are unrelated to the couple may not be the best choice because a court may find that a surrogacy contract does not terminate a woman's parental rights and grant her full custody or visitation.[65] Surrogacy brokers may be more likely to prey upon the economic vulnerabilities of women of color who will not have the resources to fight rich white American couples in court.

A second kind of technology is potentially more dangerous for poor African American women. In vitro fertilization (IVF) is the process whereby several

---

62. MARTHA A. FIELD, SURROGATE MOTHERHOOD: THE LEGAL AND HUMAN ISSUES 5-6 (1988).

63. State contract and family law, including a consideration of the best interests of the child, usually control disputes between putative parents over a child produced through surrogacy.

64.

> [S]urrogate brokers have frankly admitted that they will turn to Third World women for their stables of contract breeders . . . John Stehura, president of the Bionetics Foundation, talked about his plans for a surrogate business in Mexico that would "supply" U.S. clients: "You could devastate them [Mexican women] with money and things."
>
> In the developing world, reproductive clinics offer market and research possibilities unhampered by even the few restrictions that exist in Western contexts. The technique of sex predetermination developed out of the rationale of "female fertility out of control" in the Third World—which could be solved by reducing the number of girls born. A chain of clinics in India, Jordan, Pakistan, Egypt, Malaysia, Singapore, and Taiwan, as well as several in the U.S., has been set up by . . . a U.S. entrepreneur of sperm separation technology. Pivet, an Australian company, has established IVF clinics in Brazil, India, Malaysia, and Indonesia, particularly for sex predetermination. Such clinics have been deluged with requests from financially well-off women affected by the stigma of not having produced a son.

Janice G. Raymond, *Women as Wombs: International Traffic in Reproduction*, MS. MAGAZINE, May/June 1991, at 28, 31.

65. *See* In the Matter of Baby M, 537 A.2d 1227 (1988).

eggs are removed from a woman's ovaries and fertilized in a petri dish with the sperm of her husband. A doctor then transfers the fertilized egg into the uterus of a woman other than the egg donor. This woman is called the "gestational mother."[66] The gestational mother, unlike a surrogate mother, has no genetic connection to the child. The gestational mother is a vessel that allows infertile couples to have a child who inherits both parents' genetic make-up. Similar to the surrogate mother, the gestational mother normally agrees to relinquish all parental rights to the child and receives a fee. Because the gestational mother merely carries or incubates the fully fertilized egg of another woman and is expected to hand over the child after it is born, she becomes literally a "womb for rent."

IVF raises a number of ethical and legal issues of particular concern to African American women. Because IVF allows a woman to bear the child of a couple without having a genetic connection to the child, it creates the opportunity for African American women to give birth to completely white babies. Symbolically and historically, such an arrangement is reminiscent of the numbers of African American slave women who were forced to breastfeed white children the breast milk they created for their own children,[67] or to act as "breeders" for the master's property.[68]

A woman who gives birth to a child to whom she has no genetic connection will have little or no protection from the courts if, like Mary Beth Whitehead, the plaintiff in the *Baby M* case, she changes her mind and wants to keep the baby she has carried to term and delivered. This situation is alarming.

The case of Anna Johnson, an African American woman who gave birth to the child of a white husband and an Asian wife, highlights the conflicts. The California Court of Appeals ruled that genetics was the determining factor in parenthood.[69] In October 1989, Crispina Calvert learned of co-worker Anna Johnson's interest in serving as a surrogate mother. Eventually, Crispina involved her husband Mark Calvert in discussions with Anna Johnson. The Calverts were unable to have children because of Crispina's hysterectomy. In January 1990, Johnson signed a contract with the Calverts, which stipulated that Johnson would be impregnated with an embryo from the sperm of Mark and the egg of Crispina.[70] Johnson agreed to give birth to the child and to

---

66. FIELD, *supra* note 62, at 35. IVF has helped very few people. There were fewer than 2,000 IVF births in the U.S. by 1988, compared with more than 4,000,000 U.S. births annually. George J. Annas, *Fairy Tales Surrogate Mothers Tell*, LAW, MEDICINE & HEALTH CARE, Spring 1988, at 27 & n.2. Annas believes that IVF and surrogate motherhood are "both classist and sexist: a method to obtain children genetically related to white males by exploiting poor women." *Id.* at 27.

67. Nsiah-Jefferson, *supra* note 52, at 52.

68. *See, e.g.*, PAULA GIDDINGS, WHEN AND WHERE I ENTER 43 (1984); ANGELA Y. DAVIS, WOMEN, RACE AND CLASS 6-8 (1983); BULLWHIP DAYS 147-149 (James Mellon ed., 1990).

69. Anna J. v. Mark C., 234 Cal. App. 3d 1557, 286 Cal. Rptr. 369 (1991).

70. The Calverts used an attorney to arrange the contract, which was signed at the Center for Reproductive Health, a surrogate broker. Johnson was to be paid $2,000 at the end of the first and second trimesters, $1,000 a month before and after the birth, and $4,000 six weeks after the birth. Philip Hager, *State High Court to Rule in O.C. Surrogacy Case*, L.A. TIMES, Jan. 24, 1992, at A1.

relinquish her parental rights. On September 19, 1990, Anna Johnson gave birth to a baby boy. Two days later, the court allowed the Calverts to take the baby home without prejudice to Anna's rights.[71]

Upholding the Superior Court decision, the Court of Appeals framed the conflict between the Calverts and Johnson as a question of who is the natural mother, "the woman who nurtures the child in her womb and gives birth or an otherwise infertile woman whose egg is implanted in the woman who gives birth?"[72] In reaching the decision that Anna Johnson is not the "legal" or "natural" mother of the child because she had no genetic connection to the child, the court relied on the legislature's treatment of paternity questions.[73]

Under the California Uniform Parentage Act,[74] if a woman's husband is not the biological father of her child, he is not the legal father.[75] Using this same reasoning, the Court of Appeals found that Anna was not the genetic, biological, or natural mother of the baby that she carried to term.[76] The danger and fallacy in this type of analysis is that it fails to acknowledge that a different physiological or emotional relationship exists between a mother and child than between a father and child. A woman, unlike a man, has a physical connection to her child that goes beyond genetics. A fetus depends entirely on the mother's biology for survival. The fetus is a part of the mother. According to California courts, however, Anna Johnson's interest in the child is non-existent because her womb and body merely nurtured the fetus, her body did not produce the egg.

This case is different from the *Baby M* case.[77] In *Baby M*, Mary Beth Whitehead had a biological connection to the child she carried. Hence, she was the genetic and biological mother. Because Whitehead was the biological mother of the child—her own egg was fertilized and she gave birth to the child—maternity was never a question. The issue was whether parental rights could be legally terminated by surrogate agreements. Ultimately, the New Jersey court declared surrogate contracts illegal and unenforceable. It then weighed Whitehead's maternal rights and the father's paternal rights against the best interest of the child to determine custody.[78]

The court saw the core issue in Anna Johnson's case as the definition of

---

71. *Mother by Contract Gives Up Baby For Now*, N.Y. TIMES, Sept. 22, 1990, at 2.

72. Anna J. v. Mark C., 234 Cal. App. 3d 1557, 1562 (1991).

73. *Id.*

74. The 1975 Parentage Act sought to eliminate distinctions between legitimate and illegitimate children, and based parent and child rights on the existence of a parent-child relationship, not the marital status of the parents. CAL. CIV. CODE §§ 7000-7021 (West 1983).

75. Experts use blood tests to determine that a husband is not the biological father of the child, even if the husband and wife are cohabitating. 234 Cal. App. 3d at 1567.

76. Anna stipulated that Crispina was genetically related to the child, and blood tests confirmed that the child was not genetically related to Anna. *Id.* at 1569.

77. Matter of Baby M, 537 A.2d 1227 (N.J. 1988).

78. The court declared that the child's best interests called for awarding custody to her father, Daniel Stern. 537 A.2d at 1259.

"natural mother."[79] Johnson's attorney argued that under a California statute which states that "[t]he parent and child relationship may be established . . . [b]etween a child and the natural mother . . . by proof of her having given birth to the child . . . ,"[80] the birth mother is the natural mother.[81] The court rejected this argument, holding that the statute cannot be read to state that the birth mother is always the natural mother.[82] Having firmly established that Anna was not the "natural mother," the court found that she was not the legal mother and thus had no parental rights.

Another issue strongly pervades the Anna Johnson case: Can a Black woman ever be the "natural" mother of a white child?[83] Race issues have permeated virtually every policy issue in this country since its founding.[84] In 1903, W.E.B. DuBois said that "the problem of the Twentieth Century is the problem of the color line."[85] Given our continuing struggle with race issues, both domestically and internationally, race will also be the problem of the twenty-first century. The decision in Anna Johnson's case is painfully reminiscent of slavery and the days of the breeder woman whose feelings for her child, whether born out of love or out of rape, were disregarded when men with power over her made decisions about the child.[86]

Anna Johnson's lawyer noted the historical connection between slavery and Anna's situation:

> This case is what critics who oppose surrogacy have been warning legislators—what we are going to see is a wealthy couple like the Calverts preying on the poor, which generally translates into preying on blacks. I hope this is recognized as a civil rights issue and a classic case of exploitation and a slave contract . . .[87]

Advocates must be vigilant in their efforts on behalf of poor women. Both civil rights and women's advocates must ensure that poor women are not economically exploited through surrogacy contracts, and that their attachment

---

79. 234 Cal. App. 3d at 1562. The appellate court did not even address whether surrogate contracts were legal.

80. CAL. CIV. CODE § 7003 (West 1983).

81. 234 Cal. App. 3d at 1570-71.

82. *Id.*

83. The child in the Johnson case was of Filipino and white parentage.

84. *See generally* LEON HIGGINBOTHAM, IN THE MATTER OF COLOR (1978).

85. W.E.B. DUBOIS, THE SOULS OF BLACK FOLK xi (Penguin Books 1982).

86. *See, e.g.*, DAVIS, *supra* note 68, at 7 ("Since slave women were classified as 'breeders' as opposed to 'mothers,' their infant children could be sold away from them like calves from cows. One year after the importation of Africans was halted, a South Carolina court ruled that female slaves had no legal claims whatever on their children. Consequently, according to this ruling, children could be sold away from their mothers at any age because 'the young of slaves . . . stand on the same footing as other animals.'" (omission in original) (footnote omitted)).

87. David Behrens, *Test-Tube Baby in Tug-of-War*, NEWSDAY, Sept. 21, 1990, at 1, 23 (omission in original).

to their children is considered in any dispute over custody or visitation. The decision should not be based merely on the relative economic positions of the contending parents or on the myth that a child produced by another woman's egg has no relationship to the woman actually giving birth.

In some respects, surrogacy could be considered gainful employment. Yet it could also be considered baby-selling. Public policy advocates, medical experts, and former surrogates should discuss and debate a variety of ways to prevent economic exploitation and the potentially harmful psychological effects on a surrogate mother who is deprived of the child she bore. At the same time, they should try to make surrogate services available to all women who need them. Government regulations should be carefully developed for surrogacy. Such regulations might treat surrogate mothers like organ donors and prohibit payment for services. Surrogacy is too critical an area to allow surrogacy brokers and lawyers to control.

VI. STERILIZATION

African American women have not had genuine access to voluntary sterilization—the informed choice of a woman to forego future pregnancies by removing her uterus or cutting or blocking her fallopian tubes. Instead, African American women have often been victims of involuntary surgical procedures that strip them of their ability to reproduce.

The early birth control movement included strong factions advocating eugenics or compulsory sterilization of "unfit" persons—"[m]orons, mental defectives, epileptics, illiterates, paupers, unemployables, criminals, prostitutes and dope fiends."[88] At least twenty-six states had passed compulsory sterilization laws by 1932.[89]

During the 1970s, several instances of blatant sterilization abuse were exposed. For example, public assistance officials tricked illiterate black welfare recipients into consenting to the sterilization of their teenage daughters. Native American women under twenty-one years of age were subjected to radical hysterectomies and informed consent procedures were ignored. Doctors agreed to deliver the babies of Black Medicaid patients on the condition that the woman be sterilized. Doctors have also conditioned the performance of abortions on "consent" to sterilization.[90]

By 1982, only fifteen percent of white women were sterilized, compared to twenty-four percent of African American women, thirty-five percent of

88. DAVIS, *supra* note 68, at 214 (quoting GENA COREA, THE HIDDEN MALPRACTICE 149 (1977)). Birth control became population control, and "exterminat[ing] the Negro population" was included in the plan. DAVIS, *supra*, at 215.

89. DAVIS, *supra* note 68, at 214.

90. Nsiah-Jefferson, *supra* note 52, at 44 n.112; BLACK WOMEN'S COMMUNITY DEV. FOUND., *supra* note 52, at 80-90.

Puerto Rican women, and forty-two percent of Native American women.[91] Among Hispanic women living in the Northeast, sterilization rates as high as sixty-five percent have been reported.[92] African American women of all marital statuses were more likely than white women to use sterilization as a contraceptive method.[93] African American areas in the South have the highest rates of hysterectomy and tubal ligation in the United States.[94]

Current sterilization abuse takes many forms, and is usually not as blatantly coercive as in the past. An assessment of the voluntariness of a woman's decision to be sterilized must take into account the information available to her at the time the decision is made. Many women agree to sterilization, believing that the procedure is reversible.[95] Health professionals may contribute to this mistaken belief by describing the procedure as "tying the tubes"—an image which reinforces the belief that "what can be tied can be untied later."[96] One study reported that forty-five percent of the sterilized African American women interviewed did not realize the procedure was irreversible. Forty percent of the sample expressed regret over their decisions to be sterilized.[97]

Since the abuses of the 1970s, regulations have been adopted to ensure that informed consent is obtained for all federally-funded sterilizations.[98] Researchers complain, however, of the inadequate monitoring of these regulations. Data that is collected is not published or made publicly available.[99] As a result, the effectiveness of these regulations is unknown.

The federal government still subsidizes sterilizations for women eligible for Medicaid coverage, though it will not pay for abortions.[100] Moreover, doctors may improperly steer women with "too many" children or women whom they consider incapable of properly using contraceptives toward sterilization. Doctors may also recommend sterilization to women with fibroid tumors, even though these tumors generally do not require hysterectomies for proper treatment.[101] Doctors may also suggest unnecessary sterilizations

---

91. Vicki Alexander, *Black Women and Health*, 6 CHOICES 6, 16 (Women's Medical Ctr. 1986).

92. Nsiah-Jefferson, *supra* note 52, at 46 n.120.

93. *Id.* at 46 n.122.

94. *Id.* at 46 n.123.

95. *Id.* at 44.

96. *Id.* at 44 n.113.

97. *Id.*

98. 42 C.F.R. §§ 441.250-.259 (1991) (sterilizations); 42 C.F.R. § 441.257 (1991) (informed consent); 42 C.F.R. § 441.258 (1991) (consent form requirements).

99. Nsiah-Jefferson, *supra* note 52, at 45.

100. 42 C.F.R. §§ 441.253-.254 (1991) provide funding for sterilizations and hysterectomies while 42 CFR §§ 441.202-.203 restrict funding for abortions to situations in which a doctor certifies that the life of the mother would be endangered if the fetus were carried to term.

A recent review of the impact of Michigan's ban on state-funded abortions found that 400 more sterilizations were performed in the first year of the ban than in the previous year when state funds for Medicaid abortions were available. The study projected an increase of 1,500 sterilizations from 1988 to 1990. Medicaid sterilizations for 1990 were expected to total 6,100. Deborah Chargot, *Abortion and the Poor—Medicaid Ban Develops Cycle of Angry Mothers, Unwanted Babies*, DETROIT FREE PRESS, Aug. 5, 1990, at 1F.

101. BLACK WOMEN'S COMMUNITY DEV. FOUND., *supra* note 52, at 20.

simply to provide interns and residents with surgical experience.[102]

Before a woman can "consent" to sterilization she must have available to her all relevant information about her condition, her options, and the permanent effect of sterilization.[103] Poor women and women of color do not receive this information. In this area, as in all other matters of reproductive health, advocates should ensure that women have access to all of the available information about their condition and the possible effects and outcomes of various medical procedures before they make irreversible reproductive health decisions. Furthermore, the regulations for ensuring that women give their informed consent to sterilizations that are subsidized with federal funds should be monitored closely to prevent sterilization abuse.[104] The federal government must make the monitoring data available for scrutiny. Poor women should not be victims of unnecessary sterilization procedures, and advocates of choice should work to ensure that they are not.

VII. Workplace Toxins

The toxins and hazards to which African Americans and other people of color are exposed in the workplace damage our overall health status. Although there is a limited amount of race-specific data available, studies demonstrate that men and women of color are overrepresented in dirty and dangerous industries and jobs.[105] Clean and safe work environments are essential to the well-being of people of color.

In *United Automobile Workers v. Johnson Controls, Inc.*,[106] commonly called the "fetal protection" case, the Supreme Court decided that Title VII of the Civil Rights Act of 1964[107] prohibits an employer from excluding all women of child-bearing capacity from certain jobs even if the employer's goal is to prevent possible damage to potential or developing fetuses.[108] Johnson Controls, a battery manufacturer, excluded all women under the age of seventy

102. Nsiah-Jefferson, *supra* note 52, at 45-47 & nn.125-127.

103. Informed consent is one of the most deeply rooted traditions in American medicine. Basic tenets of professional medical ethics and established medical care standards require health care providers to obtain a patient's voluntary and fully informed consent. The Principles of Medical Ethics of the American Medical Association recognize that "the patient's right of self-decision can be effectively exercised only if the patient possesses enough information to enable an intelligent choice." GOLD, *supra* note 17, at 46 n.156, citing AMERICAN MEDICAL ASS'N, CURRENT OPINIONS OF THE COUNCIL ON ETHICAL AND JUDICIAL AFFAIRS OF THE AMERICAN MEDICAL ASS'N 31, 32 (1986).

104. In 1974, shortly after the federal informed consent guidelines were issued, a paper on the subject of sterilization labelling the guidelines as "impotent directives without any 'police powers,'" quoted the Director of the Office of Population Affairs for HEW (now HHS). The paper characterized the guidelines as "the United States government's way of saying that the poor and all those people luckless enough to have been labeled 'mentally defective' still have no rights which any physician need respect." BLACK WOMEN'S COMMUNITY DEV. FOUND., *supra* note 52, at 88 (citation omitted).

105. *See infra* notes 114-122 and accompanying text (discussing dirty and dangerous industries and jobs).

106. 111 S.Ct. 1196 (1991).

107. 42 U.S.C. §§ 2000e-2000e-17 (1988).

108. 111 S.Ct. at 1207.

(except those who provided medical proof of sterility) from jobs with a certain lead exposure level as well as from all entry-level jobs that could lead to the higher-lead jobs. As a result of the policy, the company essentially stopped hiring women for production jobs. If the policy had been upheld, as many as twenty million jobs nation-wide could have been closed to women.[109] This result would have effectively forced women to choose between their jobs and their ability to have children.

Gloyce Qualls, one of the plaintiffs in *Johnson Controls*, was demoted as a result of the fetal protection policy.[110] Two weeks after being notified of the policy, Qualls had a tubal ligation. Her salary would have been cut from $450 per week to $200 per week had she not been sterilized. While the decision protects women like Gloyce Qualls from forced sterilization in order to maintain higher paying jobs, the decision does not address the work conditions which threaten the health and safety of women and their fetuses. The *Johnson Controls* decision must therefore be viewed cautiously. Gloyce Qualls could be rendered infertile simply by doing her job.

LDF filed an amicus brief with the Supreme Court in *Johnson Controls*,[111] arguing that if the Court upheld the lower court's decision, the standards of proving an employment discrimination case would be relaxed and extended beyond fetal protection policies.[112] The belief that some individuals within a class of workers may be particularly susceptible to the harmful effects of toxic substances cannot justify categorical exclusion of that class. Title VII should not allow employers to institute policies that sacrifice women's employment opportunities merely in order to avoid cleaning dirty workplaces.

---

109. United Auto. Workers v. Johnson Controls, Inc., 886 F.2d 871, 914 (7th Cir. 1989) (Easterbrook, J., dissenting) (noting possible outcome of decision).

110. Florence Estes, *Battery Workers Pulled Together to Pursue the Right to Their Jobs*, CHI. TRIB., Apr. 28, 1991, at 3.

111. Brief for the NAACP Legal Defense and Educational Fund, Inc., and the National Black Women's Health Project, UAW v. Johnson Controls, 111 S.Ct. 1196 (1991) (No. 89-1215).

112. Title VII recognizes two forms of discrimination: disparate impact, 42 U.S.C. § 2000e-2(a)(2), and disparate treatment, 42 U.S.C. § 2000e-2(a)(1). Disparate impact claims "involve employment practices that are facially neutral in their treatment of different groups but that in fact fall more harshly on one group than another and cannot be justified by business necessity." International Brotherhood of Teamsters v. United States, 431 U.S. 324, 335 n.15 (1977). Disparate impact theories apply to employment devices such as high school requirements, pen and paper tests, and height and weight requirements. Employers may assert a business necessity defense to justify facially neutral practices that have an adverse impact on members of protected classes.

Disparate treatment theories, on the other hand, prohibit both overt or facial and covert or pretextual discrimination. Disparate treatment theories apply to contentions that an employer treats some employees less favorably than others because of their race, color, sex, religion, age, or national origin. "'Disparate treatment' . . . is the most easily understood type of discrimination." Teamsters, 431 U.S. at 335 n.15. Employers may defend against evidence of disparate treatment on the basis of sex by showing that the challenged policy or practice is a "bona fide occupational qualification reasonably necessary to the normal operation of that particular business or enterprise." 42 U.S.C. § 2000e-2(e). This is "an extremely narrow exception." Dothard v. Rawlinson, 433 U.S. 321, 334 (1977). Employers must show that "all or substantially all women would be unable to perform safely and efficiently the duties of the job involved." Weeks v. Southern Bell Tel. and Tel. Co., 408 F.2d 228, 235 (5th Cir. 1969).

By definition, an explicit classification based on sex and the capacity to have children is not gender neutral and constitutes a per se violation of the Act, or disparate treatment. Johnson Controls' policy presented a case of intentional discrimination which could be justified only by the narrow BFOQ defense.

Although the decision means that women cannot be denied access to certain blue collar jobs, the decision might also be interpreted as granting women equal access to excessively hazardous workplaces.

The *Johnson Controls* decision marks a victory against invidious discrimination based on gender and women's child-bearing capabilities, but it does not address the work conditions that place women and their fetuses at risk. In its amicus brief, LDF argued that African Americans are overrepresented in predominantly male occupations in dirty traditional industries. Policies that allow employers to focus solely on potential fetal risks do not address the broader issue of unsafe conditions for all workers. In particular, such policies do not recognize that hazardous work conditions can adversely affect men's reproductive capacity.[113] LDF argued that employers should be required to reduce hazardous exposures to safe levels for all workers or to find product substitutes.

Toxic work environments pose a particular danger to the overall health status of African American workers because of our overrepresentation in certain dirty industries and jobs. African Americans and other people of color are overrepresented in laundry and dry cleaning, tobacco manufacture, fabric mills, smelters, hospitals, and farmwork.[114] Each of these industries exposes workers to chemicals and toxic substances that have been linked to cancer. Furthermore, migrant farm workers, estimated to be seventy-five percent Chicano and twenty percent African American, have among the most dangerous and least protected jobs of all workers.[115] Their exposure to pesticides, heat, mechanical hazards, noise, and dust combine with poverty and poor medical care to lower the life expectancy of a farmworker to forty-nine years, more than twenty years less than the national average.[116]

As Davis explains in an article focusing on the workplace health and safety of African American workers:

> [I]llness, disease, and death has [sic] continued to take its toll among black workers, due to unchecked hazardous working conditions and exposure to toxic substances . . .
>
> Fifteen percent of the black work force (one to one and one-half million) are unable to work due to permanent or partial job-related disabilities. Black workers have a 37 percent greater chance than whites of suffering an occupational injury or illness. Black workers are one and one-half times more likely than whites to be severely disabled from job injuries and illness

113. The few studies of toxic hazards to male reproductive organs have concluded that such hazards exist. *See* Sandra Blakeslee, *Research on Birth Defects Shifts to Flaws in Sperm*, N.Y. TIMES, Jan. 1, 1991, at A1. *See also* Devra L. Davis, *Fathers and Fetuses*, N.Y. TIMES, Mar. 1, 1991, at A27.

114. STEPHANIE POLLACK & JO ANN GROZUCZAK, REAGAN, TOXICS AND MINORITIES 2 (1984).

115. *Id.* at 41.

116. *Id.*

> and face a 20 percent greater chance than whites of dying from job-related injuries and illnesses.[117]

The health status of African American women workers is adversely affected by the hazardous work conditions found in predominantly female occupations. Although only seventeen percent of all service occupations were filled by African Americans in 1987, African Americans, especially women, occupied an overwhelming number of the jobs involving contact with toxic substances. African Americans comprised over thirty percent of all nursing aides, orderlies and attendants; thirty percent of maids and housemen; and twenty-three percent of all private household workers.[118] African American women are over-represented among health service workers,[119] who commonly suffer from occupational diseases.[120]

Increasingly, young African American women are avoiding private household and service jobs—areas that have employed large numbers of older African American women—and moving into clerical jobs.[121] Clerical work poses dangers to the health and welfare of African American women. According to Leith Mullings, "[t]he hazards of clerical work include muscular and circulatory disorders, fatigue, and exposure to dangerous chemicals such as benzene, toluene, and other organic solvents."[122]

Any policies that permit employers to avoid their duty to provide safe and clean working environments for all workers will have doubly devastating effects on African American workers. Health indices of African Americans will continue to lag behind white Americans if employers are allowed to evade their responsibility for toxic work environments.

Advocates should not retreat from the position that a clean work

117. Morris E. Davis, *The Impact of Workplace Health and Safety on Black Workers: Assessment and Prognosis*, 31 LAB. L.J. 723, 724 (1980) (footnotes omitted).

118. BUREAU OF THE CENSUS, U.S. DEP'T OF COMMERCE, STATISTICAL ABSTRACT OF THE UNITED STATES 389 (1989).

119. While African American women comprised 5.4% of the total labor force in 1981, they constituted 24% of all nursing aides, 19% of all practical nurses, 16% of all health aides, and 16% of all other health service workers. Julianne Malveaux, Low Wage Black Women: Occupational Descriptions, Strategies for Change 93 (1984) (unpublished paper prepared for NAACP Legal Defense and Educational Fund, Inc.).

120. Leith Mullings, *Minority Women, Work, and Health*, *in* DOUBLE EXPOSURE: WOMEN'S HEALTH HAZARDS ON THE JOB AND AT HOME 127 & n.36 (Wendy Chavkin ed., 1984).

121. PHYLLIS A. WALLACE, BLACK WOMEN IN THE LABOR FORCE 24, 112 (1980). Thirty percent of African American women workers are clerical workers. Malveaux, *supra* note 119, at 77.

122. Mullings, *supra* note 120, at 128. "In offices, newsrooms and at switchboards, growing numbers of people who work hour after hour on computer keyboards are developing sometimes crippling symptoms in their hands." Jane Brody, *Computer Users' Injuries Are Often Preventable*, N.Y. TIMES, Mar. 4, 1992, at C14. One of the most common of several hand-wrist problems that beset computer operators is carpal tunnel syndrome, in which the nerve passing through the wrist becomes pinched by swollen tissues, causing numbness and tingling in the fingers at first, then crippling pain, permanent nerve damage, and loss of muscle control that can render the hand almost useless. Repetitive wrist motion injuries can be prevented with appropriately adjusted chairs and tables, proper posture, regular exercises and rest periods, and early detection. *Id. See also* Bob Baker, *Repetitive-Motion Illness on Rise, Experts Say*, L.A. TIMES, Mar. 29, 1991, at A3 (repetitive-motion illness, called "the occupational disease of the '90s," affects "wide variety of workers, from supermarket clerks to meat cutters to operators of computer terminals").

environment is necessary for all workers. Although women (as well as men) should be permitted to make their own educated and informed choices about work and reproductive risks, merely supplying women with the available information with which to make those choices is insufficient. Such information is unreliable because too few studies have been performed. Research is especially scarce regarding the effects of toxins on male reproductive systems.

Targeting the information to women alone is too narrow a focus and will perpetuate the belief that only women are responsible for poor pregnancy outcomes. Most importantly, economic concerns may deprive many women, particularly women of color, of any real choice. They may be forced to accept the best paying jobs they can find, irrespective of health concerns. Women heading families alone may not be able to survive without the health and pension benefits that accompany better paying jobs. These women may be persuaded to accept sterilization in order to work in higher paying jobs that are potentially hazardous. Gloyce Qualls' decision to undergo a tubal ligation was motivated by economic considerations. No woman should be placed in such a dilemma.

VIII. ABORTION

Women have used abortion as a method of limiting unwanted or unplanned pregnancies throughout history, regardless of the legal status of abortions. Abortions early in pregnancy were permitted under traditional common law until the middle of the 1800s. By 1900, abortion was illegal throughout the United States.[123] Historians have noted that African women enslaved in the United States aborted pregnancies rather than bearing children into slavery or giving birth to the products of slave masters' rapes.[124]

Clearly, abortions will continue, whether legal or illegal.[125] The question is whether abortions will be performed safely. Will the abortion be performed by a doctor or will it be self-induced?[126] Will it be performed in the first

123. GOLD, *supra* note 17, at 3 n.2, citing JAMES C. MOHR, ABORTION IN AMERICA (1978).

124. *See* PAULA GIDDINGS, *supra* note 68, at 46 (1984); DAVIS, *supra* note 68, at 204-05.

125. Abortion-related deaths are common in countries where abortion is still illegal, particularly in Central and South America. The Pan American Health Organization states that abortion is now Latin America's leading cause of pregnancy-related mortality. The World Health Organization reports that as many as 200,000 women die every year because of unsafe abortions in developing countries—nearly 25% of them in Latin America. They report that between 10 and 20 million women have abortions each year in Latin America and 45% of all the maternity beds in that region of the world are taken up by women who have complications of illegal abortions. Dr. Kenneth Edelin (unpublished comments, on file with author).

126. The life-threatening effects of self-induced abortions using a variety of techniques and substances have been well documented in medical literature. *See, e.g.*, A.W. Diddle & William Vandergriff, *Intravaginal Use of Potassium Permanganate as an Abortifacient: The Error in Diagnosis*, 28 OBSTETRICS AND GYNECOLOGY 155-57 (Aug. 1966); Anthony A. Deep & Ivan Jacobson, *Soap-induced Abortion: Report of 4 Cases*, 25 OBSTETRICS AND GYNECOLOGY 241-44 (Feb. 1965); John E. Moseley & Milton F. Quander, *Abortion, Chemical Peritonitis, and Pulmonary Edema Following Intrauterine Injection of Turpentine*, 24 OBSTETRICS AND GYNECOLOGY 572-74 (Oct. 1964).

twelve to fourteen weeks of pregnancy, or later, when complications are more likely to arise?[127]

For poor women, the answers to these crucial questions depend on how soon they can raise sufficient funds for the abortion.[128] Scarcity of abortion providers exacerbates this problem, especially in rural areas.[129] In 1985, eighty-two percent of all counties in the United States—home to almost one-third of all women of reproductive age—had no abortion provider.[130]

In 1973, the Supreme Court ruled in *Roe v. Wade*[131] that a woman's constitutional right to privacy includes her right to decide whether or not to terminate her pregnancy. The Court stressed, however, that a woman's right to choose abortion is not absolute. Instead, her interest must compete with the state's legitimate interest in protecting both the health of the pregnant woman and the developing life. The Court set out a framework within which states could regulate abortion. State interference with the woman's right to abort during the first trimester of pregnancy—roughly the first twelve to fourteen weeks—essentially is prohibited. States can, however, pass laws to protect the potential life of the fetus at the point of viability (the ability to survive outside the woman's body). A state's interest in protecting the fetus can be limited in cases where it is necessary to protect the life or health of the woman.

In 1980, the Supreme Court upheld the Hyde Amendment[132]—the federal government's ban on the use of federal Medicaid funds to pay for abortions for poor women (except when necessary to save the life of the mother or when the pregnancies were the result of rape or incest).[133] *Harris v. McRae* held

---

127. The timing of the abortion significantly affects women's health. More than half of all abortions are performed at or before eight weeks of pregnancy, when the procedure is the safest. The death rate from an abortion performed at or before eight weeks of pregnancy is 0.2 deaths per 100,000 procedures. The risk increases as gestation continues. Abortions performed at 11 or 12 weeks of pregnancy are three times more dangerous for the woman than abortions performed at or before eight weeks, although the rate of death from abortion at 11 or 12 weeks was only 0.6 per 100,000 procedures during 1981-1985. The nine precent of abortions performed in the second trimester accounted for 53% of all abortion deaths from 1981 to 1985. GOLD, *supra* note 17, at 29-30.

128. Nearly 60 percent of Medicaid patients, compared with 25 percent of more affluent patients, say that paying for an abortion entailed serious hardship, such as letting bills go unpaid or buying less food or clothing in order to pay for their abortion. GOLD, *supra* note 17, at 52 n.186, citing Stanley K. Henshaw & Lynn S. Wallisch, *The Medicaid Cutoff and Abortion Services for the Poor*, 16 FAMILY PLANNING PERSPECTIVES 170 (1984).

Moreover, it often takes considerable time to obtain the money needed to pay for an abortion. While not all Medicaid-eligible women are forced to delay their abortions, many abortions are delayed by as much as two to three weeks. In 1982, after the ban on federal funding was implemented, 50% of Medicaid-eligible patients at one clinic had their abortions after nine weeks of pregnancy, compared with only 37% of non-Medicaid-eligible women. *Id.*

129. Abortion services are less available in rural than in urban areas; nine out of ten nonmetropolitan counties have no facility that performs abortions. Only two percent of the nation's total abortions were performed in nonmetropolitan areas in 1985. Stanley K. Henshaw et al., *Abortion Services in the United States, 1984 and 1985*, 19 FAMILY PLANNING PERSPECTIVES, Mar.-Apr. 1987, at 63-5.

130. *Id.*

131. 410 U.S. 113 (1973).

132. Pub. L. No. 95-205, 9 Stat. 1460, § 101 (1977).

133. Harris v. McRae, 448 U.S. 297 (1980). Nine percent of women of reproductive age rely on Medicaid for their medical care. GOLD, *supra* note 17, at 51, citing SUSHEELA SINGH ET AL., THE FINANCING OF MATERNITY CARE IN THE UNITED STATES (1987).

that the Amendment did not place any governmental obstacle in the path of a woman choosing to terminate her pregnancy. Instead, the Hyde Amendment encouraged "alternative activity" deemed in the public interest, and left an indigent woman with "the same range of choice in deciding whether to obtain a medically necessary abortion as she would have had if Congress had chosen to subsidize no health care costs at all."[134]

Since *Harris v. McRae*, poor women have had a difficult time financing abortion. Only thirteen states continue to use their own revenues to provide medically necessary abortion services for their low-income residents.[135] Poor women tend to have abortions later in their pregnancies, thus increasing the health risks and the cost of the service.[136] In essence, the Court ruled that the existence of a constitutionally protected right to an abortion did not require the government to provide an access to that right. Without equal access, a right amounts to a privilege—for many African American women such a "privilege" does not exist.

In 1989, the Court's decision in *Webster v. Reproductive Health Services*[137] signaled to states that further restrictions on access to abortion are acceptable. In *Webster*, the court upheld a Missouri statute which (1) declared in its preamble that life begins at conception; (2) prohibited public employees and facilities from performing or assisting with an abortion when not necessary to save the life of the mother; (3) prohibited the use of public funds for "encouraging or counseling" about abortion; and (4) required doctors who believe a fetus may be twenty weeks to determine viability by performing certain medical tests.[138] Again, the court upheld the notion that states do not have an affirmative duty to assist women who seek to exercise the constitutional right to an abortion. Since *Webster*, a number of states have considered laws requiring parental or spousal consent or notification, expanding the information doctors are required to provide in order to have informed consent, or imposing a waiting period before the procedure can take place.[139]

In 1990, the Court upheld parental notification laws in *Hodgson v. Minnesota*[140] and *Ohio v. Akron Center for Reproductive Health,*[141]

---

134. 448 U.S. at 317.

135. Alaska, Hawaii, Maryland, New York, North Carolina, Oregon, Washington, West Virginia, California, Connecticut, Massachusetts, New Jersey, and Vermont. GOLD, *supra* note 17, at 51.

136. *See supra* note 127 (health risks increase with additional gestation time). Abortion services provided by hospitals are typically two to three times more costly than services performed in clinics and physicians' offices. GOLD, *supra* note 17, at 49.

137. 492 U.S. 490 (1989).

138. *Id.* at 501.

139. Planned Parenthood of Southeastern Pennsylvania v. Casey, 947 F.2d 682 (3d Cir. 1991), *cert. granted*, 112 S.Ct. 931 (1992), which is currently pending before the Supreme Court, may determine the continued validity of *Roe v. Wade*. The case challenges Pennsylvania's restrictive abortion laws, which require: (1) doctors to deliver a specific statement regarding the fetus before performing an abortion; (2) a 24-hour waiting period before an abortion can be performed; (3) parental consent for minors; and (4) spousal notification. *Id.*

140. 110 S.Ct. 2926 (1990).

reasoning that as long as a judicial bypass procedure was available as an alternative to parental notification, the law was constitutional.[142] For teenagers, especially poor African American ones, judicial bypass is usually no alternative at all.

> Theoretically, a young woman who can't notify both parents has the option of taking her case to a judge, regardless of her race or socioeconomic status. But to suggest that negotiating the complexities of an already overburdened legal system is a real option for a poor black teenage girl is to engage in the cruelest type of legal hair-splitting. What is the chance that the type of person least likely to be able to locate, contact, or afford an attorney can get a fair hearing when they are also those most likely to be discriminated against because of their age, sex, and race?
>
> Parental notification, which can be casually dismissed by middle-class women, literally becomes a life and death issue to less fortunate women. . . . [W]omen under 15 years of age are those who are most likely to obtain abortions later than 21 weeks. When a young woman has already had her life options limited by being poor or restricted by still-existing racial barriers, reducing her options even further increases the risk of complications of a late abortion.[143]

Women who abort after fifteen weeks of pregnancy are disproportionately under eighteen, African American, unemployed, poor enough to be Medicaid-eligible, pregnant as a result of rape or incest, or carrying a fetus diagnosed with severe defects.[144] In addition, women under fifteen years of age are those who are most likely to obtain abortions after twenty-one weeks.[145]

Time-consuming court procedures increase the risk of complications from a late abortion. More restrictive state laws are expected to be proposed. Women of color need to participate actively at the local level to prevent barriers to abortion services.

African American teens are not only more likely to have late abortions than white teens, but are also twice as likely as white teens to become pregnant in the first place.[146] Of 10,200 births to mothers under age fifteen in 1986, fifty-eight percent of the mothers were African American, thirty-nine percent

---

141. 110 S.Ct. 2972 (1990).

142. A teen who is astute enough to maneuver the judicial process can avoid parental notification requirements by showing a judge that her decision to have an abortion is a mature one made in her best interests.

143. Billye Y. Avery, Being Pro-Choice Isn't An Abstract Issue for Black Teenagers (unpublished memorandum, on file with author).

144. GOLD, *supra* note 17, at 58.

145. Avery, *supra* note 143.

146. CHILDREN'S DEFENSE FUND, *supra* note 15, at 39.

were white, and thirteen percent were Hispanic.[147] While thirty-nine out of one hundred African American teens with unintended pregnancy actually have a baby, only twenty-five out of one hundred white teens have babies.[148]

As a result, any restrictions on abortion access and low-cost abortion services provided by clinics will disproportionately affect women of color. Reproductive rights advocates should oppose such restrictions at the state as well as the federal level. For women, especially women of color, abortion is a life and death issue.[149]

IX. WOMEN'S HEALTH EQUITY ACT

In late July 1990, United States Representatives Patricia Schroeder (D-CO) and Olympia Snowe (R-ME), Co-Chairs of the Congressional Caucus for Women's Issues, introduced the Women's Health Equity Act of 1990 (WHEA).[150] This omnibus package of legislation focuses on women's health and seeks to guarantee women greater equity in medical research and the delivery of health care services. It includes twenty individual bills divided into three separate titles—research, services, and prevention.

The Act responds to the limited funding and research devoted to women's health issues.[151] For instance, the National Institute of Health (NIH), the nation's major source of funding for medical research, spends only about thirteen percent of its budget on women's health research.[152] Additionally, a study by the General Accounting Office (GAO) found that NIH had made little progress in implementing its policy designed to include women in clinical research studies and that NIH lacked appropriate mechanisms for monitoring compliance with the policy.[153] Research on diseases specific to women is inadequate. For example, breast cancer death rates increased twenty-four percent between 1979 and 1986, yet very little research has been conducted.[154] For diseases common to both sexes, virtually all research is conducted on men. Although women are the fastest growing group of those infected with AIDS and heart disease is the number one killer of women,

147. CHILDREN'S DEFENSE FUND, *supra* note 12, at 93.

148. CHILDREN'S DEFENSE FUND, *supra* note 15, at 39.

149. *See* Shirley Chisholm, *Facing the Abortion Question*, *in* BLACK WOMEN IN WHITE AMERICA 602-07 (Gerda Lerner ed., 1973). As a member of House of Representatives in 1969, Ms. Chisholm documented the harm and injuries to women of color caused by illegal abortions of women of color and stated why she changed her position to support repealing anti-abortion laws. *Id.*

150. CONGRESSIONAL CAUCUS FOR WOMEN'S ISSUES, *supra* note 53. Fifty-six members of the Congressional Caucus for Women's Issues joined Reps. Schroeder and Snowe in introducing the 1990 WHEA. The package has been endorsed by 39 national organizations. The bill has been amended and is now the Women's Health Equity Act of 1991. See Appendix for a summary of the proposed legislation.

151. Women's health issues are defined as "diseases or conditions that are unique to women, are more prevalent or more serious in women, or for which specific risk factors or interventions differ for women." CONGRESSIONAL CAUCUS FOR WOMEN'S ISSUES, *supra* note 53, at 2.

152. *Id.*

153. *Id.* at 3.

154. *Id.* at 5.

research on the particular problems of women affected by these diseases remains scarce. Only Congressional action can address this startling state of affairs.

Advocates of civil rights, health issues, and women's issues should follow this omnibus package of legislation and ensure that the general public knows such legislation is pending. The media has paid very little attention to the bills which have been introduced, though grass-roots support for these bills will be needed. Again, it is a matter of life and death for American women.

X. Conclusion

This paper has developed several specific reproductive health issues that affect African American women's definition of reproductive choice. Though all women share these health concerns, the reproductive choices of African American women in particular are too often limited by poverty. African American women genuinely support a woman's right to choose whether or not to have a baby. The larger women's movement unfortunately has focused too narrowly on the choice of *not* having a baby while ignoring the fact that many women choose to carry pregnancies to term. These women should not be penalized, nor should their children be endangered because of their economic status.

The problems and issues of African American women should not be the concern solely of African American women. Civil rights and women's groups should be responsive to the needs of the group whose interests converge with both. Yet neither group embraces African American women and our concerns. All advocates of women's rights and racial justice should join forces to ensure that the health and welfare of children born to the least economically and politically powerful group—African American women—are guaranteed. Too often, African American women have slipped through the cracks.

Appendix

Congressional Caucus for Women's Issues†

The Women's Health Equity Act of 1991

H.R. 1161, Introduced by Reps. Patricia Schroeder and Olympia Snowe
S. 514, Introduced by Sen. Barbara Mikulski
February 1991

TITLE I—RESEARCH

1. *Women's Health Research Act* (H.R. 1263, Rep. Olympia Snowe)

H.R. 1263 would permanently authorize the Office of Research on Women's Health, established by NIH [National Institute of Health] in September 1990. The office would serve as overseer and coordinator of efforts to improve women's health research at NIH, and would evaluate the representation of women among senior physicians and scientists and develop programs to increase their representation. The office would also be required to develop a plan for an intramural program in obstetrics and gynecology at NIH. That program would be instituted in the second year after the legislation's enactment. The Women's Health Research Act also authorizes the establishment of three multidisciplinary centers of excellence around the country to conduct research on women's health. H.R. 1263 would authorize $20 million in fiscal year 1992 for the Office for Women's Health Research, and $7.5 million for the centers of excellence.

2. *Clinical Trials Fairness Act* (H.R. 1160, Rep. Patricia Schroeder)

This legislation would codify NIH's policy regarding the inclusion of women and minorities in research, and extend that policy to the Alcohol, Drug Abuse, and Mental Health Administration (ADAMHA), which has been in voluntary compliance with the policy since 1987. H.R. 1160 would require the inclusion of women and minorities, where feasible and appropriate, in all clinical trials—both intramural and extramural—funded by NIH and ADAMHA. Exceptions would be permitted where inclusion of these groups

† For more information and the full text of each bill, contact the Congressional Caucus for Women's Issues, 2471 Rayburn Building, Washington, D.C. 20515, tel. (202) 225-6740.

would endanger their health or would be inappropriate to the purpose of the research.

3. *Women's Mental Health Research Act* (H.R. 1264, Rep. Olympia Snowe)

H.R. 1264 would create a new Office of Research on Women's Mental Health and Substance Abuse at the Alcohol, Drug Abuse, and Mental Health Administration (ADAMHA) charged with many of the same tasks as the Office for Women's Health Research at NIH, including coordinating efforts to improve research on women's mental health and substance abuse, supplementing existing research funds, ensuring that women are included in clinical study populations, and developing programs to increase the number of senior women scientists at ADAMHA. The Office would also make grants for the establishment of centers of excellence on women's mental health. H.R. 1264 would authorize $20 million for the new Office and $7.5 million for the centers for excellence in fiscal year 1992.

4. *Women and Alcohol Abuse* (H.R. 1000, Rep. Mary Rose Oakar)

This legislation would increase funding for the National Institute of Alcohol Abuse and Alcoholism at ADAMHA by $23 million for research on alcoholism, giving priority to problems unique to women.

5. *Breast Cancer Basic Research Act* (H.R. 381, Rep. Mary Rose Oakar)

H.R. 381 would authorize an additional $25 million earmarked specifically for basic breast cancer research at the National Cancer Institute (NCI).

6. *Contraceptive and Infertility Research Centers Act* (H.R. 2651, Reps. Patricia Schroeder and Olympia Snowe)

H.R. 2651 would provide $15 million for the development and operation of three contraceptive and two fertility research centers which would conduct clinical and other applied research; develop protocols for training physicians, scientists, nurses and other health professionals; and develop continuing education programs for such professionals. In addition, the bill creates a grant and loan repayment program to attract top-notch scientists to work at the centers.

7. *Sense of Congress Resolution on Contraceptive and Infertility Research* (H. Con. Res. 166, Reps. Patricia Schroeder and Olympia Snowe)

H. Con. Res. 166 expresses the sense of Congress that contraception and fertility research, development, and education should be a national priority, and sets specific research goals for the next 20 years. The resolution also urges Congress to require the Secretary of Health and Human Services to reestablish the Ethics Advisory Board, which was terminated in 1980. NIH currently cannot conduct research on in vitro fertilization without the express approval of the Ethics Advisory Board.

8. *Women and AIDS Research Initiative* (H.R. 1073, Rep. Constance Morella)

This legislation would authorize a $10 million Women and AIDS research initiative within NIH and ADAMHA that would support both intramural and extramural research concerning HIV transmission, development, treatment and prevention in women. The bill also authorizes $6 million to create a new program under the Community Based Clinical Research Initiative (P.L. 100-607), which provides funds for the establishment of research organizations located in community settings to provide access to clinical research for populations at high risk for HIV infection. Under the legislation, funds would be used to expand clinical trials involving AIDS treatment for women. The bill makes clear that money under this program could be used for support services, such as child care and transportation.

9. *Ovarian Cancer Research Act of 1991* (H.R. 148, Rep. Patsy Mink)

H.R. 148 would authorize $30 million a year for the National Cancer Institute on basic research to develop an early-detection test and to determine whether there is a genetic basis for ovarian cancer.

10. *Osteoporosis and Related Disorders Research* (H.R. 1212, Rep. Olympia Snowe)

H.R. 1212 authorizes an additional $62 million for federal research on osteoporosis, including basic research into the causes of the disease as well as treatments to restore bone loss or prevent further bone loss. The legislation would also provide for the establishment of an Interagency Council to promote and coordinate research and education and health promotion programs; an Advisory Panel to make recommendations about the disease; and a Resource Center to compile and disseminate information.

TITLE II—SERVICES

1. *Breast Cancer Informed Decision Act* (H.R. 382, Rep. Mary Rose Oakar)

H.R. 382 would withhold federal Medicaid funds from states that do not enact laws requiring physicians and surgeons to inform breast cancer patients of alternative effective methods of treatment for breast cancer before treatment begins.

2. *Women's Health Care Coverage Expansion Act of 1991* (H.R. 1162, Rep. Patricia Schroeder)

This legislation authorizes direct reimbursement for nurse practitioners specializing in women's health under the Medicare and Medicaid programs, in order to improve access to prenatal care and preventive gynecological care for low-income and older women.

3. *The Mickey Leland Adolescent Pregnancy Prevention and Parenthood Act of 1991* (H.R. 1398, Rep. Nancy Johnson)

H.R. 1398 would revise the Adolescent Family Life Act (AFLA) to provide $60 million for comprehensive prenatal and postpartum care, well-baby and well-child care, family planning, and family life and parenting education. The new program would also provide for preventive education and screening for sexually transmitted diseases and referral for treatment. Under H.R. 1398, both teen mothers and fathers would be eligible for counseling and referral services for employment, employment training, nutrition, substance abuse, and adoption services. Unlike the current program, H.R. 1398 would encourage parental and family involvement, but would not deny services to teenagers who choose not to consult their parents. In addition, pregnant teenagers would be informed of the availability of counseling for all legal options regarding pregnancy. No funds under the legislation could be used to pay for an abortion.

4. *Adolescent Health Demonstration Projects* (H.R. 1128, Rep. Cardiss Collins)

H.R. 1128 would provide grants for the establishment or support of adolescent health demonstration projects in secondary schools. These projects would provide health care information and services. The projects would be required to encourage family participation, to obtain the approval of the local school board before the project is implemented, and establish a community advisory committee. No funds under the legislation could be [used] to perform or to pay for an abortion.

5. *COBRA Displaced Family Amendments of 1991* (H.R. 1645, Rep. Nancy Pelosi)

This bill expands current health insurance continuation provisions under COBRA to require employers to provide health insurance coverage for widowed, divorced and legally separated spouses age 50 and above until they obtain alternate coverage or become eligible for Medicare. The bill also requires continued coverage for their dependent children until they reach age 23.

6. *The Federal Employees Family Building Act of 1991* (H.R. 927, Rep. Patricia Schroeder)

H.R. 927 requires that all federal employee health insurance plans that provide obstetrical benefits also cover medical procedures for overcoming infertility and any necessary expenses related to the adoption of a child. The legislation also allows federal employees to use sick leave for family building activities.

TITLE III—PREVENTION

1. *Medicaid Infant Mortality Amendments of 1991* (H.R. 290, Rep. Cardiss Collins)

H.R. 290 requires states to provide Medicaid coverage for all pregnant women and children under age six with incomes up to 185 percent of poverty. The House of Representatives had passed similar legislation in 1989 as part of budget reconciliation; however, Congress eventually reduced coverage to pregnant women and young children with incomes up to 133 percent of poverty.

2. *Medicaid Women's Basic Health Coverage Act* (H.R. 1129, Rep. Cardiss Collins)

This legislation would require states to provide Medicaid coverage for routine mammography and pap smear screening. Such coverage is currently optional.

3. *Breast Cancer Screening Safety Act of 1991* (H.R. 3462, Reps. Patricia Schroeder and Marilyn Lloyd)

H.R. 3462 would address concerns regarding the quality of mammography screening by requiring that all mammography screening facilities receive certification by a national accreditation body. Each facility issued a certificate must use equipment designed specifically for mammography, use only qualified individuals who meet state licensing requirements or who have been trained and certified to perform mammograms, and use only qualified radiologists who are certified to interpret the results of the mammogram. Annual inspections of certified facilities would be authorized under the legislation.

4. *The Medicare Bone Mass Measurement Coverage Act of 1991* (H.R. 1213, Rep. Olympia Snowe)

H.R. 1213 would expand Medicare reimbursement for updated technologies for diagnostic testing for osteoporosis. Only individuals at high risk of developing osteoporosis, including women with an estrogen deficiency and individuals with vertebral abnormalities, would be covered under the legislation.

5. *Women and AIDS Outreach and Prevention Act of 1991* (H.R. 1072, Rep. Constance Morella)

This legislation would authorize $10 million under the AIDS prevention program for select family planning clinics and other public health clinics that provide preventive health services for women in high risk areas to design and carry out innovative programs of outreach, referral, services, and training, including family planning, screening and treatment for sexually-transmitted diseases, and counseling and testing for the HIV virus. Clinics would also develop improved referral arrangements with agencies that serve women and their partners, including drug abuse clinics, STD clinics, and homeless shelters, and would provide appropriate follow-up services. In addition, funds would be available to train clinic personnel in dealing with persons at high risk of AIDS, sexually-transmitted diseases, and unintended pregnancy.

6. *Infertility Prevention Act of 1991* (H.R. 3461, Rep. Patricia Schroeder)

H.R. 3461 would authorize the Centers for Disease Control to provide $80 million to family planning clinics and STD clinics to provide outreach and counseling to women on the prevention and control of chlamydia and gonorrhea, to screen women for the diseases, and to provide treatment to women and their partners. Funds would also be used to develop and disseminate public information and education programs, and to collect data on the incidence of chlamydia and gonorrhea.

REPRODUCTIVE TECHNOLOGIES AND THE U.S. COURTS

KIM M. BLANKENSHIP
Yale University
BETH RUSHING
Kent State University
SUZANNE A. ONORATO
RENÉE WHITE
Yale University

This article analyzes U.S. court cases involving reproductive technologies in terms of their implications for reproductive choice, mothers' versus fathers' rights, definitions and evaluations of parenting, and the nuclear family structure. The analysis reveals that the courts have tended (a) not to recognize how social conditions shape women's reproductive choices, (b) to promote fathers' rights more than mothers' rights, (c) to ignore the social relationships that constitute childbearing and child rearing and value men's over women's biological contribution to these processes, (d) to reflect certain assumptions about the proper roles of mothers and fathers, and (e) to privilege the nuclear family. The implications of developing reproductive technology policy for an understanding of the relationships among gender, reproductive technologies, and the state are considered, and recommendations for the equitable regulation of these technologies are offered.

In the last decade, such reproductive technologies as in vitro fertilization (IVF), surrogacy, genetic engineering, and embryo transfer have offered the possibility of revolutionizing human reproduction.[1] These technologies raise some difficult questions. Will reproductive technologies expand women's reproductive choices? Or will they lead to the exploitation of poor women, encouraging them to sell their wombs? Do these technologies represent new opportunities for becoming mothers and fathers or threaten to commercialize

AUTHORS' NOTE: *A version of this article was presented at the 1991 Annual Meetings of the Society for the Study of Social Problems, Cincinnati, OH. We wish to thank two anonymous reviewers for helpful comments.*

REPRINT REQUESTS: *Kim M. Blankenship, Department of Sociology, Yale University, P.O. Box 1965, Yale Station, New Haven, CT 06520-1965.*

GENDER & SOCIETY, Vol. 7 No. 1, March 1993 8-31
© 1993 Sociologists for Women in Society

and commodify parents and children? Should these technologies be viewed as undermining family relations or extending familial options? In short, do reproductive technologies liberate or constrain men and women in their reproductive decision making, expand or control reproductive options?

How do public policies regulate reproductive technologies? Feminist scholarship reflects a complex, sometimes contradictory, view of state policies. On the one hand, feminists suspect the state of controlling women and demonstrate how public policies have perpetuated gender privilege by promoting women's dependence on men in families, defining women primarily as caregivers, and controlling their productive and reproductive labor (e.g., Eisenstein 1983; Pascal 1986; Petchesky 1984; Wilson 1977). On the other hand, feminists often hold the state responsible for promoting the public good and have shown how, under certain circumstances, public policies may liberate women or enhance gender equality (Hernes 1987; Quadagno 1990). Here we ask, do regulations regarding reproductive technologies represent an instance of state control over women's reproductive lives, or do they enhance women's reproductive freedom?

To address these questions, we focus on analysis of judicial opinion in the United States. Certainly, the courts are not the only arena in which policy relating to reproductive technologies is being formulated. State legislatures are considering or have passed laws relating to surrogacy, for example, and an analysis of the history and politics of these efforts undoubtedly would be revealing. Nevertheless, a considerable amount of policymaking regarding reproductive technologies is occurring in the courts. Furthermore, judicial opinion reveals much about developments in statutory law. While interpreting legislative mandates, judges provide considerable insight into the rationale for reproductive technology policy. Thus, although they are not the only source of information regarding reproductive technology policy, court decisions are a reasonable focus.

Our analysis of judicial opinion reveals conflict over how best to regulate reproductive technologies. In legal terms, this includes conflicts over the relationship between reproductive technologies and existing state statutes, child support, paternity rights, and adoption and custody policies. Embedded within these legal framings of the conflict, however, are debates over such issues as the role of social context in shaping reproductive choices, the relationship between mothers' and fathers' rights, the social versus biological bases of parenting, the characteristics of "good" parents, and the status of the nuclear family structure. The resolutions of these conflicts have important implications for women. After analyzing judicial opinion regarding reproductive technologies, we conclude by considering the implications of this developing policy for an understanding of the relationships among gender, reproductive technology, and the state. Finally, we offer some recommendations for the equitable regulation of reproductive technologies.

THE U.S. COURTS CONFRONT REPRODUCTIVE TECHNOLOGIES

Our analysis is based on an examination of one aspect of the national response to reproductive technologies: judicial opinion in the United States. We collected the cases analyzed here using LEXIS, an on-line data base that includes the full text of all state and federal court cases. We searched LEXIS for any case referring to reproductive technologies in general as well as to seven specific technologies: artificial insemination, in vitro fertilization, egg donation, surrogacy, embryo transfer, frozen embryo, and gamete intrafallopian transfer. After eliminating irrelevant cases, 31 remained to be examined. For each case, we analyzed the opinion of the highest court. Furthermore, we examined citations to all of these cases and, with the few exceptions noted in the text below, all remain in good standing. Thus we analyze here a comprehensive and up-to-date (as of February 1990) collection of state and federal judicial opinion.

As Table 1 indicates, reproductive technologies have entered the U.S. courts in a variety of different legal contexts. First, the courts have decided whether these technologies conflict with existing statutes. Second, they have faced questions of responsibility for child support in conjunction with reproductive technologies. Specifically, state laws establish that the husband of a woman who consents to her artificial insemination by donor (AID) is the father of the child so conceived, but whether he should pay child support upon divorce has not been clear. Third, paternity rights have been the subject of courtroom debate. Most of these cases involve either the parental rights of men who donate semen to an unmarried woman or the paternity rights regarding a child born of a surrogate arrangement. The fourth set of legal questions raised by reproductive technologies relates to adoption and custody. Finally, the courts have decided the inheritance rights of children born through AID and the right of a prisoner to donate his sperm for his wife's artificial insemination. We turn now to a detailed examination of the conflicting views on reproductive and familial relations underlying judicial resolutions of the legal questions raised by reproductive technologies.

REPRODUCTIVE CHOICE

Reproductive technologies are popularly viewed as expanding the reproductive choices available to women and men. However, feminists argue that reproductive choice is meaningful only if the conditions under which women make such choices are addressed (e.g., Gertner 1989; Petchesky 1980, 105). Feminist critics suggest a number of social and material factors that may influence women's "choice" to use reproductive technologies, such as the social stigmatization of infertility (e.g., Klein 1990; Koch 1990; Rothman 1984), the high cost of these technologies (e.g., Andrews 1989, 387; Cooper 1986), and women's lack of knowledge of the risks associated with them (e.g., Corea 1985; Klein 1990; Williams 1990). Critics also fear that because women lack economic and social power they may be coerced into using these technologies. Research suggests that male partners dominate in reproductive decision making in general (Lorber 1988) and in the decision to enter IVF programs in particular (e.g., Klein 1990, 245). Others argue that poor women and women of color may be "financially induced to serve as a breeder class for the needs of affluent white couples" (Andrews 1989, 371).

To what extent is the context of choosing accounted for in courtroom decisions? Although questions of reproductive choice have not arisen in every court case involving reproductive technologies, we can learn something from those cases in which it is a relevant matter.

Conflicts with Existing Statutes

In deciding whether surrogacy contracts violate state laws prohibiting payment in connection with adoption, the courts rarely consider how the social pressures that women face to have children or the social stigmatization of infertility may shape their decision to use these technologies. Instead, the courts tend to regard reproductive technologies as extending reproductive choice by providing a viable alternative to adoption. The courts do not consider the potential physical and psychological risks posed by these technologies, as indicated in *In the Matter of the Adoption of Baby Girl L. J.* In this case, the court contrasted adoption, which it characterized as a process fraught with "uncertainty" and "painful anxiety," with reproductive technologies, implying that the latter are neither uncertain nor anxiety provoking (p. 815). Yet, given the low success rate of IVF and embryo transfer (e.g., Klein 1990), they involve a considerable amount of uncertainty. To the extent that women are aware of their potential for psychological and physical harm (see Andrews 1989, 367-68 for a description of some of these risks), they may involve anxiety as well.

Opinions in these cases also demonstrate disagreement over the role of financial coercion in the use of reproductive technologies. The court in *Baby Girl L. J.* explains that laws prohibiting payment in connection with adoption were designed to keep baby brokers from coercing expectant mothers or parents with financial inducements to give up their child. However, both the *SPA* and the *Baby Girl L. J.* courts held that surrogacy does not violate such laws. According to the courts, adoption and surrogacy differ because surrogate arrangements are entered into before conception, whereas adoption is agreed to afterward. Financial considerations may shape a woman's choice

(Text continues on page 16)

TABLE 1: Summary of Reproductive Technology Cases in the State and Federal Courts[a]

Case Citation	*Parties/Issues Involved*	*Summary of Decision*
Conflict with existing statute(s)		
Doe v. *Keane et al.*, 658 F. Supp. 216 (1987)	Does surrogacy violate 13th and 14th Amendments of U.S. Constitution?	No claim stated under federal law.
Doe v. *Kelley*, Mich. App., 307 N.W.2d 438 (1981)	Does Michigan adoption law make surrogacy unlawful?	No: Law only prohibits payment in connection with surrogacy.
In the matter of Baby M, 537 A.2d 1227 (N.J. 1988)	Does surrogacy violate New Jersey adoption laws?	Yes: Surrogacy is entered into for purposes of adoption.
In the Matter of the Adoption of Baby Girl L. J., 505 N.Y.S.2d 813 (Sur. 1986)	Does surrogacy violate New York adoption law?	No: Law did not contemplate surrogacy.
McIntyre v. *Crouch*, 780 P.2d 239 (Or. 1989)*	Does Oregon AID law violate due process right of donor of semen to unmarried woman?	Law violates right of due process (specially concurring and dissenting opinions).
Sherwyn & Handel v. *Department of Social Services*, 218 Cal. Rptr. 778 (Cal. App. 2 Dist. 1985)	Does surrogacy violate California laws re AID and adoption?	Not a justiciable controversy.
Smith and Lifchez v. *Hartigan and Daley*, 556 F. Supp. 157 (1983)*	Does Illinois abortion law make IVF illegal?	No: Dismissed for lack of subject matter.
Surrogate Parenting Associates, Inc. v. *Kentucky* Ky., 704 S.W.2d 209 (1986)	Does surrogacy violate Kentucky adoption laws?	No violation: Surrogacy and adoption differ (opinion concurring in result only; separate dissenting opinions).
Paternity rights		
Brooks v. *Fair*, 532 N.E.2d 208 (Ohio App. 1988)	Married woman who conceived thru AID seeks to establish nonpaternity of husband.	Request denied; to rule otherwise is to contradict public policy.

C. M. vs. *C. C.* 377 A.2d 821 (N.J. 1977)	Paternity rights of donor of semen to an unmarried woman.	Paternity and visitation rights granted.
In the Interest of R. C., 775 P.2d 27 (Colo. 1989)	Parental rights of donor of semen to an unmarried woman.	Remanded: grant rights if donor known to woman and they agreed he would have parental relationship (specially concurring opinion).
Jhordan C. v. *Mary K.*, 224 Cal. Rptr. 530 (Cal.App. 1 dist. 1986)	Paternity rights of donor of semen to an unmarried woman.	Paternity claim upheld; mother gets sole custody.
McIntyre v. *Crouch*, 780 P.2d 239 (Ore. 1989)*	Parental rights of donor of semen to an unmarried woman.	Remanded: grant rights if semen donated with an agreement that man would have relationship of father (specially concurring and dissenting opinions).
Syrkowski v. *Appleyard*, 362 N.W.2d 211 (Mich. 1985)	Is father of child born to surrogate the biological father or husband of surrogate?	Remanded: Biological father has right to have paternity claim heard.
Adoption/custody		
Abajian v. *Dennett et al.*, 15 Misc. 2d 260 (N.Y. 1958)	Does wife have right to deny former husband custody and visitation of children she alleges were conceived through AID?	No: Wife stopped from asserting children conceived by AID.
Davis v. *Davis*, Circuit Court for Blount CountyTeam at Maryville Equity Division, September 21, 1989	Who should gain custody of seven frozen embryos when biological parents divorce?	Custody granted to biological mother to preserve embryos for implantation.
In the Matter of Baby M, 537 A.2d 1227 (N.J. 1988)	Custody of and parental rights re child born of surrogate arrangement to biological mother or biological father.	Biological mother is legal mother; biological father gets sole custody; biological mother's visitation rights determined on remand.

(continued)

TABLE 1: Continued

Case Citation	*Parties/Issues Involved*	*Summary of Decision*
Adoption/custody		
In the Matter of the Adoption of Anonymous, 74 Misc. 2d 99 (N.Y. 1973)	Husband of woman who conceived through AID during their marriage refuses to consent to woman's second husband's petition to adopt child.	Petition denied: First husband is child's father.
In the Matter of the Adoption of Baby Girl L. J., 505 N.Y.S.2d 813 (Sur. 1986)	Custody of child born of surrogate arrangement to biological mother or biological father?	Custody granted to biological father and his spouse.
In the Matter of Adoption of Reams (1989) Ohio App. LEXIS 4538	Who has the right to adopt child born of surrogate arrangement between a childless couple and a married woman upon couple's divorce: infertile woman or her husband?[b]	Reversed in part and remanded: Husband has right to have request for adoption heard.
Strand v. *Strand*, 190 Misc. 786 (N.Y. 1948)	Does former husband who consented to AID have visitation rights re child?	Yes: Child considered semiadopted by husband.
York and Adler-York v. *Howard W. Jones, Jr., M.D., et al.*, 717 F. Supp. 421 (1989)*	Who should gain possession of frozen prezygotes, biological parents or doctors/medical center performing technique?	Prezygotes property of biological parents.
Child support		
Anonymous v. *Anonymous*[c] 542 N.Y.S. 2d 586 (1989)*	Must husband of woman who allegedly conceived through AID pay child support?	No: Should get a chance to contest allegation of paternity.
Anonymous v. *Anonymous*, 41 Misc. 2d 886 (N.Y. 1964)	Must husband of woman conceiving through AID provide child support?	Yes: Consent implies contract to support child.

Gursky v. *Gursky*, 39 Misc. 2d 1083 (N.Y. 1963)	Must husband of woman conceiving through AID provide child support?	Yes: Although child not legitimate, he gave consent and must support child.
In re Baby Doe, 353 S.E.2d 877 (S.C. 1987)	Must husband of woman conceiving through AID pay child support?	Yes: Actions imply consent, is child's father.
In re Marriage of Adams, 528 N.E.2d 1075 (Ill. App.2 Dist. 1988)	Must husband of woman conceiving through AID provide child support?	Yes: Gave consent, is father of child (opinion concurring in part and dissenting in part).
In the Matter of Karin T. v. *Michael T.*, 127 Misc. 2d 14 (1985)	Must "husband" of woman conceiving through AID provide child support?[d]	Yes: Consented to AID, must support children.
K. S. v. G.S., N.J. Super.Ch., 440 A.2d 64 (1981)	Must husband of woman conceiving through AID provide child support?	Yes: Gave consent, is father of child.
People v. *Sorensen*, 66 Cal. Rptr. 7 (1968)	Must husband of woman conceiving through AID provide child support?	Yes: He is lawful father of child.
R. S. v. *R. S.*, 670 P.2d 923 (Kan. App. 1983)	Must husband of woman conceiving through AID provide child support?	Yes: Gave consent, is father of child.

a. In addition to the cases cited here, we found a number of cases that addressed the question of whether reproductive technologies are or should be covered by medical insurance, for example, *Kinzie* v. *Physician's Liability Insurance Co. Okl.*, 750 P.2d 1140 (1987); *Settles* v. *Golden Rule Insurance Co. and Jim Toyne Insurance, Inc.*, 715 F. Supp. 1021 (1989); *McKinnon* v. *Blue Cross-Blue Shield of Alabama, et al.* 691 F. Supp. 1314 (1988); *Michael* v. .*Metropolitan Life Insurance Company*, 631 F. Supp. 451 (1986); *Reilly* v. *Blue Cross and Blue Shield United of Wisconsin*, 846 F.2d 416 (1988). These cases certainly are relevant to the issue of the accessibility of reproductive technologies. However, we have not analyzed them in depth (and, hence, they are not included in this table) because the underlying issues they involve—for example, questions of what constitutes a medically necessary procedure—are not directly relevant to this article.
b. This case is further complicated by the fact that conception with sperm of the husband of the contracting couple failed. The child in question was conceived through the artificial insemination of the surrogate with the sperm of another man, also not her husband. She then turned the child over to the contracting couple in exchange for $10,000. This means that the child in question was not biologically related to either member of the childless couple.
c. Although this case is listed in the table because it is potentially relevant, the opinion provides very little information to analyze.
d. In this case, the "husband" is a woman who put herself forward as a man.
* Hard copy unavailable; citation in text refers to LEXIS copy of the case.

to give up her child for adoption, but the "essential consideration [with regard to surrogacy] is to assist a person or couple who desperately want a child but are unable to conceive one in the customary manner" (*SPA*, p. 216). The implication is that women cannot be coerced into conceiving a child but they can be coerced into giving one up. This interpretation is reinforced by the fact that most courts have decided that surrogate contracts are voidable. According to the *SPA* majority, a surrogate mother's consent given before five days following birth is not legally binding: This "preserve[s] to the mother her right of choice regardless of decisions made before the birth of the child" (p. 213). However, much as providing formal legal rights to women does little to promote their substantive equality (e.g., MacKinnon 1989, 215-34; Williams 1982), extending reproductive choice by providing for the voidability of surrogate contracts does not necessarily expand their reproductive options. Many women cannot exercise the choice to void the surrogate contract because the financial need that influenced their initial decision to agree to such a contract makes it impossible for them to keep the child. Furthermore, any woman who opts for this choice, forfeits her right to any money and must fight for the custody of the child.

Only one court to take up the surrogacy and payment question appears to have been concerned with the extent to which a woman's economic situation may influence her choice to become a surrogate.[2] Chief Justice Wilenz spends more than five pages in the *Baby M* opinion discussing the potential for financial coercion in surrogate arrangements, thereby declaring surrogate contracts illegal (pp. 1246-50). According to the court, "Under the [surrogate] contract, the natural mother is irrevocably committed before she knows the strength of her bond with her child. She never makes a totally voluntary, informed decision, for quite clearly any decision prior to the baby's birth is, in the most important sense, uninformed, and any decision after that, compelled by a pre-existing contractual commitment, the threat of a lawsuit, and the inducement of a $10,000 payment, is less than totally voluntary" (pp. 1248-49). Yet, in spite of this incisive analysis, the *Baby M* court refuses to consider how circumstances affect choices when it addresses the issue of custody. The court sympathizes with Mary Beth Whitehead by suggesting that most mothers would have put up a struggle to keep their child (pp. 1258-59). Still, it refuses to consider how her actions were shaped by the existence of the surrogacy contract in reaching its custody decision.

Paternity Rights

Most state laws stipulate that the donor of semen for the artificial insemination of a woman not his wife is not the father of any child so conceived. However, the courts have been called on to decide whether there are any circumstances under which donors ought to be accorded paternity rights. Opinions in these cases demonstrate that women who resist the medicalization of reproduction and unmarried women may be less able than other women to "choose" to use reproductive technologies.

Statutes in at least four states – Connecticut, Georgia, Oklahoma, and Oregon – make self-insemination illegal. An Idaho statute requires that insemination be performed by or under the supervision of a physician. And a California statute provides different rights to the parties involved, depending on who performs the insemination. That such laws restrict the reproductive choices of some women is clearly revealed in the case of *Jhordan C.* v. *Mary K.* At issue was California Civil Code Section 7002 (b), which stipulates that "the donor of semen provided to a licensed physician for use in artificial insemination of a woman other than the donor's wife is treated in law as if he were not the natural father of a child thereby conceived."

Mary sought to bear a child through AID and raise it with her close friend Victoria. They chose Jhordan as the donor, and Mary inseminated herself with his semen and gave birth to Devin in March 1980. In December, Jhordan sought paternity and visitation rights. Mary and Victoria, in turn, sought joint legal custody. The court ruled that Jhordan was Devin's legal father, stressing that, because a doctor did not inseminate Mary, the law protecting her from a paternity claim did not apply. It further explained that there were "two sound justifications" for the physician requirement (p. 534). The first was health considerations: According to the court, a doctor can obtain a complete medical history of the donor and screen him for hereditary or communicable diseases. In reality, however, these screenings are rarely done (Andrews 1984, 56), calling into question whether health considerations explain the physician requirement. Second, the court argued that requiring physician involvement creates a "formal, documented structure" for the relationship between donor and recipient, thereby forestalling misunderstandings regarding parental status. Yet the court's suggestion that a doctor would be able to prevent such "misunderstandings," a task that seems more in a lawyer's domain, is puzzling, given that the courts have generally defined the boundaries of professional expertise quite narrowly. And the law requires only the semen to exchange hands with a physician. It is hard to see how this will create the formal documented structure referred to above, or how it guarantees donor screening. Furthermore, Section 7004 of the California Civil Code, covering *married women* who conceive through AID with their husband's consent, does *not* require physician involvement.

The court expressly denies that the physician requirement interferes with the right to procreative choice, because it neither prevents self-insemination nor in "any other way prevents women from artificially conceiving children under circumstances of their own choice" (p. 537). Yet California's law certainly shapes choice: Women who choose to inseminate themselves do so at the risk of losing certain legal protections, and only unmarried women face this risk. Even when the courts have not interpreted the physician requirement so stringently, their actions have still affected women's reproductive choices. For example, in *McIntyre* v. *Crouch* the court ruled that, in Oregon, an unmarried woman who conceives through AID without involving a doctor commits a misdemeanor and may have to pay a fine, but she still is protected from a paternity claim. (Ultimately, the court *did* grant the donor the right to a hearing on his paternity claim.)

Child Support

Child support cases involving reproductive technologies have been the one instance in which court actions appear to recognize how social context shapes reproductive choices. Typically, these cases involve husbands who consented to the AID of their wives but who, upon divorce from their wives, deny responsibility for support of the children so conceived. Because state laws specify that the husband under such circumstances is the child's natural father, these decisions revolve around the question of what constitutes consent and how that affects the initial decision to conceive. Furthermore, although some state laws stipulate that a husband must provide his consent in writing, the courts have held that the absence of written consent does not bar them from inquiring into the circumstances surrounding the decision to use AID. If the husband's conduct indicates knowledge of the procedure and failure to object actively, he is presumed to have consented and is held responsible for child support. Underlying this focus on the legal question of consent is a concern with the effect of a husband's actions on reproductive decisions. In one of the earliest and most frequently cited cases of its kind, the *Gursky* court writes:

> There is nothing in the record to indicate that the wife would have undergone artificial insemination in the absence of the husband's consent. Hence it is reasonable to presume that she was induced so to act and thus changed her position to her detriment in reliance upon the husband's expressed wishes. (Pp. 1088-89)

Although the *Gursky* court expressly considered whether the AID would have occurred in the absence of the husband's consent, other courts have simply *presumed* that without her husband's participation and consent, a wife would not conceive through AID (e.g., *People* v. *Sorensen*; *Anonymous* v. *Anonymous* 1964). Thus, in child support cases, the courts have recognized that social context affects a woman's reproductive choices. In all but one of these cases, the husband was held responsible for child support.

PARENTING

Often reproductive technologies are heralded for extending options for parenthood by enabling couples to have biologically related children when one partner is infertile or enabling women to have children when they are unwilling or unable to conceive through intercourse. Feminist critics are concerned, however, that reproductive technologies may extend fathers' rights at the expense of mothers' rights (e.g., Hanmer 1983; Smart 1987). They also charge that these technologies degrade motherhood by emphasizing biological over social definitions of parenting (e.g., Rothman 1989). And critics charge that these technologies have become an occasion on which to impose definitions of good mothers that few women can meet (Harrison 1987; Rothman 1987, 1989). Indeed, Morgan (1985) cautions that, in the course of regulating these technologies, "maternity and motherhood will be made and re-made male" (p. 232). In regulating reproductive technologies, how have the courts viewed mothers' and fathers' rights? How do they define and evaluate parenting?

Mothers' Versus Fathers' Rights

Court actions have expanded the parental options of infertile men more than infertile women. As previously indicated, men need only consent to the AID of their wives to be named as father on a child's birth certificate. This suggests that AID is more for men to become fathers than it is for women to become mothers. Otherwise, we would expect that husbands would have to adopt children so conceived. The frequently cited opinion in *People* v. *Sorensen* all but confirms this interpretation of AID: "[T]he word father is construed to include a husband who, unable to accomplish his objective of creating a child by using his own semen, purchases semen from a donor and uses it to inseminate his wife to achieve his purpose" (p. 11). Here the court denies both women's desire for children and their involvement in "creating a child."

Conflicts with existing statutes. Rulings in cases involving conflicts between surrogacy contracts and existing state laws have the effect of extending to an infertile woman's *husband* the opportunity to become a father. Some courts have ruled that an infertile woman is not a party to a surrogate contract. Both the *SPA* and *Baby Girl L. J.* courts held that surrogacy contracts do not violate laws prohibiting payment in connection with adoption because they are *not* entered into for the purposes of adoption; that is, they are not entered into to provide an infertile woman the opportunity to become a mother. Even when the courts have found adoption to be the intent of a surrogate contract, they have not expanded the options for infertile women to become mothers. Chief Justice Wilenz, in the *Baby M* case, opined that the "purpose and effect of the surrogacy contract [is] to give the father the exclusive right to the child by destroying the rights of the mother" (p. 1247). Yet the court ruling in this case, that surrogacy violates New Jersey policy, prevented Mrs. Stern from adopting Baby M. even as it awarded Mr. Stern custody of the child. Whether surrogacy contracts are upheld or voided by the courts, motherhood for infertile women has not been the result.

The courts have also expanded men's options for becoming fathers in resolving conflicts between surrogacy arrangements and state laws declaring the husband of a married woman who conceives through AID the child's father. In such cases, they have either ruled or implied that they would rule in favor of the sperm donor. It is worth noting, however, that, in each of these cases, the surrogate's husband has supported the donor's paternity claim. We do not know how the courts would resolve competing disputes for paternal rights between two men because this has never been at issue.

Paternity rights. The one instance in which public policy seems to expand women's options for motherhood involves state statutes that specify that the donor of semen to a woman not his wife has no paternity rights. According to the courts, these laws afford "unmarried . . . women a statutory vehicle for obtaining semen for artificial insemination without fear that the donor may claim paternity" (*Jhordan C.* v. *Mary K.*, 534); see also, *McIntyre* v. *Crouch*; *In the Interest of R. C.*). Yet, in every instance in which a donor of semen to an unmarried woman has contested his nonpaternity status, he has won the case.

Legal commentators suggest that, in such conflicts between unmarried women and donors, the intent of both donor and recipient should be considered (e.g., Andrews 1986, 668). Yet the importance the courts have placed on any alleged agreement between donor and recipient regarding the donor's parental rights has varied considerably. In *Jhordan*, the court held that, because a licensed physician was not involved in the insemination, an unmarried woman was not protected from a paternity claim. Any relationship the donor had with the child was irrelevant to this part of the court's determination. Similarly, the court in *C. M.* v. *C. C.* noted that C. M. testified, and C. C. denied, that he "assumed he would act toward the child in the same manner as most fathers act toward their children" (p. 822). But the court focused on whether C. M. was the "natural" father of the child. The court reasoned that, if conception had taken place through intercourse, there would be "no question that the 'donor' would be the father. . . . The issue becomes whether a man is any less a father because he provides the semen by a method different from that normally used" (p. 824). The court ruled he was not. Furthermore, the court held that because C. M. was known to C. C., "[he] cannot be considered to be any less a father because he is not married to [her]" (p. 824). The court briefly returned to the question of any preconception agreement in concluding its opinion, maintaining that the evidence supported C. M.'s contention that he intended to assume the responsibilities of fatherhood. But that evidence did not consist of any agreement *between* the two. It consisted of the fact that "no one else was in a position to become the father" (p. 824).

Even when the *McIntyre* court ruled that a donor must do more than know an unmarried woman, but must also have "grasped the opportunity to participate in the rearing of his child" (p. 14), its ruling provided no criteria for evaluating evidence of a donor taking such initiative. It implied, however, that filing a suit for paternity might itself indicate an effort to grasp an opportunity for parenthood.

Only the court in *In the Interest of R. C.* placed considerable weight on the preconception intentions of the parties and indicated what would constitute evidence of such intent (the recipient's promise that the donor would be treated as a father, the donor's purchase of baby equipment and opening of a college trust fund, and his assistance in the birth and with child rearing on a daily basis). However, the court would have disallowed any preconception agreement between a donor and a married woman whose husband consented to her AID. Court actions in most of these cases involving disputes between semen donors and unmarried women, then, have had the effect of promoting men's desires to become fathers at the expense of the parental rights of unmarried women.

Adoption/custody. What have the courts done when men *do not* desire to be fathers? In *Davis* v. *Davis*, when the biological parents of frozen embryos separated, the court granted Mrs. Davis custody of the frozen embryos for future implantation, even though Mr. Davis claimed that if she used the frozen embryos without his consent he would be "raped of his reproductive rights" and forced into parenthood (p. 83). Yet it did not grant her custody to expand her parental options but because it was in the "best interests of the children, in vitro, that they be made available for implantation" (p. 20).

Child support. In child support cases, too, where the husband of a woman who conceives through AID disputes his child support obligations on the grounds that he is not the child's father, the courts have almost invariably ruled that he is a father and obliged to support the children. Such actions have the effect of promoting fatherhood against men's desires. But they have been justified in the name of the best interests of the child, the nuclear family, and the financial integrity of states, for, if men do not support their children, states may have to.

Incarcerated men. Court actions have restricted incarcerated men's parental options. The court in *Goodwin* v. *Turner* ruled that prison officials are not required to assist a man in artificially inseminating his wife even when he pays for it. According to the court, any privacy interest an individual may have in becoming a parent through artificial insemination is "necessarily infringed on by the nature of incarceration" (p. 1455). To the extent that men of color are more likely to be incarcerated than white men, this decision is potentially discriminatory by race.

Social Versus Biological Bases of Parenting

Feminists argue that reproductive technologies degrade motherhood and turn it into something to be purchased in the market (Rothman 1989). At the heart of this argument is the concern that reproductive technologies foster an emphasis on the biological over the social bases of parenting. Once motherhood is viewed as a biological process, it can be divided up into different pieces and separately sold to the highest bidder. Court rulings reflect disagreement over whether motherhood and fatherhood are social or biological relationships.

Conflict with existing statues. Some courts have equated AID and surrogacy (e.g., *Baby Girl L. J.* and *SPA*). In so doing, they not only define parents in terms of their genetic contribution but exalt men's and ignore women's role in childbearing. As the courts use the terms, both AID and surrogacy involve a donor who provides the genetic material that a spouse cannot. But only in surrogacy is the spouse unable to contribute the genetic material *and* either unable or unwilling to carry the baby to term. Thus the female donor in a surrogacy contract provides both an egg and her body for nine months for the purposes of gestating that egg. By equating AID and surrogacy, the courts exalt the biological contribution of egg and sperm and ignore the role of gestation, part of a woman's biological contribution, in the development of a baby.[3] Furthermore, gestation involves a social relationship as well as a biological one (Rothman 1989). Thus, although donors in both surrogacy and AID arrangements must "give up" a child that contains their genetic material, only in surrogacy arrangements is a donor required to give up a child to which she has been intimately related for nine months.

Adoption/custody. The *Davis* case presents a clear example of the courts defining parenting in biological terms and thereby undervaluing women's contribution to birth. Here, the court adopts the logic of the best interests of the child to decide the status of seven frozen embryos. To do this, the court relies on the testimony of one expert witness, Dr. Jerome LeJeune, who argues that "as soon as he [sic] has been conceived, a man is a man [sic]" (p. B18). And it rejects the arguments of the other three experts that the embryos only represent *potential* human life and therefore should not be treated as children. In this way, it appears not to recognize that the "man" to which Dr. LeJeune refers only becomes a "man" after nine months in a woman's body.

In contrast to *Davis*, the *Baby M* court, called on to decide whether the biological father or the surrogate mother should receive custody of the baby, resists equating men's and women's contributions to childbearing.[4] Chief Justice Wilenz also acknowledges that gestation can be both a biological and a social process. Accordingly, he disallows surrogate contracts in part because they require a woman to commit herself to giving up a child "before she knows the strength of her bond with [it]" (p. 1248). And he concludes that when a father and mother are separated at the birth of the child and disagree about custody, the child should only be taken from its mother under extreme circumstances, because of "[t]he probable bond between mother and child . . . along with the likelihood, in most cases, of a significantly lesser, if any, bond with the father" (p. 1261). Furthermore, he refuses to equate AID and surrogacy. Still, he fails to recognize the full extent of women's contribution to reproduction by equating AID and egg donation. For, when an egg is donated, the infertile woman participates in reproduction: She is pregnant for nine months and gives birth to the child. When semen is donated, an infertile man does nothing but consent to the procedure.

Child support. If court actions have tended to exalt biology in one set of circumstances, in conflicts over child support, they clearly recognize the importance of social relationships in parenting. As indicated above, the courts almost invariably have found husbands who consent to the AID of their wives obliged to support any children so conceived. Clearly, the husbands in such instances are not biologically related to the children, yet the courts define them as fathers. Indeed, in one instance, a court declared a woman (who put herself forward as a man, married another woman, and consented to her "wife's" AID) to be the father of two children (*Karin T.* v. *Michael T.*). According to this court, even though a parent is technically one who brings forth offspring, a spouse can do this by consenting to AID (even if "he" is another woman).

In some instances, the courts cite a variety of other evidence to support the view that a husband has established a social and therefore parental relationship with a child, from putting it forth publicly as his own, to declaring it a dependent on his tax forms, to changing its diapers and feeding it. In other instances, however, the courts have assumed that the consent of husbands to their wives' AID alone signifies their desire to develop a social relationship with that child (e.g., *K. S.* v. *G. S.*).

Evaluating Mothers and Fathers

Judicial opinion relating to reproductive technologies contains judgments regarding "good" and by implication "bad" parenting. Class shapes these evaluations. For example, the *Baby M* court comes close to equating good parents with membership in the middle or upper class. Among the qualities speaking favorably for the Sterns the court notes: The household is stable; finances are adequate; and the Sterns have a willingness to learn, listen, and encourage the child's independence that is enhanced by their professional training (p. 1259). Similarly, the court in *C. M.* v. *C. C.* grants C. M.'s paternity claim in these terms: "C.M. has shown a genuine interest in the child; he is a teacher and educationally able to aid his development, and is financially capable of contributing to his support" (p. 825). Here the court all but equates interest in the child with class status. But gender also matters in courtroom evaluations of parenting, as we argue below.

Conflicts with existing statutes. That judicial opinion frequently reinforces gender stereotypes is revealed in cases involving conflicts between surrogacy and existing state laws. For example, the courts seem to agree that good mothers put the interests of others above their own. They disagree, however, in their interpretation of how and when women demonstrate this quality, as is revealed clearly in the *SPA* opinions. The majority views surrogacy as an act of altruism (p. 212). In contrast, Justice Vance's dissenting opinion implies that surrogacy is an act of profound selfishness by women "who will readily, for a price, allow themselves to be used as human incubators and who are willing to sell, for a price, all of their parental rights in a child thus born" (p. 214).

Child support. If good mothers are selfless, good fathers provide for their children; so judicial opinion regarding child support assumes. For, as we have seen, the courts almost invariably have held men who consent to their wives' artificial insemination responsible for supporting any children so conceived. Even when couples have tried to abide by different roles, the courts have promoted this view. Thus a court has declared "immaterial" a wife's agreement with her husband that she wanted no child support from him in holding her husband responsible for providing it (*People* v. *Sorensen*, 12).

Paternity. In conflicts over paternity rights, the courts have demanded less of men seeking to establish a social relationship with a child than they have of women. This works to the disadvantage of unmarried women, especially if their sexuality is in question. For example, in the *Jhordan* case, the court permitted Jhordan's claim to paternity on the grounds that he developed "a social relationship with Mary and Devin as the child's father"

(p. 537). What precisely did he do? He maintained contact with Mary during the pregnancy (she attended a Christmas party at his workplace, and he visited her "several" times at her workplace); he collected some baby equipment and started a trust fund; and he visited the day after Devin's birth and once a month thereafter. In contrast, the court denied joint custody with Mary's friend Victoria even though Victoria, too, had established a social relationship with Devin and Mary. She took Mary to her appointments, went to birthing classes, and assisted in the delivery. After Devin's birth, she spent at least two days a week with him, talked to him on the phone every day she did not see him, participated in making decisions about his daily care and development, and took vacations with him and Mary.

FAMILY STRUCTURE

Reproductive technologies are popularly portrayed as enabling diverse family arrangements. Feminist critics argue, however, that they are being used to reinforce patriarchal family structures and relations and thereby reinforce women's subordination (e.g., Morgan 1985; Spallone 1986; Spallone and Steinberg 1987).

Our analysis of judicial opinion reveals a firm commitment to the traditional nuclear family structure both on the part of the courts and in the statutes they discuss. For example, several courts have indicated that AID laws that confer paternity on the consenting husband of a woman who conceives through AID express public policy favoring the nuclear family (e.g., *C. M.* v. *C. C.*, 824; *R. S.* v. *R. S.*, 927; *In the Matter of the Estate of Gordon*, 825). Furthermore, the courts have almost always held these husbands responsible for child support. To do otherwise, they argue, would be to controvert public policy favoring the legitimization of children born to married women. Sometimes these decisions are rationalized in the best-interests-of-the-child logic, but they are also justified in terms of a policy preference for traditional family arrangements.

Conflicts with Existing Statutes

A preference for the nuclear family structure underlies conflict in judicial opinion over surrogacy. Some courts suggest that surrogacy would promote traditional families by enabling infertile couples to have children. The *Baby M* court, however, finds that surrogate contracts violate New Jersey policy because they "guarantee permanent separation of [a] child from its natural parents" and thereby contradict a long-standing policy that, "to the extent possible, children should remain with and be brought up by both of their natural parents" (pp. 1246-47). The courts may disagree over whether surrogacy would threaten the nuclear family structure, but they do not differ in their preference for this family form.

Paternity Rights

Do reproductive technologies enable alternative family arrangements by allowing single women and men to become parents? Whether the courts would support alternative family arrangements for men has not been an explicit issue. The *SPA* court, in deciding that surrogate arrangements did not conflict with adoption law, wrote that there is no requirement that the biological father of a child born of a surrogate arrangement be married (p. 211). On the other hand, the court in *In the Interest of R. C.* held that it would not honor any preconception agreements between donors and married women whose husbands had consented to their AID, thereby restricting the familial options of unmarried men.

The right of unmarried women to bear children and establish families without a father has been a direct issue before the courts. Some state laws that permit the use of reproductive technologies by unmarried women do so in ways that can constrain their familial options. California, as we saw, permits the AID of unmarried women, but unmarried recipients are protected from paternity actions only when the semen is provided to a licensed physician. Furthermore, a donor assertion of paternity subsequent to AID is precluded for married women in California, regardless of physician involvement. The *Jhordan* court, in addressing the equal protection implications of this, responded, "In the case of a married woman, the marital relationship invokes a long-recognized social policy of preserving the integrity of marriage. . . . No such concerns arise where there is no marriage at all. Equal protection is not violated by providing that certain benefits or legal rights arise only out of the marital relationship" (p. 536). Here the court admits a preference for traditional families but denies that this disadvantages women who opt for alternative family arrangements. Similarly, in *C. M.* v. *C. C.* the court argues, "[We] have consistently shown a policy favoring the requirement that a child be provided with a father as well as a mother," in support of its decision to grant C. M.'s paternity request (p. 824). Yet, even as it admits its preference for one family form, the court denies that it restricts the familial options of unmarried women by stating that "the court takes no position as to the propriety of the use of artificial insemination between unmarried persons but must be concerned with the best interests of the child." These interests are "two parents wherever possible" (p. 825). And, as we have seen, one court held that preconception agreements are relevant between an unmarried woman and a donor but irrelevant between married women and donors. In so doing, it restricted the possibility of establishing alternative families in which children might be raised by more than two adults.

If the courts have expressed a preference for particular family arrangements, they continue to insist on the sanctity of family autonomy. Indeed, they concur that whether and how to have children is a private decision. Yet, at the same time, court actions have implications for such decisions. The question of family autonomy arises expressly in the *Jhordan* case. Mary and Victoria claimed that they and the child Devin constituted a family unit and that giving Jhordan legal rights infringed on their right to family autonomy. The court rejected this definition of family and insisted that Jhordan was not only Devin's father but a member of the family. Furthermore, it denied that it upheld Jhordan's paternity claim because Mary was unmarried and seeking custody with another woman: "[W]e do not express any judicial preference toward traditional notions of family structure or toward providing a father where a single woman has chosen to bear a child" (p. 537). Instead, it ruled against her because a licensed physician had not performed the insemination and because Jhordan had become a family member. Yet we have seen that the licensed physician requirement in California applies only to unmarried women and that Jhordan's contributions pale in comparison to those of Victoria, who lost her bid for joint custody. A preference for traditional family arrangements lies at the heart of this decision. Furthermore, in preferring one family form, the court's actions place others at a disadvantage.

EVALUATING DEVELOPING POLICY: WOMEN, REPRODUCTIVE TECHNOLOGY, AND THE STATE

Our analysis confirms the social nature of reproductive technologies. Whatever reproductive options they may offer women and men, these options are not inherent in the technologies themselves but are the product of the struggles over them. The courts participate in these struggles and, in so doing, participate in the control of women's reproductive lives. Court decisions have tended not to recognize the extent to which social conditions shape women's reproductive choices. They have done more to promote fathers' rights than mothers' rights. They have often ignored the social relationships that constitute childbearing and child rearing and valued men's over women's biological contribution to these processes. Court decisions are also based on certain assumptions about the proper roles of mothers and fathers. Finally, far from promoting alternative family forms, court decisions reflect an unwavering commitment to the nuclear family structure.

Still, the courts have not acted monolithically. They have recognized the influence of social conditions on reproductive choices in child custody cases. They also have recognized that parenting involves social as well as biological relationships both in child custody cases and in cases involving the paternity rights of donors to unmarried women. This suggests to us the need to analyze more systematically the conditions under which the courts will act in ways compatible with women's interests. Although our analysis does not directly address this question, it does have implications for it. On the one hand, it appears that the courts recognize the social conditions shaping reproductive choices and the social relationships that constitute parenting when doing so promotes traditional family arrangements. Hence they uphold men's financial responsibility for children born to their wives through AID because to do otherwise would be to contradict public policy favoring the legitimization of children. However, they acknowledge that a husband's consent forms part of the context in which a woman decides to engage in AID and signifies his desire to develop a social relationship with the child as its father. Similarly, they confirm the paternity rights of sperm donors to unmarried women—even in the face of laws that stipulate donors have no such rights—by suggesting that these laws were not meant for men who sought to develop a social relationship with the child so conceived. But their actions in this regard have the effects of restricting unmarried women's ability to develop alternative family forms and of promoting at least the ideology, if not the reality, of the two-parent family.

The decision of the New Jersey Supreme Court in the *Baby M* case represents another instance in which a court has adopted perspectives more consistent with women's interests. Chief Justice Wilenz recognized the role of economic conditions in a woman's decision to become a surrogate and the subsequent potential for surrogacy to exploit women. He also acknowledged both the importance and the social nature of gestation. Did the *Baby M* court rule against surrogacy out of fear that it undermined the nuclear family? Or was the decision influenced by the tremendous amount of feminist attention the case received? We stress that our analysis can only speculate on these kinds of questions. Clearly, there is need for further research here. Ultimately, this research can shed more light on how gender shapes state actions and how state actions, in turn, construct gender relations.

Our analysis also confirms that reproductive technologies and efforts to regulate them derive their meaning in a social context. And it is difficult to make nonexploitative policy in a world characterized by exploitation. On the one hand, policymakers can regulate reproductive technologies more equitably. They should oppose surrogate contracts involving money and support the voidability of surrogate arrangements because, under present conditions, surrogacy has a far greater potential to oppress women than to open opportunities to them. Second, no exceptions should be made to state laws preventing donors of semen for the artificial insemination of women who are not their wives from claiming legal rights as fathers. It would still be possible for men and women to have informal agreements regarding paternal rights.

But donors will have to risk the possibility of being denied these parental rights, with no legal recourse. Currently, this seems the only way both to protect the reproductive and familial autonomy of unmarried women and to acknowledge their contribution to reproduction. Third, women must be informed of the risks associated with reproductive technologies, and these risks must be recognized in legal disputes. Fourth, women's contribution—both biological and social—to childbearing must receive greater recognition: The woman who labors for nine months with and gives birth to a child should have maternal rights if she so desires. Fifth, unmarried women should receive the same legal protections as married women. Sixth, doctors should be required to perform tests to protect both recipients of semen and fetuses, but doctors should not be required to participate in artificial insemination, and protections should not be contingent on that participation. Finally, neither class status nor adherence to traditional gender roles should be a primary basis on which to evaluate parenting.

Thus there are a variety of ways by which policymakers can do more to prevent reproductive technologies from oppressing women. But these alone will not end the inequality and exploitation associated with reproduction. This requires a much larger struggle for better and greater access to health care; safer and healthier workplaces and communities; recognition of the value of women's work, including mothering, and more and better-paying jobs for all people; comprehensive parental leave policies; high-quality and affordable day care; and fundamental reforms in the welfare system.

NOTES

1. Although *reproductive technologies* refers to a range of technologies available for manipulating reproduction, we focus here on what Stanworth (1990) refers to as "conceptive technologies": technologies aimed at people who cannot or do not want to conceive children through intercourse.

2. The surrogacy and payment question arose in the case of *Sherwyn & Handel*, but the court refused to rule on it, arguing that the case did not present a justiciable controversy. The court in *Doe* v. *Kelley* held that adoption laws do make payment in connection with surrogacy unlawful, although they do not prohibit surrogacy altogether. But the court is not concerned with the potential for coercion of such payment. Instead, it views these prohibitions as reasonable government regulation of the status of children.

3. Surrogacy arrangements generally fail to recognize women's contribution to birth because they often stipulate that if the mother miscarries after her fifth month she is entitled only to a small portion of the total fee. The surrogate involved in the *Doe* v. *Keane* case, for example, miscarried in her fifth month and, by the contract, was to receive only 10 percent of the total payment. Eventually, however, she arranged to receive 70 percent.

4. The *McIntyre* court also suggested that different rights ought to be accorded the different contributions of men and women (though ultimately it ruled against the woman).

REFERENCES

Andrews, Lori B. 1984. The stork market: The law of the new reproductive technologies. *American Bar Association Journal* 10:50-56.

———. 1986. Legal aspects of assisted reproduction. *Annals of the New York Academy of Science* 54:668-74.

———. 1989. Position paper: Alternative modes of reproduction. In *Reproductive laws for the 1990s*, edited by Sherrill Cohen and Nadine Taub. Clifton, NJ: Humana Press.

Corea, Gena. 1985. *The mother machine*. New York: Harper & Row.

Cooper, Glinda. 1986. An analysis of the costs of infertility treatment. *American Journal of Public Health* 76:1018-19.

Eisenstein, Zillah. 1983. The state, the patriarchal family and working mothers. In *Families, politics, and public policy: A feminist dialogue on women and the state*, edited by Irene Diamond. New York: Longman.

Gertner, Nancy. 1989. Position paper: Interference with reproductive choice. In *Reproductive laws for the 1990s*, edited by Sherrill Cohen and Nadine Taub. Clifton, NJ: Humana Press.

Hanmer, Jalna. 1983. Reproductive technology: The future for women. In *Machina ex dea*, edited by Joan Rothschild. New York: Pergamon.

Harrison, Michele. 1987. Social construction of Mary Beth Whitehead. *Gender & Society* 1:300-11.

Hernes, Helga. 1987. *Welfare state and women power*. Oslo: Norwegian University Press.

Klein, Renate D. 1990. IVF research: A question of feminist ethics. *Issues in Reproductive and Genetic Engineering* 3:243-51.

Koch, Lene. 1990. IVF – an irrational choice? *Issues in Reproductive and Genetic Engineering* 3:235-42.

Lorber, Judith. 1988. In vitro fertilization and gender politics. In *Embryos, ethics, and women's rights*, edited by Elaine H. Baruch. New York: Harrington Park Press.

MacKinnon, Catharine A. 1989. *Toward a feminist theory of the state*. Cambridge: Harvard University Press.

Morgan, Derek. 1985. Making motherhood male: Surrogacy and the moral economy of women. *Journal of Law and Society* 12:219-38.

Pascal, Gillian. 1986. *Social policy: A feminist analysis*. London: Tavistock.

Petchesky, Rosalind Pollack. 1980. Reproductive freedom: Beyond a woman's right to choose. In *Women: Sex and sexuality*, edited by Catharine R. Stimpson and Ethel Spector Person. Chicago: University of Chicago Press.

———. 1984. *Abortion and woman's choice: The state, sexuality, and reproductive freedom*. Boston: Northeastern University Press.

Quadagno, Jill S. 1990. Race, class and gender in the U.S. welfare state: Nixon's failed family assistance plan. *American Sociological Review* 49:632-47.

Rothman, Barbara Katz. 1984. The meanings of choice in reproductive technologies. In *Test-tube women: What future for motherhood*, edited by Rita Arditti, Renate Duelli Klein, and Shelly Minden. London: Pandora.

———. 1987. Comment on Harrison: The commodification of motherhood. *Gender & Society* 1:312-16.

———. 1989. *Recreating motherhood: Ideology and technology in a patriarchal society*. New York: Norton.

Smart, Carol. 1987. "There is of course the distinction dictated by nature": Law and the problem of paternity. In *Reproductive technologies: Gender, motherhood and medicine*, edited by Michelle Stanworth. Minneapolis: University of Minnesota Press.

Spallone, Patricia. 1986. The Warnock Report: The politics of reproductive technology. *Women's Studies International Forum* 9:543-50.

Spallone, Patricia, and Deborah Steinberg. 1987. International report. In *Made to order: The myth of reproductive and genetic progress*, edited by Patricia Spallone and Deborah Steinberg. Oxford: Pergamon.

Stanworth, Michelle. 1990. Birth pangs: Conceptive technologies and the threat to motherhood. In *Conflicts in feminism*, edited by Marianne Hirsch and Evelyn Fox Keller. New York: Routledge.

Williams, Linda S. 1990. Wanting children badly: A study of Canadian women seeking in vitro fertilization and their husbands. *Issues in Reproductive and Genetic Engineering* 3:229-34.

Williams, Wendy. 1982. The equality crisis: Some reflections on culture, courts, and feminism. *Women's Rights Law Reporter* 7:175-200.

Wilson, Elizabeth. 1977. *Women and the welfare state*. London: Tavistock.

Kim M. Blankenship is an Assistant Professor of Sociology at Yale University. Her research and publications have focused on the interrelations among gender, race, class, and law and social policy. In particular, she is interested in antidiscrimination policy, reproductive technologies, and AIDS.

Beth Rushing is an Assistant Professor in the Department of Sociology at Kent State University. Her current research focuses on the effects of employment, marriage, and parenthood on health, as well as analyses of state legislation regarding reproductive technologies.

Suzanne A. Onorato received her Ph.D. in sociology from Duke University. Her research has focused on various women's health issues, including the "new" reproductive technologies, contraception, and premenstrual syndrome (PMS). She currently is the executive director of Y-Me of the New England Breast Cancer Support Program and is a visiting lecturer in the Sociology Department at Yale University.

Renée White is a doctoral candidate in sociology at Yale University in New Haven, Connecticut. She is a member of Students of Color for Reproductive Health and is co-chair of the Black Graduate Network. Her dissertation examines the socioeconomic and cultural influences on Black adolescents' sexuality and AIDS knowledge.

Adolescent Pregnancy and Public Policy

DEBORAH L. RHODE

Although teenage pregnancy has recently emerged as a major social "problem," its frequency is by no means a recent social phenomenon. The appropriate age for sexual relations and parenthood has always been a matter of cultural definition, and in the United States it has varied considerably across time, region, class, race, ethnicity, and gender. Over the past two centuries, the age at which childbirth is biologically possible has declined, while the period of adolescents' economic dependence has increased. One result of these changes has been a growing cultural conflict over reproductive choices during the transition to adulthood. During the past two decades, the rising birth rate among unmarried teenagers has prompted increasing national concern but no coherent policy. Much of the problem stems from disputes over the nature of the problem. Is the primary issue morality, fertility, or poverty? What choices should adolescents make in sexual relationships, what role should the state play in shaping the choices available, and who should decide those questions?

This article explores public policies on adolescent pregnancy against their broader historical, legal, and socioeconomic backgrounds. Its central premise is that such policies have frequently misdescribed the

---

DEBORAH L. RHODE is a professor of law at Stanford University and the former director of Stanford's Institute for Research on Women and Gender. Her recent books include *The Politics of Pregnancy: Adolescent Sexuality and Public Policy*, *Theoretical Perspectives on Sexual Difference*, and *Justice and Gender*.

Political Science Quarterly Volume 108 Number 4 1993-94 635

problem and misled as to the solution. Too often, decision makers have located the problem at the individual level and faulted teens who want "too much too soon" in sexual relationships. Insufficient attention has focused on the societal level, on structures that offer female adolescents "too little too late": too little reason to stay in school, too little assistance in birth control, too little opportunity for childcare, health services, vocational training or decent jobs, and too little understanding of the responsibilities of single parenthood. Both legislative and judicial decision making have proceeded without adequate information on the experience of women or the consequences of various public policies.[1]

To understand the limitations of current frameworks, some historical background is useful. Since specific policies toward adolescent pregnancy are a recent development, their evolution must be understood in terms of broader reproductive, family, and regulatory patterns. An overview of those patterns suggests certain dynamics that are particularly relevant to contemporary debates: the variation in cultural norms concerning teenage sexuality and the class and racial biases in public policy responses.

An Historical Overview

Historical material on adolescent sexual activity is limited, in part because only in the last century has adolescence been recognized as a distinct developmental phase. However, research on premarital intercourse, premarital pregnancy, and out-of-wedlock births provides some indirect evidence of cultural norms.[2]

Until the nineteenth century, American attitudes toward early childbearing were relatively permissive. Pregnancy among very young adolescents was rare, since the average age of menstruation was 15 or older. Available evidence reveals no strong prejudice against teenage fertility when the couple was willing to marry and could become economically independent. Nor was premarital sexual intercourse among

[1] Sharon Thompson, "Search for Tomorrow: On Feminism and the Reconstruction of Teen Romance" in Carole S. Vance, ed., *Pleasure and Danger* (Boston: Routledge and Kegan Paul, 1984), 350.

[2] Susan E. Harari and Maris A. Vinovskis, "Adolescent Sexuality, Pregnancy, and Childbearing in the Past" in Annette Lawson and Deborah L. Rhode, *The Politics of Pregnancy: Adolescent Sexuality and Public Policy* (New Haven: Yale University Press, 1993); John Demos and Virgilio Demos, "Adolescence in Historical Perspective," *Journal of Marriage and the Family* 31 (1969): 632.

adolescents seen as especially problematic. Although many colonial communities made efforts to punish fornicators of all ages, intimacy during courtship met with somewhat greater tolerance, as the practice of bundling in New England and Middle Atlantic regions suggests. Because courting couples often lived at considerable distance from each other, in housing that had little heat, privacy, or extra space, they were often permitted to share the same bed as long as they remained fully clothed or kept a bundling board between them. Sexual intercourse was not expected to occur, but if it did the parties were expected to marry. The assumption was that a woman would go from her father's to her husband's household without any intervening period of independence. Bastardy was a serious offense in a culture that made families the central social and economic unit. The mother—and father if he could be identified—was subject to criminal penalties and social opprobrium, while the child experienced similar stigmas and various legal disabilities. Until the mid-eighteenth century rates of illegitimacy remained low, around 3 percent of all births, although rates of premarital pregnancy varied over time and region; an estimated one in three brides was pregnant among seventeenth-century Chesapeake immigrants and one in ten among New England residents.[3]

During the latter part of the eighteenth century, premarital pregnancies and out-of-wedlock pregnant births began to rise. The increase in part reflected greater geographical mobility and the breakdown of stable communities during the war and postwar periods. Formal enforcement of bastardy and fornication laws ceased, and informal sanctions weakened. For some young women, pregnancy also became a means of insuring that they rather than their parents controlled the choice of husband. By the late eighteenth century, an estimated 30 percent of brides were pregnant. Most premarital intercourse took place as part of courtship, but if pregnancy occurred and marriage did not, the costs were borne disproportionately by women. Paternity could be difficult to prove and fathers difficult to locate. Special hard-

[3] Maris A. Vinovskis, *An "Epidemic" of Adolescent Pregnancy?* (New York: Oxford University Press, 1988), 3-21, 32; John D'Emilio and Estelle B. Freedman, *Intimate Matters: A History of Sexuality in America* (New York: Harper and Row, 1987), 10-28, 32-33, 43-46. For sanctions against fornication, see ibid., 13; Michael D. Freeman and Christina M. Lyon, *Cohabitation without Marriage: An Essay in Law and Social Policy* (Aldershot, England: Gower, 1983); Edmund S. Morgan, "The Puritans and Sex" in Michael Gordon, ed., *The American Family in Social-Historical Perspective*, 3rd ed. (New York: St. Martin's Press, 1983), 311.

ships fell on indentured servants who were impregnated by already married masters.[4]

For women of color, sexual norms evolved out of different cultural traditions and, for blacks, out of different sociolegal constraints imposed by slavery. Many Native American tribes condoned premarital intercourse and among some a first birth usually preceded marriage. Under bondage, blacks had no formal right to marry and their African heritage often provided no sanctions against premarital sexual activity. In a plantation culture, early fertility increased women's value. Accordingly, many slave communities did not stigmatize premarital intercourse or childbirth, although couples typically preferred stable unions when possible.[5]

In the mid-nineteenth century, the growth of religious revival movements and moral reform societies, together with the culture idealization of domesticity, more actively discouraged adolescent sexual relations. Conventional medical wisdom, which assumed that early intercourse drained "vital life forces," also contributed to that trend. Rates of premarital pregnancy declined to 10 percent and in white middle-class communities a strong double standard prevailed. The increasing importance of education, particularly for males, counseled against early marriage, while for most young women pregnancy without marriage was disastrous; the inadequacy of work and welfare options for mothers made single parenthood difficult, if not impossible. Gender discrimination in employment was widespread, and until the Depression, public assistance programs provided meager subsidies, while frequently excluding unmarried or otherwise "unfit" mothers from coverage. The social stigma against "fallen" women added to these difficulties and

[4] Henry B. Parkes, "Morals and Law Enforcement in Colonial New England," *New England Quarterly* 5 (1932): 431, 442–43; Vinovskis, "*Epidemic*," 13; D'Emilio and Freedman, *Intimate Matters*, 32–33, 43, 73. For a discussion of sexual images of women, see ibid., 43–46; Laurel Thatcher Ulrich, *Good Wives: Image and Reality in the Lives of Women in Northern New England, 1650–1750* (New York: Knopf, 1982).

[5] D'Emilio and Freedman, *Intimate Matters*, 8, 13, 65, 87, 97–100; James Axtell, *The Indian Peoples of Eastern America: A Documentary History of the Sexes* (New York: Oxford University Press, 1989), 71–72; John Upton Terrell and Donna M. Terrell, *Indian Women of the Western Morning: Their Life in Early America* (New York: Dial Press, 1974), 116–17, 121; Herbert G. Gutman, *The Black Family in Slavery and Freedom, 1750–1925* (New York: Pantheon, 1976), 295–96; Eugene D. Genovese, *Roll, Jordan, Roll: The World the Slaves Made* (New York: Pantheon, 1974), 461, 465–66; Daniel Scott Smith, "The Long Cycle in American Illegitimacy and Prenuptial Pregnancy" in Peter Laslett, Karla Oosterveen, and Richard M. Smith, eds., *Bastardy and Its Comparative History* (Cambridge, MA: Harvard University Press, 1980), 37; Dorothy Sterling, ed., *We Are Your Sisters: Black Women in the Nineteenth Century* (New York: W. W. Norton, 1984), 32–33. See also sources cited in fn. 7 below.

the law did little to cushion their impact. Judicial determination of paternity and enforcement of child support was available in theory but often inaccessible in practice, given the financial, procedural, and psychological burdens of litigation. For most female adolescents, marriage was the only respectable solution to an early pregnancy. However, at least from an economic standpoint, that solution generally did not represent a major sacrifice, since few women expected to develop the educational and employment skills that would permit financial independence.[6]

Among some groups of women, the sanctions against premarital sexual activity were less pronounced. Young urban working-class women were less subject to community or family oversight and accounted for most of the rise of premarital pregnancy rates, to around 23 percent at the close of the nineteenth century. Although relatively well-to-do blacks assimilated white middle-class norms of chastity, many lower-class and rural black communities condoned premarital intercourse and out-of-wedlock births. Particularly among tenant farming families, child labor was a crucial economic asset, and early fertility often enhanced rather than diminished a woman's standing. Such attitudes frequently persisted among migrants. Within these communities, young single mothers received support from kinship networks that mitigated the costs of nonmarital childbearing.[7]

[6] The nineteenth-century cult of domesticity is reviewed in Barbara Welter, "The Cult of True Womanhood, 1820–1960," *American Quarterly*, 18 (1966): 151. For religious revivalism and the growth of moral reform societies, see Carroll Smith-Rosenberg, *Disorderly Conduct: Visions of Gender in Victorian America* (New York: Knopf, 1985), 109–28. For medical wisdom, see Edward Clarke, *Sex in Education: Or a Fair Chance for Girls* (Boston: Houghton, Mifflin, 1873), 104, 137; Barbara Ehrenreich and Deidre English, *For Her Own Good: 150 Years of the Experts' Advice to Women* (Garden City, NY: Anchor, 1978), 113–18; Vinovskis, "*Epidemic*," 10, D'Emilio and Freedman, *Intimate Matters*, 178–84; Harari and Vinovskis, "Adolescent Sexuality." For social welfare policies, see Michael B . Katz, *In the Shadow of the Poor House: A Social History of Welfare in America* (New York: Basic Books, 1986); Mimi Abramovitz, *Regulating the Lives of Women: Social Welfare Policy from Colonial Times to the Present* (Boston: South End Press, 1988), 145; Jacqueline Jones, *Labor of Love, Labor of Sorrow: Black Women, Work, and the Family from Slavery to the Present* (New York: Basic Books, 1985), 266–67, 307–08. Discrimination in employment is reviewed in Alice Kessler-Harris, *Out to Work* (New York: Oxford University Press, 1982); Deborah L. Rhode, *Justice and Gender: Sex Discrimination and the Law* (Cambridge, MA: Harvard University Press, 1989), 38–48. As Constance Nathanson notes, the sanctions for illegitimacy varied somewhat, depending on whether the circumstances suggested "innocence betrayed," mental deficiency, hereditary weaknesses, or original sin. Despite these variations, however, there was "scarcely any other way [in which] a woman [could] lose status so completely," as Constance A. Nathanson, quoting Kingsley Davis, writes in *Dangerous Passage: The Social Control Sexuality in Women's Adolescence* (Philadelphia: Temple University Press, 1991), 104.

[7] Kathy Lee Peiss, *Cheap Amusements: Working Women and Leisure in Turn-of-the-Century New York* (Philadelphia: Temple University Press, 1986), 62, 70–71, 108–10; Christine Stansell,

In the culture generally, rates of premarital sexual intercourse jumped sharply during the 1920s and then remained stable until the 1960s. A variety of forces contributed to the liberalization of sexual mores: urbanization, wartime experiences, media images, and rising opportunities for unchaperoned activities in college, automobiles, and recreation centers such as dance halls. Alfred Kinsey's 1948 study reported that about half of all women and 90 percent of men had engaged in premarital sexual intercourse. But until the late 1960s, most women, particularly white middle-class women, viewed such intimacy as a prelude to marriage, and the dominant culture retained strong sanctions against out-of-wedlock births. Adolescents received mixed messages; the popularization of Freudian psychology and the sexual imagery flooding the media made intimate relations seem more acceptable, while teen advice manuals, parental pressures, and socioeconomic constraints gave the opposite signal. Polls during the 1950s and early 1960s revealed that only a quarter of all Americans approved of premarital intercourse. Most adults were equally opposed to sex education programs, which were commonly accused of "teaching vice to little children." The dominant message was the one popularized by a leading guide, *Facts of Life and Love for Teens*: "anything beyond a good night kiss could lead to trouble."[8]

That trouble was experienced disproportionately by women. Outside of certain black communities, young women who became pregnant often faced grim alternatives: undesired marriages, illegal and unsafe abortions, social ostracism, and exclusion from educational programs. Some temporary support and vocational assistance were available through homes for unwed mothers. Adoption agencies also of-

---

City of Women: Sex and Class in New York (New York: Knopf, 1986), 178–186; D'Emilio and Freedman, *Intimate Matters*, 199–200, 272-73, 299–300; Charles S. Johnson, *Shadow of the Plantation* (Chicago: University of Chicago Press, 1934), 49–63. For contemporary discussion of these patterns, see Carol B. Stack, *All Our Kin: Strategies for Survival in a Black Community* (New York: Harper and Row, 1974), 108–15; Leon Dash, *When Children Want Children: The Urban Crisis of Teenage Childbearing* (New York: William Morrow, 1989), 238–62.

[8] D'Emilio and Freedman, *Intimate Matters*, 256, 262–66, 333; Joseph F. Kett, *Rites of Passage: Adolescence in America, 1790 to the Present* (New York: Basic Books, 1977), 258–60; Alfred C. Kinsey, Wardell B. Pomeroy, Clyde E. Martin, and Paul H. Gebhard, *Sexual Behavior in the Human Female* (Philadelphia: W. B. Saunders, 1948), 321–26; B. K. Singh, "Trends in Attitudes Toward Premarital Sexual Relations," *Journal of Marriage and the Family* 42 (1980): 387; David B. Tyack and Elisabeth Hansot, *Learning Together: A History of Coeducation in American Schools* (New Haven: Yale University Press, 1990), 223 (discussing instruction in "vice"); Patricia J. Campbell, *Sex Education Books for Young Adults, 1892–1979* (New York: R. R. Bowker, 1979), 93 (discussing conduct beyond a kiss).

fered placement alternatives for their infants, an option that about 90 percent of all single mothers accepted, generally under varying degrees of pressure. However, services for unwed parents were highly inadequate and could not mitigate the public stigma, psychological trauma, and educational and economic disadvantages that often accompanied nonmarital births. Such costs were exacerbated by a legal system that treated "promiscuous" female (but not male) adolescents as delinquent, and visited additional penalties on illegitimate children, such as exclusion from insurance and governmental entitlement programs.[9]

As this overview suggests, American norms for adolescent sexual expression and childbearing reveal strong variations but certain common themes. During some eras and within some subgroups, early premarital intercourse has not been viewed as especially problematic, provided that any resulting childbirth remained within marital unions or, in many black communities, within a kinship network. Yet for much of American history, the dominant white, middle-class culture has also imposed substantial costs on premarital sexual activity. Those costs have been paid largely by women.

The Evaluation of Reproductive Rights Law

Legal doctrine reflected and reinforced these broader cultural patterns. Until the nineteenth century, there was little formal regulation of birth-control practices. Folk wisdom regarding contraceptive techniques was passed down across generations and during the antebellum period some devices began to be advertised publicly. Relatively few married couples, however, appear to have practiced birth control. Effective contraception often required male cooperation or technical knowledge not yet available, and social norms encouraged high birth rates. For unmarried women, the only reliable means of limiting fertility was abortion. Under early common-law provisions, it was permissible until "quick-

[9] Karen DeCrow, *Sexist Justice* (New York: Random House, 1974), 268–79; Herma Hill Kay, *Text, Cases and Materials on Sex-Based Discrimination* (St. Paul, MN: West, 1988), 375–81. Maris Vinovskis, "An 'Epidemic' of Adolescent Pregnancy? Some Historical Considerations," *Journal of Family History* 6 (1981): 205 (discussing adoption rates); Jane Gross, "Anti-Abortion Revival: Homes for the Unwed," *New York Times*, 23 July 1989; Harry Krause, *Illegitimacy: Law and Social Policy* (Indianapolis: Bobbs-Merrill, 1971); Edward Vincent, "Illegitimacy," *International Encyclopedia of Social Science* (1988), 7, 85, 88; Rickie Solinger, *Wake Up Little Susie: Single Pregnancy and Race before Roe v. Wade* (New York: Routledge, 1992), 168–86. See also sources cited in fn. 8.

ening," the point at which a fetus moved in the woman's body and, according to ecclesiastical doctrine, acquired a soul.[10]

During the nineteenth century, the same ideological forces that made premarital sexual activity less acceptable made birth control less accessible. In 1873, Congress passed the Comstock Law, which prohibited dissemination of information about abortion or contraception. Over the next several decades, all but one state made abortion a felony, and many raised the age at which females could legally consent to sexual activity. Such restrictions responded to various concerns. Moral reformers worried that any separation of sex from procreation would result in increased veneral disease, "psychological derangement," and social instability. In addition, the decline in fertility rates among the "better classes" during the late nineteenth and early twentieth centuries sparked fears of "race suicide." Many physicians shared these concerns, and also found restrictions on abortion and contraceptive services to be an effective means of undercutting their competitors—midwives and other lay practitioners who provided such services. Although the organized medical profession generally presented its campaign against birth control in moral and spiritual terms, nineteenth-century religious leaders were relatively uninterested in the effort.[11]

Yet neither did the newly emerging feminist movement actively oppose the physicians' campaign, despite the significance of reproductive

[10] Linda Gordon, *Woman's Body, Woman's Right: A Social History of Birth Control in America* (New York: Grossman, 1976), 45-53, 69-71. Plantation economies benefited from high fertility rates, but some slaves nonetheless resorted to folk methods of birth control, especially when they were subjected to coerced intercourse with white owners or overseers. See Jessie M. Rodrique, "The Black Community and the Birth Control Movement" in Kathy Peiss and Christine Simmons with Robert A. Padgug, eds., *Passion and Power: Sexuality in History* (Philadelphia: Temple University Press, 1989), 138. See also Rosalind Pollack Petchesky, *Abortion and Woman's Choice: The State, Sexuality, and Reproductive Freedom* (New York: Longman, 1990), 45-50; James C. Mohr, *Abortion in America: The Origins and Evolution of National Policy, 1800-1900* (New York: Oxford University Press, 1978), 18-43.

[11] John P. Harper, "Be Fruitful and Multiply: Origins of Legal Restrictions on Planned Parenthood in Nineteenth-Century America" in Carol Berkin and Mary Beth Norton, eds., *Women of America: A History* (Boston: Houghton Mifflin, 1979), 246-65; James Reed, *The Birth Control Movement and American Society: From Private Vice to Public Virtue* (New York: Basic Books, 1978), 40-41, 188 (quoting Alexander Skene); William Henry Chafe, *The American Woman: Her Changing Sexual, Economic, and Political Roles, 1920-1970* (New York: Oxford University Press, 1972), 100; Sheila M. Rothman, *Woman's Proper Place: A History of Changing Ideals and Practices, 1870 to the Present* (New York: Basic Books, 1978), 46; Nancy Woloch, *Women and the American Experience* (New York: Knopf, 1984), 394; Kristin Luker, *Abortion and the Politics of Motherhood* (Berkeley: University of California Press, 1984), 27-28; Gordon, *Woman's Body*, 59-60; Reva Siegel, "Reasoning from the Body: A Historical Perspective on Abortion Regulation and Questions of Equal Protection," *Stanford Law Review* 44 (1992): 261.

issues for women's independence. To some movement leaders, these issues seemed "too narrow . . . and too sordid." For others, they were too threatening. Contraception appeared to reduce the risks of extramarital sex and thus to jeopardize traditional domestic relationships. For the vast majority of late nineteenth- and early twentieth-century women, marriage and motherhood were the best source of economic security and social status. To most white middle-class wives, any threat to the family was a threat not worth provoking, particularly if the major gain was to license sexual activity, which many had experienced as a duty rather than a pleasure. Although most leaders of the women's rights movement were committed to "voluntary motherhood," they wished to ensure that it remained within marital relationships and that birth control occurred through abstinence rather than through other, "degrading," methods. Feminists recognized unwanted pregnancies as a "problem," but assumed that the solution was to eliminate unwanted sex.[12]

As a result, the initial campaign for reproductive freedom remained largely independent of the organized women's movement. Moreover, that campaign often lost sight of egalitarian principles and concrete concerns of many groups, including young, low-income, and minority women, who most needed assistance. In attempting to broaden their movement's appeal, most early family planning activists emphasized eugenics and supported policies giving physicians control over the distribution of birth control materials. Since few doctors were willing to provide such materials to unmarried minors, and low-income and minority women often lacked access to physicians' services, this approach left many sexually active couples without protection. Yet most leaders, including Margaret Sanger, disavowed any support for abortion and presented their preventive approach as a sufficient alternative.[13]

Appeals to eugenics helped fuel the campaign for sterilization practices that resulted in involuntary fertility control for thousands of slightly retarded or impoverished women, particularly among minority groups. During the 1890s, the development of reasonably safe surgical

[12] Carrie Chapman Catt, quoted in Gordon, *Woman's Body,* 109–11, 238; D'Emilio and Freedman, *Intimate Matters*, 57–73, 80–81; Reed, *Birth Control Movement*, 132.

[13] Ellen Chesler, *Woman of Valor: Margaret Sanger and the Birth Control Movement in America* (New York: Simon and Schuster, 1992); David M. Kennedy, *Birth Control in America: The Career of Margaret Sanger* (New Haven: Yale University Press, 1970), 115–21; Angela Y. Davis, *Women, Race & Class* (New York: Vintage Books, 1981), 210–19; Petchesky, *Abortion*, 87–93.

techniques laid the foundations for a wave of compulsory sterilization laws, which received Supreme Court blessing in 1927. In *Buck v. Bell*, the Court permitted the sterilization of Carrie Buck, an institutionalized and assertedly "feeble-minded" daughter of a feeble-minded mother, after she had given birth to an illegitimate feeble-minded child. Writing for the majority, Justice Oliver Wendell Holmes concluded that "three generations of imbeciles are enough." Subsequent evidence exposed the risks of such decision making. Carrie Buck does not appear to have been mentally deficient; she was institutionalized to hide a pregnancy resulting from rape by one of her foster relatives.[14]

In the late 1930s and 1940s, the eugenic sterilization movement began to decline, partly in response to scientific evidence undermining its premises and to the lessons of Nazi Germany. Without overruling *Buck*, the Supreme Court cast doubt on its reasoning and limited its application in a subsequent decision that invalidated mandatory sterilization laws for habitual criminals. However, family planners, public hospital administrators, and welfare officials continued to condition assistance on "consent" to sterilization by poor and minority women, who were assumed to lack competence or motivation for other birth-control techniques. Coercive and uninformed sterilization remained a major problem until the 1980s, when well-publicized litigation and the promulgation of federal regulations finally began to curb abuses. Yet proposals to coerce certain groups of women to allow implants of a long-term contraceptive, Norplant, have rekindled earlier concerns. In light of this legacy, birth-control strategies have often provoked suspicion or resistance within minority communities, and the issue continues to raise broader questions of nationalism, separatism, and cultural survival.[15]

[14] See Thomas M. Shapiro, *Population Control Politics: Women, Sterilization, and Reproductive Choice* (Philadelphia: Temple University Press, 1985); Petchesky, *Abortion*, 130, 180; *Buck v. Bell*, 274 U.S. 200, 205, 207 (1927); J. Ralph Lindgren and Nadine Taub, *The Law of Sex Discrimination* (St. Paul, MN: West, 1988), 413 (discussing Stephen Jay Gould's analysis of *Buck v. Bell*).

[15] Petchesky, *Abortion*, 130, 159, 179–80; Lindgren and Taub, *Law of Sex Discrimination*, 413–18; *Skinner v. Oklahoma*, 316 U.S. 535 (1942); Hyman Rodman, Betty Sarvis, and Joy Walker Bonar, *The Abortion Question* (New York: Columbia University Press, 1987), 74 (describing medical services conditioned on agreement to sterilization); Robert G. Weisbord, *Genocide? Birth Control and the Black American* (Westport, CT: Greenwood Press, 1975); Patricia J. Williams, *The Alchemy of Race and Rights* (Cambridge, MA: Harvard University Press, 1991), 218; Philip Reilly, *The Surgical Solution: A History of Involuntary Sterilization in the United States* (Baltimore: Johns Hopkins University Press, 1991); Rodrique, "Black Community"; *People v. Johnson* (Tulare Supr. Ct. No. 29290, 2 January 1991) imposes the implant of Norplant as a condition of probation for black woman convicted of child abuse; Faye Wattleton, "Perspective on Race and Poverty: Using Birth Control as Coercion," *Los Angeles Times*, 13 January 1991 (describing the proposal to use

Throughout most of the twentieth century, the reproductive rights movement has reflected class and racial biases. The movement's earliest legal victories involved reinterpretations of federal statutes to permit prescription and distribution of birth-control materials where permissible under state law (for example, in order to prevent disease). The result was that women who could afford a sympathetic private physician could usually obtain contraceptives. Those with sufficient resources were also sometimes able to establish a physical or psychological justification for a legal abortion; alternatively, they could afford a relatively safe illegal one, or travel to a more permissive jurisdiction. For the young, poor, uneducated, or unsophisticated, the options were far more limited. Despite efforts by some community organizations to provide assistance for low-income and minority groups, misconceptions about conception as well as deaths from self-induced abortions were common.[16]

Legal restraints on birth control gradually weakened in response to various cultural forces: liberalization of sexual mores, efforts to prevent veneral disease, women's rising educational opportunities and workforce participation, and increasing economic pressures to curtail fertility. After 1960, the availability of an oral contraceptive helped liberalize public attitudes and practices, and concerns about global overpopulation had similar effects.

Beginning in the mid-1960s, the Supreme Court issued a series of decisions that both responded and contributed to these trends. In 1965, a majority of justices interpreted the due process clause to protect private use of contraceptives by married couples. Decisions in 1972 and 1977 extended that right to unmarried minors. Although the justices' reasoning differed, a majority found constitutional grounding for a limited right of privacy in matters related to procreation. According to the Court's opinion in *Eisenstadt v. Baird*, the state's interest in discouraging promiscuity could not justify bans on contraceptive sales to minors since it would be "unreasonable" to infer "pregnancy and the birth of an unwanted child [as an appropriate] punishment for fornication." Due process and equal protection concerns also con-

---

Norplant to reduce the underclass); Tamar Lewin, "A Plan to Pay Welfare Mothers for Birth Control," *New York Times*, 9 February 1991 (describing Kansas state legislator's proposal to condition welfare benefits on agreement to use Norplant).

[16] Rodman, Sarvis, and Boner, *Abortion Question*, 154; Davis, *Class*, 210; Luker, *Abortion*, 36–39; Petchesky, *Abortion*, 45–50; Solinger, *Wake Up Little Susie*, 4–5. See Discussion in the text accompanying fn. 34–38.

vinced lower courts to begin striking down public school policies that excluded unwed mothers but not unwed fathers.[17]

During the 1960s, many of the same forces underlying liberalized contraception policy also encouraged abortion-law reform. As improvements in medical technology reduced the circumstances in which abortion was necessary to preserve a mother's life, the rigidity of existing statutes became more apparent. Women's increased employment and sexual activity also increased the number of unwanted pregnancies. Physicians faced growing pressure to evade both the letter and spirit of the law, and that pressure was reflected in abortion rates. By the 1960s, most estimates indicated that about a million abortions were performed annually, few of them legally; approximately one abortion occurred for every three to four live births. Abortion was the most frequent form of criminal activity after gambling and narcotics violations, and its perpetrators were also its victims. Procedures could be painful and dangerous if performed hastily or by unskilled practitioners. Somewhere between one and ten thousand women died each year as a result of botched abortions, and more suffered permanent physical or psychological injuries. Predictably, those most at risk were young, poor, and minority women. These human costs prompted increasing reform activity.[18]

Changes in public sentiment were apparent in state legislative initiatives, in national opinion surveys, and ultimately in Supreme Court rulings. In 1973, the Supreme Court issued its landmark decision *Roe v. Wade*. There, a majority of justices concluded that an adult woman had fundamental privacy interests entitling her to terminate a pregnancy without restriction during its first trimester. That ruling set off some of the most strident and sustained criticism in the Court's history. In its aftermath, the justices remained enmeshed in controversies over maternal health and funding regulations designed to restrict abortions.[19]

[17] *Griswold v. Connecticut*, 381 U.S. 479 (1965); *Eisenstadt v. Baird*, 405 U.S. 438, 448 (1972); *Carey v. Population Services International*, 431 U.S. 678 (1977); *Ordway v. Hargraves*, 323 F. Supp. 1155 (D. Mass. 1971); *Shull v. The Columbus Municipal Separate School District*, 338 F. Supp. 1376 (N.D. Miss. 1972); *Farley v. Reinhard* (No. 15569, N.D. Ga., 22 Sept. 1971) in *Clearinghouse Review* 5 (1972): 620; Barbara A. Babcock, Ann E. Freedman, Eleanor Holmes Norton, and Susan E. Ross, *Sex Discrimination and the Law: Cases and Remedies* (Boston: Little, Brown, 1975), 1033.

[18] Luker, *Abortion*, 55–76; Frederick S. Jaffe, Barbara L. Lindheim and Philip R. Lee, *Abortion Politics: Private Morality and Public Policy* (New York: McGraw-Hill, 1981), 22–24.

[19] *Roe v. Wade*, 410 U.S. 113 (1973).

Beginning in the late 1970s, the Supreme Court issued a number of decisions upholding some of the most substantial restrictions. In *Maher v. Roe* and *Harris v. McRae*, a majority of justices sustained denial of public funds for abortion even for women entitled to childbirth subsidies. Subsequent decisions in *Webster v. Reproductive Health Services* (1989) and *Planned Parenthood of Southeastern Pennsylvania v. Casey* permitted further constraints on reproductive choice. *Webster* upheld bans on abortion in public facilities and permitted requirements of costly tests to determine fetal viability in the second and third trimesters of pregnancy. *Planned Parenthood v. Casey* struck down spousal notification provisions as an undue burden on reproductive choice but sustained other requirements, including pre-abortion waiting periods, provision of information concerning fetal development, reporting procedures for abortion facilities, and parental consent for minors. Although the Court declined invitations to overrule *Roe* and return abortion issues to state legislatures, it clearly signaled greater tolerance for state restrictions on reproductive rights. Since adolescents are disproportionately likely to lack private funds for abortions and to delay the decision to abort until after the first trimester, they are particularly vulnerable to such restrictions.[20]

In upholding limitations on reproductive choice, the Court has been strikingly insensitive to the needs of young and poor women. Underlying the majority's analysis is a revived form of the much-discredited distinction between benefits and burdens. According to the court in *Harris v. McRae*, the government has

> place[d] no obstacles—absolute or otherwise—in the pregnant woman's path to an abortion. . . . By making childbirth a more attractive alternative, the state may have influenced a woman's decision but it has imposed no restriction on access to abortion that was not already there. . . . The financial constraints that restrict an indigent woman's ability to enjoy the full range of constitutionally protected freedom of choice are the product not of governmental restrictions on access to abortions, but rather of her indigency.[21]

This reasoning is problematic on several levels. The notion that the state has played no role in creating economic constraints on choice suggests an extraordinary degree of myopia: welfare, education, em-

[20] *Maher v. Roe*, 432 U.S. 464 (1977); *Harris v. McRae*, 448 U.S. 297 (1980); *Webster v. Reproductive Health Services*, 492 U.S. 490 (1989); *Planned Parenthood of Southeastern Pennsylvania v. Casey* 112 Sup. Ct. 2791 (1992); Jaffe et al., *Abortion Politics*, 143–45.

[21] *Harris v. McRae*, 448 U.S. at 316. See also *Maher v. Roe*, 432 U.S. at 474.

ployment, housing, tax, and innumerable other public policies always affect income distribution. It is of course true that indigency limits the exercise of many fundamental rights. Yet it is also the case that the right at stake was as significant as others for which the Court has required state assistance, such as court costs in divorce proceedings or free transcripts for defendants in criminal cases. Moreover, the usual justification for the government's failure to provide such assistance—conserving scarce resources—has been unavailable in abortion-funding cases. When these cases arose, the average cost of subsidizing childbirth was close to ten times that of underwriting abortions. Since many indigents' children would also require continuing welfare support, the statutes in question have been inexplicable as revenue-saving measures. Rather, the government has sought to accomplish indirectly what it cannot do directly: to coerce women into completing unwanted pregnancies.[22]

Federal regulations have sought to impose further restrictions on reproductive choice by barring provision of information about abortion in government-funded health programs. In 1991, in *Rust v. Sullivan*, the Supreme Court sustained those regulations on reasoning similar to that in *Harris*. Under the *Rust* majority's analysis, it was irrelevant that indigent women would be unable to obtain abortion-related services; those individuals were in no worse position than if the government chose not to provide any assistance at all. Yet as dissenting justices noted, such a decision ran counter to well established precedents holding that federal funds could not be conditioned on content-based restrictions of speech.[23]

The targets of these abortion-related restrictions are the women least able to protect themselves through conventional political channels and least able to support an additional child. Although most indigent women have continued to receive subsidies after federal cutbacks through state or private sources, limitations on funding and lack of information about its availability impose significant barriers. Young, unsophisticated teenagers have been among those most at risk. States with adolescent pregnancy initiatives generally prohibit the use of public funds for abortion and impose other limitations on access. The result, as

[22] Jaffe et al., *Abortion Politics*, 143–45. See also dissenting opinions in *Harris v. McRae* and *Maher v. Roe*.

[23] The regulations affect Title X-funded facilities, which serve some five million low-income women. *Rust v. Sullivan*, 111 S. Ct. 1759 (1991).

subsequent discussion suggests, is to constrain reproductive choices and increase their physical risks and psychological trauma.[24]

When questioned about the justice of the government's selective attempt to coerce childbirth, President Jimmy Carter replied that "there are many things in life that are not fair." Why the state should amplify that unfairness remains unanswered. Moreover, the political conflicts that have constrained abortion decision making have been equally apparent in policies specifically directed toward adolescent pregnancy. And the costs have been equally substantial.[25]

The Emergence of Adolescent Pregnancy as a Social Problem

The late 1960s and early 1970s witnessed the first perception of teenage pregnancy as a major social problem. The increase in public concern was not a function of increases in adolescent fertility. In fact, rates of childbirth among females ages 15–19 dropped 45 percent between 1957 and 1983, largely because of liberalization of contraceptive and abortion policy. Although the birth rate began to rise again in the late 1980s, it still remained substantially lower than in the decade before the *Roe* decision. When teenage childbearing became a focus of attention, the reasons had less to do with its frequency than with a cluster of volatile issues involving sexuality, abortion, family values, and welfare policy.[26]

During the past quarter-century, although fewer teens were having children, more were having sex, more were becoming pregnant, and more were becoming single mothers. By the late 1980s, the United

[24] Jaffe et al., *Abortion Politics*, 143–46; Stanley K. Henshaw, Asta N. Kenney, Debra Somberg, and Jennifer Van Vort, *Teenage Pregnancy in the United States: The Scope of the Problem and State Responses* (New York: Alan Guttmacher Institute, 1989), 56; Brief for the National Abortion Rights Action League et al., in *Thornburgh v. American College of Obstetricians and Gynecologists* (1985), 14 (and sources cited therein); Laura Nsiah-Jefferson, "Reproductive Laws, Women of Color, and Low-Income Women" in Sherrill Cohen and Nadine Taub, eds., *Reproductive Laws for the 1990s* (Clifton, NJ: Humana Press, 1988), 23; Dorothy L. Roberts, "The Future of Reproductive Choice for Poor Women and Women of Color," *Women's Rights Law Reporter* 12 (1990): 59, 63n35.

[25] Jimmy Carter, quoted in Sylvia Law, "Rethinking Sex and the Constitution," *University of Pennsylvania Law Review*, 132 (1984): 1016n219.

[26] Vinovskis, "*Epidemic*," 25; Elise F. Jones et al., *Teenage Pregnancy in Industrialized Countries* (New Haven: Yale University Press, 1986), 37; National Center for Health Statistics, *Advanced Report of Final Natality* (1985) (in 1985 there were 51 births per thousand teens; in 1970; there were 68.3); National Center for Health Statistics, *Advanced Report on Final Natality Statistics, 1988* (1990) (from 1986 to 1988, the birthrate climbed 10 percent among 15-, 16-, and 17-year-olds).

States had the highest teen pregnancy rate in the developed world. Approximately 45 percent of adolescent females were sexually active before marriage, an increase of over 15 percent since 1971, and substantial numbers used contraceptives intermittently or ineffectively. The result was a million teen pregnancies each year, and about four-fifths of these were unintentional. An estimated four out of every ten American women were becoming pregnant at least once before age 20, a rate that has doubled since 1950. About half of that group were carrying their pregnancy to term (20 percent of all adolescents), and half of those who did so were unmarried. By the late 1980s, estimates indicated that less than 5 percent of all unmarried teens and less than 1 percent of blacks were placing their children for adoption, a dramatic shift from earlier eras. The rate of teenage childbirth among blacks has become almost two and one-half times greater than among whites; among Hispanics, the rates have become almost twice as great. Over a quarter of single black women have had at least one child by the age of 18. Although teens under age 15 have accounted for only 2 percent of all adolescent mothers, their rate of pregnancy has been increasing.[27]

Although such patterns are widely assumed to constitute a social problem, there is no consensus on its cause or cure. At the risk of some oversimplification, it makes sense to distinguish two dominant positions in public debate. Most liberals "begin with the premise that teenagers should not have babies [while most] conservatives begin with the premise that single teenagers should not have sex." For conservatives, the problem involves primarily moral and fiscal concerns: premarital sexuality is not only objectionable in itself, it promotes other objectionable practices, such as abortion and the destruction of fetal "life," or nonmarital childbearing and the erosion of traditional values

[27] Jones et al., "Teenage Pregnancy in Developed Countries: Determinants and Policy Implications," *Family Planning Perspectives* 17 (1985): 53, 56–57; Janet Benshoff and Harriet Pilpel, "Minors' Rights to Confidential Abortions: The Evolving Legal Scene" in J. Douglas Butler and David F. Walbert, eds., *Abortion, Medicine and the Law* (New York: Facts on File, 1986), 137–44; Cheryl D. Hayes, ed., *Risking the Future: Adolescent Sexuality, Pregnancy and Childbearing* (Washington, DC: National Academy Press, 1987), 1, 15, 50, 52–67; Claire Brindis and Rita J. Jeremy, *Adolescent Pregnancy and Parenting in California: A Strategic Plan for Action* (San Francisco: Center for Population and Reproductive Health Policy, University of California, 1988), 32–41 (discussing increase in sexually active teens and divergence in estimated adoption rates of .4 percent for black teens and 12.2 percent of whites); Gross, "Anti-Abortion Revival" (quoting the president of The National Committee for Adoption), A-25; Children's Defense Fund, *Teenage Pregnancy: An Advocate's Guide to the Numbers* (Washington, DC: Children's Defense Fund, 1988), 11, 22; "Young Teen-agers Have More Babies," *New York Times*, 31 August 1991.

and financial self-sufficiency. For liberals, the problem involves primarily health and socioeconomic status: single parenthood is linked with disrupted education, reduced employment opportunities, and an increased likelihood of poverty for mothers, as well as heightened medical risks and developmental difficulties for their children.[28]

Each of these definitions of the problem is itself problematic on both descriptive and prescriptive levels. Most conservative and liberal accounts distort the dynamic they seek to counteract. Both constituencies have overstated the adverse consequences of adolescent pregnancy and understated the barriers to addressing it. Such distortions are in some sense endemic to the American political process. The most effective way of getting an issue onto the policy agenda is usually to paint it as a crisis that can be addressed without major political conflict or financial expenditure. Yet the construction of the issue best suited to attracting public notice and building coalitions for reform is often ill suited to generating adequate policy responses. This is clearly the case with adolescent pregnancy. Conflicting definitions of the problem have resulted in political compromises that are inadequate to serve societal needs.[29]

From conservatives' perspective, recent increases in teenage sexuality, pregnancy, and nonmarital childbirth are both a cause and consequence of deeper social difficulties. As growing numbers of adolescents reject conventional family patterns, more adults have viewed the situation with subdued "moral panic." In their view, the problem is immorality, not poverty, and stems from cultural permissiveness, a decline in parental authority, and a weakening of community sanctions against illegitimacy.[30]

[28] For conservative objections, see Carole E. Joffe, *The Regulation of Sexuality: Experiences of Family Planning Workers* (Philadelphia: Temple University Press, 1986), 45 (quoting Cunnaught Marshner); and sources cited in n. 30. For liberal perspectives see Dorothy I. Height, "What Must Be Done about Children Having Children," *Ebony*, March 1985, 78; Joycellyn Elders, Jennifer Hui, and Steff Padilla, "Adolescent Pregnancy: Does the Nation Really Care?" *Berkeley Women's Law Journal* (1989–1990): 170; Children's Defense Fund, *Teenage Pregnancy*. For discussion of the disadvantages connected with teenage pregnancy, see ibid. and Hayes, *Risking the Future*, 123–9.

[29] Evelyn L. Brodkin, "Teen Pregnancy and the Dilemmas of Social Policy Making" in Margaret K. Rosenheim and Mark F. Testa, ed., *Early Parenthood and the Transition to Adulthood* (New Brunswick, NJ: Rutgers University Press, forthcoming).

[30] See Nathanson, *Dangerous Passage*, 16 (arguing that moral panic is particularly likely when normative violations are compounded by normative conflict over the legitimacy of traditional roles); Sonia L. Nazario, "Abortion Foes Pose Threat to the Funding of Family Planning," *Wall Street Journal*, 8 March 1990 (quoting Tom Phillips, president of Catholics Serving the Lord: "Whenever you increase accessibility, you increase use . . . [and] the problems that go along with use," and Joe Scheidler, founder of the Chicago Pro-Life Action League, claiming that clinics promote a "contraceptive mentality" in which "people are just having sex for fun").

That position gained increasing public attention during the 1992 presidential campaign. Vice President Dan Quayle attracted substantial media coverage, first by calling for social sanctions against unmarried women (including a leading television-series character) who bear children and call it "just another lifestyle choice." He subsequently faced questions about how he would reconcile his absolutist antiabortion commitments with family values if his own daughter told him that she had decided to have an abortion. Initially he responded that he would "counsel her and talk to her, and support her whatever decision she made," although he hoped she would not choose an abortion. After widespread outcry from all points on the political spectrum, he backpedaled slightly. Both he and his wife, Marilyn Quayle, made it clear that if their daughter were to become pregnant now, at age 13, "she'll take the child to term." However, he reaffirmed his willingness to support whatever decision an adult daughter made. This compromise hardly mollified many commentators, who noted the hypocrisy of his support for a choice by his own family member that he would deny to women generally. Others noted the irony of his double standards, which are reflected in current abortion law. In the "Quayle world, an 18-year-old gets to be supported whatever her decision. A 13-year-old gets to have a baby. This is no joke. It's the foolishness that passes now for public policy."[31]

From conservatives' perspective, however, these policies are still too permissive. Public subsidies for birth-control programs are held accountable for legitimating conduct they should seek to prevent. According to New Right legislators such as former Senator Jeremiah Denton, author of the federal Adolescent and Family Life Act, "the most effective oral contraceptive yet devised is the word 'no.'" Under conservatives' analyses, any outreach program designed to prevent pregnancy rather than sex appears counterproductive; it encourages the activity that creates the problem.[32]

[31] Kevin Sack, "Quayle Insists Abortion Remarks Don't Signal Change in His View," *New York Times*, 24 July 1992; Ellen Goodman, "Abortion Double Standard," *San Francisco Chronicle*, 30 July 1992. Then presidential candidate Bill Clinton, who supports both choice and parental consent provisions, more deftly avoided confrontation with that double standard. When asked a similar question about what he would do if his 12-year-old daughter were pregnant, Clinton responded, "I wouldn't talk to the press about it." Sack, above (quoting Clinton).

[32] Petchesky, *Abortion*, 270 (quoting Orrin Hatch and Jeremiah Denton); Allan C. Carlson, *Family Questions: Reflections on the American Social Crises* (New Brunswick, NJ: Transaction, 1988), 102–03 (blaming family planning clinics for increased sexual activity); Bob Gould, "Fear of Feminism: The Right Wing and Family Values," *Social Justice* 17 (1990): 141, 143.

Most available research, however, fails to support such claims. Studies of contraceptive and abortion services generally find no evidence that their availability has increased sexual activity. The vast majority of adolescents seek assistance only after they have engaged in sexual intercourse. Comparative data underscore the point. Many European countries have rates of adolescent sexual activity equivalent to those in the United States, but substantially higher rates of contraceptive use and lower rates of adolescent pregnancy and abortion.[33]

Although liberals tend to be more permissive toward teenage sexual activity, they are typically no less judgmental than conservatives about teenage childbirth. Both groups share assumptions about the experience of adolescent pregnancy that rest on dubious factual premises. Most public debate conflates unintended with unwanted pregnancy and ignores data indicating that adolescent attitudes fall across a spectrum. The frequent caricature of children having children is highly distorted: two thirds of all births to teenagers are to 18- and 19-year-olds, and only a tiny percentage are to mothers under 16. The equally common assumption that early pregnancy constitutes a direct path to poverty is equally exaggerated. The leading longitudinal study by Furstenberg, Brooks-Gunn, and Morgan, involving predominantly black adolescent mothers in Baltimore, found that the majority was eventually able to obtain a high school education, secure full-time employment, and avoid welfare dependency. About two thirds of the children have completed or are close to completing comparable levels of schooling and only a quarter have become teenage parents themselves.[34]

[33] Melody G. Embree and Tracy A. Dobson, "Parental Involvement in Adolescent Abortion Decisions: A Legal and Psychological Critique," *Law and Inequality* 10 (1991): 53, 58; Brigid Rentoul, "Cognitus Interruptus: The Courts and Minors' Access to Contraceptives," *Yale Law and Policy Review* 5 (1986): 212, 231–32nn99–100; Eve Paul and Dana Klassa, "Minors' Right to Confidential Contraceptive Services: The Limits of State Action," *Women's Rights Law Reporter* 10 (1987): 45, 45n11; Harrell R. Rodgers, Jr., *Poor Women, Poor Families: The Economic Plight of America's Female-Headed Households* (Armonk, NY: M. E. Sharpe, 1986), 92–93; Ann Harper, "Teenage Sexuality and Public Policy: An Agenda for Gender Education" in Irene Diamond, ed., *Families, Politics, and Public Policy: A Feminist Dialogue on Women and the State* (New York: Longman, 1983), 220; Hayes, *Risking the Future*, 153, 208; Jones et al., "Teenage Pregnancy," 57–58.

[34] See fn. 28 and Diana M. Pearce, "'Children Having Children': Teenage Pregnancy and Public Policy from the Woman's Perspective" in Lawson and Rhode, *Politics of Pregnancy*. Compare Elders, Hui, and Padilla, "Adolescent Pregnancy," 170–71 (arguing that adolescent pregnancy eliminates "any glimmer of hope for [teenagers'] future," and that a "poor teenager with a baby is captive to a slavery the 13th Amendment did not anticipate") with Frank F. Furstenberg, Jr., Jeanne Brooks-Gunn, and S. Philip Morgan, *Adolescent Mothers in Later Life* (Cambridge, England) Cambridge University Press, 1987), and Frank Furstenberg, Jr., Mary Elizabeth Hughes, and Jeanne Brooks-Gunn, "The Next Generation: Children of Teenage Mothers Grow Up" in Rosenheim and Testa, *Early Parenthood*. See also Sarah McCue Horwitz, Lorraine V. Klerman, H. Sung Kuo, and James

So, too, many factors commonly assumed to be consequences of early pregnancy—educational difficulties, low self-esteem, poverty, and unemployment—appear to be partial causes as well. For example, although adolescent mothers are more likely to drop out of school than peers of similar socioeconomic backgrounds, it is unclear how much of this difference is due to childbearing and how much to lower academic commitment and competence. Recent studies suggest that most young mothers leave school before becoming pregnant rather than the converse and that mothers who give birth while in school are as likely to graduate as their peers. If such findings are confirmed, then labeling early parenthood the paramount problem will misrepresent the appropriate solution.[35]

A growing number of scholars has also suggested that early pregnancies are more adaptive for economically and racially subordinate groups than is commonly acknowledged. According to researchers such as Arline Geronimus, Mary Edwards, and Mark Testa, teenagers who are no longer in school may find advantages to having their most intensive parenting demands during a period when they have fewest employment opportunities and greatest access to free childcare assistance from their relatives. Since young black mothers are likely to live in an extended family, they often have better access to networks of kin support and childcare than older mothers. From a physiological standpoint, early motherhood is far less problematic than conventional accounts imply. Most health risks currently associated with teenage childbirth are attributable to socioeconomic status and the failure to obtain adequate prenatal care. If such care were available, some analysts believe that childbearing by poor women between ages 16 and 19 would

F. Jekel, "Intergenerational Transmission of School Age Parenthood," *Family Planning Perspectives* (1991): 168 (reporting the longitudinal study finding that most children of teenage parents had not become teenage parents themselves and noting that studies finding correlation of early parenthood between mothers and offspring failed to control for relevant demographic variables); Sarah McCue Horwitz, Lorraine V. Klerman, H. Sung Kuo, and James F. Jekel, "School-age Mothers: Predictors of Long-Term Educational and Economic Outcomes," *Pediatrics* 87 (1991): 862.

[35] For discussion of interrelationships and causal ambiguities, see Frank Furstenberg, Jr. and Jeanne Brooks-Gunn, "Teenage Childbearing: Causes, Consequences, and Remedies" in Linda N. Aiken and David M. Mechanic, eds., *Applications of Social Science to Clinical Medicine and Health Policy* (New Brunswick, NJ: Rutgers University Press, 1986), 307, 316–17. See also Claire Brindis, "Antecedents and Consequences: The Need for Diverse Strategies in Adolescent Pregnancy Prevention" in Lawson and Rhode, *Politics of Pregnancy*. For school drop-out findings, see Dawn Upchurch and James McCarthy, "The Timing of First Birth and High School Completion," *American Sociological Review* 55 (1990): 224; Randall J. Olsen and George Farkas, "Endogenous Covariates in Duration Models and the Effect of Adolescent Childbirth on Schooling," *Journal of Human Resources* 19 (1989): 39.

pose fewer risks to mothers or newborns than later childbearing because many medical problems associated with poverty increase with age.[36]

Such claims have become matters of considerable dispute. Some commentators challenge the conclusions regarding health difficulties and claim that if risks to infants after the first few months of birth are considered, early childrearing carries significant disadvantages. Recent research also suggests that a substantial number of young mothers remain outside of support networks and find family childcare inadequate. Even if, as the Baltimore study suggests, most adolescent mothers are eventually able to achieve financial independence, a significant number experiences enduring poverty and many face prolonged periods of severe hardships. So, too, children of teenage mothers have disproportionate difficulties, such as high rates of delinquency and low rates of educational attainment. Although it is unclear how much of their problem is attributable to socioeconomic variables and how much to parentage age and capacity, at least some evidence suggests that young adolescents are less well equipped for certain childbearing demands. In any event, the existence of such controversies points up the need for qualifications largely missing in current policy debates.[37]

Similar qualifications are necessary to assess conventional solutions for the adolescent pregnancy "problem." Recent data call into question the widespread assumption that economically and racially subordinate groups gain significant benefits from deferring parenthood until marriage or the completion of education. Divorce rates for such young

[36] Arline T. Geronimus, "On Teenage Childbearing and Neonatal Mortality in the United States," *Population and Development Review* 13 (1987): 245, 253, 260–62; Mary G. Edwards, "Teenage Childbearing: Redefining the Problem for Public Policy" (paper presented to the American Political Science Association, 30 August 1990); Mark Testa, "Racial Variation in the Early Life Course of Adolescent Welfare Mothers" in Rosenheim and Testa, *Early Parenthood*; Catherine Riessman and Constance Nathanson, "The Management of Reproduction: Social Construction of Risk and Responsibility" in Aiken and Mechanic, *Applications of Social Science*, 251, 262–63.

[37] Karen Pittman, *Special Report: A Rebuttal of Two Controversial Teen Pregnancy Studies* (Washington, DC: Children's Defense Fund, May 1990) (criticizing Geronimus's study for considering only neonatal morality while excluding risks to older infants and for relying on dated studies of caregiving networks); Cathy Trost, "Census Survey of Child Care Increases Concern about How Much Poor Can Pay," *Wall Street Journal*, 15 August 1990 (noting studies documenting many low-income women's preference for nonfamily childcare); Furstenberg, Brooks-Gunn, and Morgan, *Adolescent Mothers*; Furstenberg et al., "The Next Generation"; Anne C. Peterson and Lisa J. Crockett, "Adolescent Sexuality, Pregnancy and Childrearing: Developmental Perspectives" in Rosenheim and Testa, *Early Parenthood* (noting ambiguities concerning parental competence); Irwin Garfinkel and Sara McLanahan, *Single Mothers and Their Children: A New American Dilemma* (Washington, DC: Urban Institute, 1986) (discussing parental competence).

couples are substantial, and various factors, including relatively high rates of unemployment, violence, suicide, and substance abuse, have reduced the pool of eligible male partners. Although the extent to which such factors explain current racial differences in family formation remains subject to debate, it is clear that marriage is not an adequate solution for many pregnant adolescents.[38]

Moreover, because women of color remain concentrated in occupations that offer low pay and little return for completing secondary education, they often have relatively little to lose from early childbearing. Thus, some longitudinal studies have found little differences in the labor market position of black women who gave birth in their teens and those who did not. Similar findings have emerged from research comparing adolescents who became mothers with their sisters who delayed childbearing: deferral of parenthood and additional schooling did not significantly increase income or reduce welfare dependency. By contrast, a recent study of largely black Baltimore teenagers found that those who obtained abortions did better economically and educationally than adolescents who gave birth, even after controlling for various demographic and motivational factors.[39]

Although such conflicting data underscore the need for further research, one general point attracts widespread consensus. Childbirth patterns are responsive to socioeconomic and cultural forces that vary considerably across class, race, and ethnicity. Expectations about future opportunities play an important role in shaping collective norms and individual choices at both conscious and preconscious levels.

The point of this discussion is not, as Arline Geronimus emphasizes, that "in the best of all possible worlds, teen childbearing would be personally or socially optimal." It is, however, to suggest that "we are very far from the best of all possible worlds, and that there are serious flaws in the logic of current policy approaches intended to get us there."

[38] See fn. 39, below; William J. Wilson, *The Truly Disadvantaged* (Chicago: University of Chicago Press, 1987); Margaret C. Simms, "Adolescent Pregnancy among Blacks in the United States: Why Is It a Policy Issue?" in Lawson and Rhode, *Politics of Pregnancy*; Maris Vinovskis, "Teenage Pregnancy and the Underclass," *Public Interest* 93 (1988): 87.

[39] Compare Elaine McCrate, "Labor Market Segmentation and Relative Black/White Teenage Birth Rates," *Review of Black Political Economy* (1990): 37; Shelly Lundberg and Robert D. Plotneck, "Teenage Childbearing and Adult Wages," Discussion Paper 90–24 (Washington, DC: Institute for Economic Resources, 1989) (finding that premarital birth reduces wages for white but not black women); Arline T. Geronimus and Sanders Korenman, "The Socioeconomic Consequences of Teen Childbearing Reconsidered," Research Report 90-190 (Ann Arbor: University of Michigan Population Studies Center, Sept. 1990); and A. Steven Holmes, "Teenage Study Hints Gain for Those Having Abortion," *New York Times*, 25 January 1990.

As the following discussion makes clear, conflict and confusion over the nature of the adolescent pregnancy "crisis" has seriously compromised societal responses.[40]

Adolescent Pregnancy as a Policy Issue

The central problem in formulating public policies on adolescent pregnancy lies in too narrow a definition of what the problem is. According to the most comprehensive cross-cultural data from Western industrialized nations, two key factors accounting for the high teenage fertility rate in the United States are the culture's "ambivalent, sometimes puritanical attitudes regarding sex" and the existence of a large population subgroup that is economically deprived. At the ideological level, the problem in formulating policy lies in cultural conflict about sexual expression, family authority, and reproductive choice. At the material level, the problem reflects an unwillingness to translate societal concerns into financial commitments and to invest in the necessary research and support structures for populations at risk.[41]

Much of the difficulty in designing adolescent pregnancy policy reflects deeper cultural ambivalence. Few if any societies exhibit a more perverse combination of permissiveness and prudishness in their treatment of sexual issues. A majority of Americans no longer view premarital intercourse as wrong, and the media bombards teens with scenes of sexual involvement and innuendo. Sex is often presented as a goal in and of itself, with little discussion of its risks or of responsible contraceptive behavior. Female adolescents remain subject to double standards that make spontaneous intercourse seem acceptable but suggest that adequate preparation is evidence of promiscuity. Conventional norms present casual sex as unacceptable but declare teenagers too young for serious relationships. On major television networks, sex sells everything from automobiles to laundry detergent but advertisements for birth control remain almost taboo. Our cultural imagery links masculinity with sexual prowess and femininity with sexual attractiveness. Yet we expect decision makers to pay homage to traditional values and accept policies that preach abstinence and restrict adolescent access to birth control.[42]

[40] Geronimus, "On Teenage Childbearing," 273n14.

[41] Jones et al., *Teenage Pregnancy in Industrialized Countries*, 230.

[42] Ibid., 58–62, 240; Brindis and Jeremy, *Adolescent Pregnancy and Parenting in California*, 45–48; Patricia Voydanoff and Brenda W. Donnelly, *Adolescent Sexuality and Pregnancy* (Newbury Park, CA: Sage, 1990), 22–27; see fns. 48–50, below.

This cultural ambivalence over teenage sexual activity has been apparent in the increasing federal, state, and local policies that address the issue. In 1978 Congress passed the first national legislation specifically focusing on teenage pregnancy—the Adolescent Health, Services, and Pregnancy Prevention Act. In essence, the act amended a prior statute to make federally funded contraceptive services (excluding abortion) available to adolescents. As reviews of that legislation make clear, congressional debate was ahistorical, uninformed, and more attentive to political pressures than to adolescent needs. By supporting contraception as an alternative to abortion, the act was able to unite pro-choice and pro-life constituencies. However, it provided such limited services for such a limited percentage of teens that its impact remained minimal. In addition, the statute's inadequate provision for evaluation prevented informed judgments about which subsidized services were most cost-effective.[43]

Three years later, Congress passed the Adolescent Family Life Act (AFLA), which reflected a similar political compromise. The legislation secured support for pilot demonstration projects and their evaluation, but funding limitations again prevented adequate analysis of program impact. Moreover, the act focused on encouraging "chastity" and "sexual self-discipline" rather than on providing contraceptive services, and banned any use of federal funds for abortion counseling. Many programs receiving AFLA funds have been religious. Their instruction has included distribution of pamphlets on "Reasons to Wait," which remind readers that "God wants us to be pure," and advise them to "pretend Jesus was on [their] date." Despite evidence of widespread church-state entanglement, a 1988 Supreme Court decision sustained the act. Funding for AFLA programs has remained at low levels, and its limited availability reflects an uneasy political truce. Federal policy remains caught in the crossfire between liberals, who favor government-subsidized birth-control services for adolescents, and conservatives, who argue that providing such assistance undermines parental authority and legitimates sexual promiscuity.[44]

[43] Pub. L. No. 95-626. Tit. VI, 92 Stat. 3595–3601; Vinovskis, "*Epidemic*," 32–39.

[44] Pub. L. No. 97-35, 95 Stat. 578, 42 U.S.C. paras. 300z ff; Maris Vinovskis, "The Use and Misuse of Social Science Analysis in Federal Adolescent Pregnancy Policy," *Social Science Research Institute* (DeKalb: Northern Illinois University Distinguished Lecture Series, November 1989); *Congressional Quarterly Weekly Report*, 27 June 1981, 1153; Hayes, *Risking the Future*, 22–25; Appellee's Document No. 73, Docket No. 87-253, *Bowen v. Kendrick*, 487 U.S. 589 (1988). See also fns. 28, 30, above; Vinovskis, "*Epidemic*," 198–294.

Similar political controversies have occurred at state and local levels and have resulted in similar policy constraints. During the 1980s, over two-thirds of the states developed initiatives regarding adolescent pregnancy, but only a few targeted contraception for significant efforts. Although about three-fourths of all high schools and junior high schools offered some form of sex education, only about a fifth identified adolescent pregnancy as one of their concerns. To avoid antagonizing vocal conservative minorities, many local decision makers have vetoed both contraceptive counseling and school-based clinic services. By the early 1990s, only about one hundred clinics were operating nationwide, and only about two dozen schools had condom distribution programs: over half of all sexually active teens lacked adequate reproductive health services. The cost, inaccessible locations, and nonconfidentiality of existing programs have imposed substantial barriers to contraceptive choice.[45]

A related limitation in existing policies involves federal inattention to male attitudes and behavior. Program design has often proceeded as if pregnancy occurs via the stork. Too little focus has centered on men who pressure their partners to have sex or to have a baby that will demonstrate their own virility but who are unable or unwilling to support a child. Particularly in low-income communities, rising rates of unemployment, school failure, drug use, and criminal activity have compromised opportunities for responsible parenting. Although federal policy has begun to give more attention to fathers, its scope has been severely constrained, and only a few states with adolescent pregnancy initiatives address such needs.[46]

[45] Vinovskis, "*Epidemic*," 194–203; Hayes, *Risking the Future*, 143–45; Elders, Hui, and Padilla, "Adolescent Pregnancy," 177; Henshaw, Kenney, Somberg, and Van Vort, *Teenage Pregnancy*, 46, 65; Theodora Downs and Susan Solonka, *Background Briefing Report, Evolving State Policies on Teen Pregnancy and Parenthood: What More Can the Feds Do to Help* (Washington, DC: Family Impact Seminar, American Association for Marriage and Family Therapy, 1990); Jacqueline Darroch Forrest and Jane Silverman, "What Public School Teachers Teach about Preventing Pregnancy, AIDS and Sexually Transmitted Diseases," *Family Planning Perspectives* 21 (1989): 65 (noting that on the average, schools devote less than two hours a year to contraceptive instruction, and many withhold all information concerning birth control); Carole Joffe, "Sexual Politics and the Teenage Pregnancy Prevention Worker in the United States" in Lawson and Rhode, *Politics of Pregnancy*; Tamar Lewin, "Studies on Teenage Sex Cloud Condom Debate," *New York Times*, 8 February 1991. See also fns. 43–44, above, and 46–51, below.

[46] Children's Defense Fund, *Adolescent and Young-Adult Fathers: Problems and Solutions* (Washington, DC: Children's Defense Fund, 1988); 2–6, 16–18; Children's Defense Fund, *What about the Boys: Teenage Pregnancy Prevention Strategies* (Washington, DC: Children's Defense Fund, 1988); Catherine Chilman, "Feminist Issues in Teenage Parenting," *Child Welfare* 64 (1985): 225; Brindis and Jeremy, *Adolescent Pregnancy and Parenting in California*, 133–34; Vinovskis, "*Epidemic*," 166–68. For discussion of the disjuncture between norms supporting parental responsibility and the

The cost of these policy limitations is considerable. Restrictions on contraceptive services discourage contraception, not sex. Many teens grossly underestimate the risk of unprotected sexual activity and overestimate the risks of various forms of contraception. Given the increasing rates of unintended pregnancies, AIDS, and drug-addicted offspring among teenage populations, federal refusal to provide adequate outreach services carries enormous personal and social costs.[47]

Parental consent requirements in birth-control programs have reflected similar political compromises, and adolescents have paid the price. A 1981 regulation by the Department of Health and Human Services would have required parental notification for teens seeking contraceptives in all federally funded clinics. Although lower federal courts enjoined enforcement of that "squeal rule," subsequent surveys indicated that about a fifth of hospitals and well over a third of private physicians voluntarily have required parental consent before prescribing contraception. Many states and localities have similarly mandated parental involvement for minors seeking abortions, while over a third of hospitals and about a fifth of surveyed clinics have voluntarily adopted comparable rules. Such restrictions reflect another attempt to mediate between conservatives, who oppose any abortion or provision of birth control to adolescents, and liberals, who favor such assistance.[48]

---

capacity of low-income fathers to provide assistance, see Mercer Sullivan, *The Male Role in Teenage Pregnancy and Parenting: New Directions for Public Policy* (New York: Vera Institute for Justice, 1990); Mercer Sullivan and the Study Group on the Male Role in Teenage Pregnancy and Parenting, "Absent Fathers in the Inner City," *Annals of the American Academy of Political and Social Science* 501 (1989): 48; Henshaw, Kenney, Somberg, and Van Vort, *Teenage Pregnancy*, 51.

[47] See text accompanying fn. 33; Embree and Dobson, "Parental Involvement," 58; Catharine Chilman, *Adolescent Sexuality in a Changing American Society: Social and Psychological Perspectives* (Washington, DC: Dept. & Publications No. [NIH] 79-1426, 1979, 2nd ed. 1983); Kristin Luker, *Taking Chances: Abortion and the Decision Not to Contracept* (Berkeley: University of California Press, 1975); Kristin A. Moore, Margaret C. Simms, and Charles L. Betsey, *Choice and Circumstance: Racial Differences in Adolescent Sexuality and Fertility* (New Brunswick, NJ: Transaction, 1986), 31-32; Dash, *When Children Want Children*, 68, 115, 127, 136. Also see fn. 58, below.

[48] 48 Fed. Reg. 3600, 3614 (1983); Rentoul, "Cognitus Interruptus"; *Planned Parenthood Federation of America v. Schweiker*, 559 F. Supp. 658 (D.D.C.), aff'd, 712 F.2d 650 (D.C. Cir. 1983); *State of New York v. Heckler*, 719 F.2d 1191 (2d Cir. 1983); American Civil Liberties Union of Northern California, *Minor's Rights to Abortion: The Challenge to California's Parental Consent Laws* (San Francisco: American Civil Liberties Union of Northern California, 1991). See also fn. 52, below; American Civil Liberties Union, Reproductive Freedom Project, *Parental Notice Laws: Their Catastrophic Impact on Teenagers' Right to Abortion* (New York: American Civil Liberties Union Foundation, 1986); Benshoof and Pilpel, "Minors Rights," 152n60; Hayes, *Risking the Future*, 155. By the late 1980s, about twenty states had parental involvement mandates, but over half were not in force because of judicial or administrative determinations of unconstitutionality. See ACLU *Parental Notice Laws*.

Litigation challenging parental notification and consent policies has also resulted in compromises, and they make little long-term sense. In a series of cases during the late 1970s and early 1980s, the Supreme Court held that states could require parental consent or notification for abortion services to minors as long as adjudicative procedures were available to bypass such requirements under specified circumstances. To a majority of justices, parental involvement rules were justified by the "peculiar vulnerability" of adolescence and the importance of preserving family ties. However, a minor should be able to avoid consent and notice requirements by establishing in court either that she is sufficiently mature and well informed to make an independent decision concerning abortion or that, even if she is immature, the abortion would be in her best interest.[49]

In 1990, a sharply divided Court sustained further restrictions on adolescents' abortion rights. A majority of justices in *Hodgson v. Minnesota* upheld a requirement that both parents be notified before an adolescent's abortion if a judicial procedure is available to bypass that requirement under the circumstances previously identified. In *Ohio v. Akron Reproductive Services*, the Court sustained a legislative scheme that required physicians personally to notify one parent, placed heightened burdens of proof on minors seeking judicial bypass of the notice requirement, imposed delays of up to twenty-two days for bypass applicants, and failed to insure confidentiality. In upholding the reasonableness of these regulations, the majority concluded: "It is both rational and fair for the State to conclude that, in most instances, the family will strive to give a lonely or even terrified minor advice that is both compassionate and mature. . . . It would deny all dignity to the family to say that the State cannot take this reasonable step in regulating its health professions to ensure that, in most cases, a young woman will receive guidance and understanding from a parent.[50]

On a symbolic level, this resolution is understandable. As one California legislator noted, most decision makers "are parents too." And most parents feel that they should be involved in matters that call for mature guidance. A notice-plus-bypass procedure affirms the value of parental involvement but also recognizes that in some instances such

[49] *Planned Parenthood Association of Kansas City v. Ashcroft*, 462 U.S. 476 (1983); *H. L. v. Matheson*, 450 U.S. 398 (1981); *Bellotti v. Baird*, 443 U.S. 622, 634-35, 637-39, 643-44 (1979); *Planned Parenthood of Central Missouri v. Danforth*, 428 U.S. 52 (1976).

[50] *Hodgson v. Minnesota*, 497 U.S. 417 (1990); *Ohio v. Akron Reproductive Services*, 497 U.S. 502 (1990).

involvement may result in punitive or otherwise counterproductive measures. Yet on a practical level, this compromise carries significant costs, and the law has offered a poor forum for assessing them. Legislative and judicial decision making generally has proceeded without demanding or assembling adequate data on the effects of mandatory parental involvement. Do such requirements result in greater communication on matters involving sexuality and birth control? What are the consequences for teenagers? How many experience better parental relations, how many suffer increased physical risks from self-help or delayed decisions to abort, how many decide not to abort, and what are the effects of that decision? Finally, how well are legal procedures likely to function? How many minors evade consent requirements by seeking abortions out of state? Do petitions to bypass parental involvement result in thoughtful exercise of judicial discretion? Or should the resources consumed by courts and counsel in bypass cases be directed to more productive approaches, such as pregnancy prevention?[51]

Although initial legislative and judicial decision making proceeded without systematic evidence on these points, subsequent research has provided partial answers. Yet the data available have had disturbingly little policy impact. For example, parental knowledge does not correlate with more effective contraceptive practices on the part of adolescents. About half of all adolescents seek birth control services without informing their parents, and about half that group (a quarter of all teens) will not seek such assistance if parents must consent or receive notification. Only 2 percent of surveyed adolescents indicate that policies requiring parental involvement would cause them to restrict their sexual activity. Statutory notice requirements do not significantly increase the likelihood that teenagers will consult their parents on birth control matters.[52]

[51] Mark E. Rust, "Old Enough to Conceive, Old Enough to Abort?" *California Lawyer* 49 (March 1988): 30, 35 (quoting Maxine Waters); Robert M. Mnookin, "*Bellotti v. Baird:* A Hard Case" in Robert H. Mnookin, ed., *In the Interests of Children* (New York: W. H. Freeman 1985), 149. For a critique of the Supreme Court's approach to adolescent pregnancy issues, see Ruth Colker, "An Equal Protection Analysis of United States Reproductive Health Policy: Gender, Race, Age, and Class," *Duke Law Journal* (1991): 324.

[52] Mnookin, "*Bellotti v. Baird,*" 149; Asta M. Kennedy, Jacqueline D. Forrest, and Aida Torres, "Storm over Washington: The Parental Notification Proposal," *Family Planning Perspectives* 14 (1982): 185, 190; Susan F. Newcomer and J. Richard Udry, "Parent-Child Communication and Adolescent Sexual Behavior," *Family Planning Perspectives* 17 (1985): 169, 189; Blum et al., "The Impact of a Parental Notification Law on Adolescent Abortion Decision-Making," *American Journal of Public Health* 77 (1987): 619–20 (showing that in Minnesota, which has a notification law, 65

Nor have judicial bypass procedures usefully contributed to adolescent decision making. Of some 1,300 Massachusetts abortion cases involving petitions to bypass parental consent, courts found the adolescent to be mature in 90 percent of these cases and in all but five of the remainder held that abortion was in her best interest. Four of those cases were either overturned on appeal or resulted in abortions authorized by another judge. In the single case where the court refused to permit termination of the pregnancy, the petitioner accomplished that objective in another state. Although the frequency of adolescent abortion has declined in Massachusetts since the implementation of consent requirements, almost all the decline appears attributable to an increase in out-of-state abortions.[53]

Studies of other state notification procedures similarly find that all but a tiny percentage of bypass petitions are granted, but that the procedure itself imposes substantial barriers, particularly for poor, minority, and very young teenagers. As is true with other abortion restrictions, the individuals most adversely affected by parental involvement rules are those least able to bear the costs of early childbirth. Adolescents have often been deterred by the lack of information about legal assistance, the risks of public exposure, the costs of multiple trips to distant locations, and the fear of courtroom interrogations. That fear is understandable given the judicial questions some minors encounter, such as: How will you feel about having a dead child? Are you aware that abortion can jeopardize your future fertility? Would your parents be willing and able to raise your baby? Notice and consent requirements increase the risk of physical abuse, psychological trauma, medically threatening delays, and unwanted childbirths. In effect, such rules compound the "peculiar vulnerability" they are designed to address.[54]

---

percent of parents were informed; in Wisconsin, which does not, 62 percent were informed). See studies cited in Benshoof and Pilpel, "Minor's Rights," 144–45; Nanette Dembitz, "The Supreme Court and a Minor's Abortion Decision," *Columbia Law Review* 80 (1980): 1251, 1255–58; Theresa M. Walker, "California's Parental Consent Statute: A Constitutional Challenge," *Hastings Law Journal* 40 (1988): 169, 200. See also Embree and Dobson, "Parental Involvement," 69–70 (noting few differences between adult and adolescent decision making on birth control).

[53] Mnookin, "*Bellotti v. Baird*," 239–40; Virginia G. Cartoof and Lorraine V. Klerman, "Parental Consent for Abortion: Impact of the Massachusetts Law," *American Journal of Public Health* 76 (1986): 397.

[54] In Minneosta, only nine bypass petitions were denied between 1981 and 1986. See *Hodgson v. Minnesota*, 497 U.S. 497, n21 (1990). See ACLU, *Parental Notice Laws*, 9–15; Patricia Donovan, "Judging Teenagers: How Minors Fare When They Seek Court-Authorized Abortions," *Family Planning Perspectives* 15 (1983): 259–60, 265, and "Your Parents or the Judge: Massachusetts' New Abortion Consent Law," *Family Planning Perspectives* 13 (1981): 224; *Hodgson v. Minnesota*, 648 F. Supp. 756, 763–64 (D.C. 1988), *rev'd*, 110 S. Ct. 2926 (1990); *American Academy of Pediatrics*

Those involved in administering consent regulations express little support for their requirements. Criteria for assessing teenagers' maturity and best interests are arbitary and inconsistent, while judicial cross-examination often increases the guilt and misery that accompanies unwanted pregnancies. As the dissent in *Hodgson* recognized, and the results in bypass cases reflect, it is almost impossible to identify circumstances in which a minor is too immature to make an abortion decision for herself but in which her best interest lies in having a child she does not want. Given the increased medical and socioeconomic risks associated with carrying a fetus to term, on what basis could courts legitimately seek to compel childbirth? In case studies of Massachusetts and Minnesota regulations, participants in the legal process overwhelmingly agreed that resources now spent on obtaining judicial rubber stamps of bypass petitions should be allocated to more productive strategies.[55]

This is not to understate the value of fostering better communication between parents and children concerning responsible sexual behavior. It is rather to suggest that involuntary notification is an ineffective means of fostering such communication. Virtually every major professional study has concluded that compulsory parental involvement ill serves adolescent needs and family relationships. A preferable strategy would be voluntary parental outreach programs. Such programs seek to improve family communication and adolescent decisions making skills while avoiding the notice or consent requirements that deter teenagers from seeking assistance. Yet all too often legislative, judicial, and administrative analyses have proceeded without consideration of alternative strategies for parental involvement. As with other teenage pregnancy initiatives, policy makers have offered compromises more responsive to symbolic politics than to practical needs.[56]

---

v. Van de Kamp, (Cal D. Ct. 1990), trial transcript, 273–74, 282–84, Brief for Respondents, 19–21; *Bellotti v. Baird*, 443 U.S. 634–35; Embree and Dobson, "Parental Involvement," 76 (noting that the small number of abortion providers and judicial decision makers in some communities multiplies barriers).

[55] For experts' views see Mnookin, "*Bellotti v. Baird*," 258; ACLU, *Parental Notice Laws*, 1, 13, 17; Donovan, "Judging Teenagers," 265–66; *Hodgson v. Minnesota*, 110 S. Ct. 2957–58 (Marshall, J., dissenting in part). For comparisons of the increased medical, psychological, and socioeconomic costs associated with childbirth in comparison with abortion, see sources cited in Walker, "California's Parental Consent Statute," 196–97; Gary B. Melton and Anita J. Pliner, "Adolescent Abortion: A Psycholegal Analysis" in Gary B. Melton, ed., *Adolescent Abortion: Psychological and Legal Issues* (Lincoln: University of Nebraska Press, 1986). For discussion of inconsistency and trauma resulting from judicial decision making, see sources cited in fns. 53–54.

[56] Kennedy, Forrest and Torres, "Storm over Washington," 195; ACLU, *Parental Notice Laws*, 1, 8, 17.

REFORMULATING THE PROBLEM/RETHINKING THE RESPONSES

In formulating more effective public policies, a critical first step is to convince decision makers and their constituents to rethink the problem that teenage pregnancy presents. No significant progress can occur as long as Americans view the issue in terms of individual rather than societal responsibilities and insist on policies that reflect traditional family values rather than contemporary adolescent needs. The alternative framework proposed here begins from different premises and demands different resource priorities. Its objectives are to enhance adolescents' capacity for informed and uncoerced choices and to expand the choices available. These objectives in turn require policy initiatives that not only increase birth-control information and assistance, but also respond more effectively to education, employment, health, and welfare needs.

A more promising policy approach will depend on more systematic research. Despite the recent attention to the adolescent pregnancy "crisis," we know surprisingly little about certain key issues. What factors predict effective use of contraceptives among sexually active teens? What strategies are most successful in promoting responsible sexual behavior and how do they differ across gender, class, age, race, and ethnicity? How much of the disadvantage for young mothers and their children is attributable to age and how much to other demographic and socioeconomic factors? What individual coping strategies and social programs are most effective in reducing that disadvantage?[57]

Yet despite these gaps in current understanding, we know more than enough about the inadequacies of current frameworks to identify more promising alternatives. The most obvious inadequacy involves birth-control information and assistance. Widespread ignorance among teenagers concerning the substantial risks of unprotected sexual activity and the relatively low risks of contraceptive use demonstrate the need for fuller information. Increasing rates of AIDS among adolescents further underscore the urgency of that need. Yet since existing educational programs have had little measurable impact on contraceptive behavior, more innovative campaigns are necessary from multiple sources, including schools, community organizations, and the media.[58]

[57] Riessman and Nathanson, "Management of Reproduction," 259; Furstenberg and Brooks-Gunn, "Teenage Childbearing," 319; Hayes, *Risking the Future*, 231–59; Furstenberg, Hughes, and Brooks-Gunn, "The Next Generation."

[58] See sources cited in fn. 47; Andrew W. Boxer, "Adolescent Pregnancy and Parenthood in the Transition to Adulthood" in Rosenheim and Testa, *Early Parenthood*; Furstenberg and Brooks-

Such campaigns should include greater attention to decision making skills for both sexes. Encouraging men to exercise greater responsibility in sexual relationships from an early age is a crucial part of reproductive freedom for women. More emphasis should also center on empowering female adolescents to resist pressures for unwanted sexual intercourse or parenthood. Many teenage women report engaging in sexual activity that they do not find pleasurable, and deciding to have a baby more because of peer and family influence than because of affirmative desires to rear a child. All too often, single motherhood becomes an ill-considered means to other ends: a way to assert independence, to punish parents, to gain prestige, or to win approval from their child's father. Better education, mentoring, peer counseling, and community outreach strategies should be available to help young women assess and assert their own best interests. Especially for young teenagers, abstinence can be a feminist strategy, but not on the terms conservatives usually offer. Rather, the objective should be a revived but reformulated understanding of what the nineteenth-century women's movement labeled "voluntary motherhood."[59]

To make reproductive choices truly voluntary will also require more accessible birth control. As noted earlier, a majority of sexually active teens lack adequate reproductive health services, and many are deterred by the cost, distance, nonconfidentiality, and stigma associated with existing clinics. To reduce unplanned pregnancies, we need to reduce the barriers to reproductive planning, such as parental notification and consent requirements, harassment of clinic users, and shortages of free or low-cost services. Comprehensive health programs located in or near schools are among the most promising outreach strategies, for they can avoid singling out participants who request contraceptive assistance and can offer accessible follow-up services. Development and distribution of alternative birth-control technologies, such as RU 486, should become higher public priorities.[60]

Gunn, "Teenage Childbearing," 321-22; Voydanoff and Donnelly, *Adolescent Sexuality*, 96; Douglas Kirby, *Sexuality Education: An Evaluation of Programs and Their Effects* (Santa Cruz, CA: Network Publications, 1984), 95-100.

[59] Joffe, *Regulation of Sexuality*, 157-58; Dash, *When Children Want Children*, 121-25, 216, 220, 265; see Sara Ruddick, "Procreative Choice for Adolescent Women" in Lawson and Rhode, *Politics of Pregnancy*; Elijah Anderson, "Sex Codes and Family Life Among Poor Inner-City Youths," *Annals of the American Academy of Political and Social Science* 501 (1989): 59.

[60] Brindis and Jeremy, *Adolescent Pregnancy*, 104-05; Jones et al., *Teenage Pregnancy in Industrialized Countries*, 236-38. See, generally, Hayes, *Risking the Future*. For discussion of the inadequacy of contraceptive alternatives, see Cohen and Taub, *Reproductive Laws for the 1990s*.

Yet if we are serious about not just removing the obstacles to choice but expanding the choices available, further initiatives are necessary. First and most fundamental, we must find ways to improve school programs and reduce dropout rates. Particularly in low income communities, the current educational structure seems irrelevant to immediate needs and promises little economic security. Female adolescents who lack confidence in their own abilities and future prospects often see little to lose from early pregnancy and childbearing. Low expectations about educational and vocational opportunities readily become self-fulfilling prophecies. Young women who leave school are disproportionately likely to become pregnant, to have multiple pregnancies, and to experience poverty and welfare dependency.[61]

Breaking this cycle will require changing adolescents' aspirations and experience. More programs are needed to expand life-planning options, build self-confidence, improve school performance, increase literacy, and strengthen vocational skills. Additional support is necessary for scholarships and bilingual assistance. These initiatives must be coordinated with other opportunities for vocational training, job placement, and higher education, so that participants can see ways to achieve status and self-worth apart from early parenthood. We cannot respond effectively to teenage motherhood without responding also to the broader range of social problems that make motherhood seem to be a teenager's best option. Nor can we address the inadequacy of paternal support without also addressing the inadequacy of unemployment, education, and drug treatment opportunities.[62]

[61] Jewell Taylor Gibbs, "The Social Costs of Teenage Pregnancy and Parenting in the Black Community: Implications for Public Policy" in Rosenheim and Testa, *Early Parenthood*; Wilson, *Truly Disadvantaged*; n. 35, above and accompanying text; Brindis and Jeremy, *Adolescent Pregnancy*, 48–49; Hayes, *Risking the Future*, 125–28; Denise F. Polit and Janet R. Kahn, "Early Subsequent Pregnancy among Economically Disadvantaged Teenage Mothers," *American Journal of Public Health* 76 (1986): 167.

[62] Gibbs, "Social Costs"; Brindis and Jeremy, *Adolescent Pregnancy*, 85, 88–89; Marion Wright Edelman, *Families in Peril: An Agenda for Social Change* (Cambridge, MA: Harvard University Press, 1987), 58–60; Wilson, *Truly Disadvantaged*; William Wilson and Kathryn M. Neckerman, "Poverty and Family Structure: The Widening Gap between Evidence and Public Policy Issues" in Sheldon H. Danzig and Daniel H. Weinberg, eds., *Fighting Poverty: What Works and What Doesn't*? (Cambridge, MA: Harvard University Press, 1986), 232, 252–59; Tom Joe, "The Other Side of Black Female-Headed Families: The Status of Adult Black Men," *Family Planning Perspectives* 19 (1984): 74; Sullivan, *The Male Role*; Sullivan and Study Group, "Absent Fathers"; Hortensia Amaro and Barry Zuckerman, "Psychoactive Substance Use and Adolescent Pregnancy Compounded Risk among Inner City Adolescent Mothers" in Mary Ellen Colten and Susan Gore, *Adolescent Stress: Causes and Consequences* (New York: Aldine de Gruyter, 1991); Center for Population Options, *Teenage Pregnancy and Too-Early Childbirth: Public Costs and Personal Consequences* (Washington, DC: Center for Population Options, 1991).

The importance of some of these initiatives has attracted growing attention. Recent changes in federal welfare regulations require adolescent AFDC recipients to obtain job training and skills instruction if childcare is available, and to participate in continuing education if they lack a high school degree. Although the effectiveness of these requirements has yet to be assessed, past experience with similar vocational programs suggests grounds for caution. Participants typically register minor income gains in comparison with program cost. So, too, recent efforts to increase paternal child support for AFDC families are likely to be of limited effectiveness, given most fathers' inadequate financial resources. Significant progress for adolescent parents is likely to require broader macroeconomic interventions and antipoverty initiatives, such as increases in minimum wage, job-creation strategies, and expanded health, housing, childcare, family counseling, and income-transfer policies.[63]

More emphasis must also center on expanding prenatal services and coordinating responses to various risk-taking behaviors, including drug and alcohol abuse, as well as unprotected sexual activity. The necessity of such initiatives is underscored by the increasing number of infants who are deformed or disabled as a result of maternal addictions. Recent efforts to prosecute addicted mothers rather than provide adequate treatment programs again point up the risks in placing blame at the individual rather than societal level. It is neither just nor effective to condemn teenage mothers who make tragic choices while ignoring the constraints on choices that are available.[64]

All these policy initiatives must be formulated with greater community involvement and more responsiveness to variations within and

[63] David Ellwood, "Conclusion" in David T. Ellwood and Phoebe H. Cottingham, eds., *Welfare Policy for the 1990s* (Cambridge, MA: Harvard University Press, 1989), 269, 272 (noting that supported work programs that cost over $13,000 per participant produced earnings gains of only $1,500); Gary Walker, "Comment" in Ellwood and Cottingham, *Welfare Policy*, 141; Gina Adams, Karen Pittman, and Raymond O'Brien, "Adolescent and Young Adult Fathers: Problems and Solutions" in Lawson and Rhode, *Politics of Pregnancy*; Michael B. Katz, *The Undeserving Poor: From the War on Poverty to the War on Welfare* (New York: Pantheon Press, 1989); Wilson, *Truly Disadvantaged*; Diana M. Pearce, "The Feminization of Poverty: A Second Look" (Washington, DC: Institute for Women's Policy Research, 1989); Erik Eckholm, "Forcing Welfare Fathers to Pay Up Hits Stumbling Block: Lack of Jobs," *New York Times*, 20 July 1992.

[64] Roberts, "Future of Reproductive Choice," 64n39 (describing gross inadequacy of treatment facilities and the growing trend toward prosecution of addicted mothers). See Lisa Maher, "Criminalizing Pregnancy: The Downside of a Kinder, Gentler Nation?" *Social Justice* 17 (Fall 1990): 111; Amaro and Zuckerman, "Psychoactive Substance Use"; Dorothy Roberts, "Punishing Drug Addicts Who Have Babies: Women of Color, Equality, and the Right of Privacy," *Harvard Law Review* 104 (1991): 1419.

among ethnic groups. Third-generation Puerto Rican women have different concerns from those of recent emigrees from Mexico, and blacks from urban ghettos face different constraints from those of teenagers from middle-class suburbs or impoverished rural communities. Public policies need to take fuller account of culturally specific attitudes and kinship support structures affecting early childbirth. No single set of interventions will be equally effective with all subgroups; rather, we need multiple strategies that directly enlist families at risk.[65]

The policy alternatives identified here do not constitute a modest agenda. But neither can we expect major progress until we translate our rhetorical concerns into resource commitments. Coherent policies on teenage pregnancy require a redirection of adult priorities.*

[65] Children's Defense Fund, *Teenage Pregnancy*; Linda M. Burton and Carol B. Stack, "Conscripting Kin: Reflections on Family, Generation, and Culture" in Lawson and Rhode, *Politics of Pregnancy*; Nsiah-Jefferson, "Reproductive Laws," 23; Brindis, "Antecedents and Consequences"; Edelman, *Families in Peril*; Francis Ianni, *The Search for Structure: A Report on American Youth Today* (New York: Free Press, 1989).

* The comments of Annette Lawson and Michael Wald and the research assistance of Joan Krause and Wendy Olsen are appreciated. This article is adapted and updated from Annette Lawson and Deborah L. Rhode, *The Politics of Pregnancy: Adolescent Sexuality and Public Policy* (New Haven: Yale University Press, 1993).

Keeping Abortion Clinics Open: The Importance of *Ragsdale v. Turnock* in the Post-*Casey* Era

Susan Gluck Mezey Raymond Tatalovich Michael Walsh
Loyola University Chicago

In 1973, Roe v. Wade *constitutionalized a woman's right to an abortion. But, while* Roe *removed most legal obstacles to abortion, it did not address the limited availability of abortion services in the nation. The case examined here,* Ragsdale v. Turnock, *revolved around an Illinois statute that imposed far-reaching restrictions on abortion clinics, the site of most U.S. abortions since* Roe. *The crucial role of clinics in providing abortion services explains why the dispute represented by* Ragsdale *had the potential for an enormous impact on legalized abortion in the United States. Because of the number of women affected, the* Ragsdale *litigation could have led to the most significant judicial ruling since* Roe. *The suit resulted in a settlement in which plaintiffs secured the right to a legal clinic abortion during the first 18 weeks of pregnancy. Although the case was settled to the satisfaction of pro-choice advocates, a similar law today might well survive constitutional scrutiny.*

Roe v. Wade constitutionalized a woman's right to an abortion, striking down restrictive laws throughout the nation. But while *Roe* removed most legal obstacles to abortion, it did not speak to the limited availability of abortion services in the nation, and did not mandate that hospitals begin to provide such services more readily (Rosenberg, 1991). Decades later, there exists a maldistribution of abortion services that worsens each year: There are no known abortion providers in 83 percent of the 3,135 counties in the U.S. (Henshaw and Van Vort, 1988, p. 106). This situation has arisen because, as Jaffe, Lindheim, and Lee (1981, p. 32), pointed out: "The medical mainstream, in effect, copped out. By doing so, it has produced great disparities in the availability of abortion services in many parts of the country. This has led to the creation of an entirely new system of specialized clinics, which now account for most U.S. abortions."

Since *Roe,* many states have imposed various constraints on abortion services, including restrictions on public funding, and "right to conscience" laws permitting health care workers to refuse to participate in abortions. The greatest threat to the survival of *Roe*, however, would be a successful attack on the very existence of abortion clinics. The case examined here revolved around a statute with that potential. Although the case was settled to the satisfaction of pro-choice advocates, a similar law today might well survive constitutional scrutiny.

CLINIC LITIGATION

On June 28, 1985, Dr. Richard M. Ragsdale, director of the Northern Illinois Women's Center (NIWC), brought a class action suit in federal district court, asking the court to enjoin the enforcement of three Illinois statutes that prevented physicians from performing—and women from obtaining—abortions in the first and early-second trimesters. The laws, the Medical Practice Act (MPA) of 1973, the Ambulatory Surgical Treatment Center Act (ASTCA) of 1973, and the Health Facilities Planning Act (HFPA) of 1974, contain both general and abortion-specific provisions. The suit charged that the three statutes would essentially require all abortions to be performed in hospitals or their "functional equivalent[s]."

The case arose in February, 1985 when the Northern Illinois Women's Center, located in the Rockford Medical Arts Building, was informed that its lease would not be renewed. The building's owner told Ragsdale that the estimated expense ($240,000) of conforming the Northern Illinois Women's Center's offices to the standards of the ASTCA was too high. Ragsdale unsuccessfully sought to obtain new facilities.

The Medical Practice Act specifies licensing requirements for Illinois physicians and prohibits the practice of medicine by unlicensed practitioners. Section 16 of the act authorizes the Illinois Department of Public Health (IDPH) to revoke or suspend medical licenses under certain circumstances. Section 16(1), the only part of the act challenged in the lawsuit, specifically refers to abortion by allowing Illinois' health department to "revoke or suspend the license of any physician who performs an 'elective abortion' in any place other than an ASTC [ambulatory surgical treatment center], a hospital, or a facility run by the state, the federal government, or a state university or college." This section prevented physicians from performing abortions in their offices, and required them to comply with both the ASTCA and the HFPA.

The Ambulatory Surgical Treatment Center Act set forth standards of care in ambulatory surgical treatment centers as well as codes pertaining to construction and maintenance of the premises. Section 3 of the new act defined an ambulatory surgical treatment center as "any institution, place or building devoted primarily to the maintenance and operation of facilities for the performance of surgical procedures or any facility in which a medical or surgical procedure is utilized to terminate a pregnancy, irrespective of whether the facility is devoted primarily to this purpose." Thus, the act applies to facilities that primarily perform surgical procedures as well as to facilities in which any abortions are performed. Clinics, such as Ragsdale's, which primarily performed first trimester abortions, were subject to the act's rigorous standards.

Extensive regulations pertaining to patient care, equipment, supplies, and physical plant facilities accompanied the Ambulatory Surgical Treatment Center Act. Subpart G, devoted entirely to abortions, restricted ambulatory surgical treatment centers to performing first trimester abortions (the first 12 weeks of gestation beginning at ovulation) and created a number of requirements for abortions that did not pertain to other medical/surgical procedures. It specified Ambulatory Surgical Treatment Center Act provisions for patient care, and included counseling and examination requirements, pre- and post-operative care rules, and reporting procedures. The act's regulations also mandated the size of examination rooms, hallways and corridors, and contained rules regarding the air conditioning, heating, and ventilation systems, essentially requiring ambulatory surgical treatment centers to become mini-hospitals.

The Health Facilities Planning Act, enacted more than a year after the other two laws, required all "health care facilities" including ambulatory surgical treatment centers, to obtain permits before acquiring major medical equipment, building a new facility or altering an existing one. Securing a permit, known as a certificate of need, was based on presenting a comprehensive application to the Illinois Health Facilities Planning Board, attending a public hearing on the application, and obtaining board approval which required consideration of a number of factors, including the board's view of where the public interest lay.

In his suit, Ragsdale contended that the Health Facilities Planning Act, like the Ambulatory Surgical Treatment Center Act, would lead to diminished abortion services by creating a "government-imposed forum" for abortion opponents that allowed them to erect obstacles to individuals who desired to build, renovate, or relocate any facility providing abortions. He described the Health Facilities Planning Act as part of a "interlocking whole" impeding women from securing early abortions in Illinois by restricting the procedures to hospitals. In response, the state argued that the act was intended primarily to produce a comprehensive health care delivery system, covering all kinds of health care facilities, services and equipment; its secondary purpose was to limit rising health care costs by precluding the construction and duplication of unnecessary facilities.

DISTRICT COURT GRANTS INJUNCTION

In granting the injunction against Section 16(1) of the Medical Practices Act, as well as the Ambulatory Surgical Treatment Center Act and the Health Facilities Planning Act and their regulations, District Court Judge John Nordberg found that clinics performing abortions would be forced to raise their fees if required to comply with the ASTCA's building and personnel regulations. Such increases, he said, would inflate the price of an abortion, placing it beyond the reach of some women.

Nordberg also agreed that Ragsdale had shown that the ASTCA and HFPA, particularly the latter's certificate of need process with its attendant public hearings, would make it virtually impossible to open a new facility or add abortion services to an existing facility. Ragsdale's attempt to obtain a new site for his clinic, noted the judge, had resulted in a public hearing with more than a thousand people creating a "riotous" atmosphere. Forty-eight hours after the hearing—despite a formal agreement between Ragsdale and his proposed future landlord—the site was withdrawn.

Based on this evidence, Nordberg concluded that the challenged laws and regulations "place a very real and a very heavy burden" on a woman's right to choose to terminate a pregnancy, that enforcement would increase costs and limit the availability of abortions. Women would have to travel further, spend more money, and risk a delayed abortion because of these three Illinois laws, he said.

The state argued that the plaintiffs could not show irreparable injury, necessary to obtain an injunction, because sections of the statutes and regulations were not being enforced. The court dismissed the claim because the state had never formally notified Illinois physicians of its intent not to enforce the laws nor had Illinois' public health department ever attempted to amend the regulations or statutes. The mere existence of the regulations and statutes created a chilling effect on abortion providers, Judge Nordberg ruled.

Citing *Roe* and its progeny, particularly the 1983 decision *City of Akron v. Akron Center for Reproductive Health*, the court then assessed the likelihood that the plaintiffs would succeed based on the merits of the case. As established in *Roe* and *Akron*, the court must view abortion regulations with strict scrutiny; that is, the state must offer a compelling reason to justify regulations restricting the right of abortion. In *Roe*, the Supreme Court held that the state's interest in a woman's health becomes compelling approximately at the end of the first trimester. The court held that, before the end of the first trimester, the decision to abort was between the woman and her physician, and the woman must be allowed to obtain an abortion without "interference by the state."

This position was modified somewhat by the ruling in *Akron* (1983) that permitted states to regulate first trimester abortions as long as there was "no significant impact" on the woman's right to choose an abortion and the regulations were "justified by important state health objectives."

Utilizing the trimester approach established in *Roe*, modified by *Akron*, Judge Nordberg pointed out that the challenged statutes and regulations pertained to all abortions without distinction among trimesters. He applied strict scrutiny to the challenged laws and regulations, noting that all regulations, including those expressed in general terms, that burden a woman's right to terminate a pregnancy during the first trimester must be subjected to the compelling state interest test.

Against this backdrop, Nordberg found that the plaintiffs established a likelihood of success on the merits of the case by "overwhelmingly demonstrat[ing]" that the statutes burdened the woman's first and early-second trimester abortion rights. The state, he held, had "failed to produce any evidence at all of a compelling or even rational basis for the challenged statutes and regulations." Equating a "medical necessity" standard with the compelling state interest test, he held, that while the state had shown that some regulations are "consistent with accepted medical practice," there was no evidence that the laws were justified by medical necessity. He added in a footnote that the regulations could even be seen as "medically detrimental" to women seeking abortions because they created more imposing, formidable surroundings for the abortion procedure.

THE SEVENTH CIRCUIT AFFIRMS DISTRICT COURT

In a 2-1 vote, a panel of the Seventh Circuit Court of Appeals affirmed, for the most part, the preliminary injunction. Speaking for himself and Judge William Bauer, Judge Jesse Eschbach devoted most of his opinion to reviewing the district court's finding that plaintiffs would likely succeed on the merits of the case because the challenged statutes and regulations violated a woman's right to privacy established in *Roe v. Wade*.

Over the objections of the state, the circuit court concluded that the lower court's "medical necessity" standard was appropriate for both trimesters. Agreeing with the district court, the appellate court required the state to show a compelling interest in the regulations applying to both first and second trimester abortions. Citing *Roe* and *Akron* (1983), Eschbach explained that regulations affecting first trimester abortions must be insignificant and must be "justified by important state health objectives." This is "consistent," he said with Seventh Circuit rulings that regulations with more than a "*de minimis* impact on the abortion decision," must be justified by a "compelling" reason; that is, the state must demonstrate that the "burden is not undue or unjustifiable." Although acknowledging that second trimester regulations may be judged by a more lenient standard, he applied the more stringent standard here because the regulations and statutes did not distinguish between trimesters.

The state argued that, because the regulations pertained to all surgical procedures and did not single out abortions, the court should not apply the stricter standards normally used to judge abortion regulations but should utilize the more deferential standard normally applied to health regulations. The court disagreed, saying it believed that the Ambulatory Surgical Treatment Center Act was primarily intended to regulate abortion clinics and was only extended to other outpatient surgical clinics to save it from unconstitutionality.[1]

Eschbach reasoned that the law was designed to limit access to abortions because it made any place where an abortion was performed an ambulatory surgical treatment center, thereby subjecting it to the act's regulations. He added that the state cannot merely characterize all abortions as surgery and apply the same standards to abortions as it would to major surgical procedures.

Having established the proper standard of review, the circuit court affirmed the lower court's findings that the regulations would have a significant impact on the woman's abortion decision by raising the cost and limiting the availability of abortions. Moreover, the circuit court held, they would subject a woman to the added stress of having an abortion in a "hospital-like facility" rather than in the relative comfort of her doctor's office.

Because there was no evidence that women's health was being compromised by physicians performing abortions in their offices, the court concluded that the state could not require abortion facilities, including the offices of doctors whose primary practice was performing abortions, to be licensed as ambulatory surgical treatment centers. With the licensing requirement no longer applicable to clinics performing abortions, the court ruled that the substantive rules of the act no longer applied.

In the alternative, the court addressed many of the challenged act's regulations and found the entire scheme unconstitutional. The court was particularly concerned with the Subpart G regulations; with one or two exceptions, it found them invalid. Conceding that some of the other regulations (outside Subpart G) might be constitutional, the court nevertheless felt that the challenged regulations were sufficiently unconstitutional to render the scheme unconstitutional as a whole.

The court particularly singled out the counseling requirements of Subpart G as conflicting with Supreme Court case law. It also found the physical plant, equipment, and staffing specifications to be unrelated to the safety of abortions and medically unjustified. In short, the court held that the plaintiffs showed that the regulations imposed a substantial burden, and the state had not demonstrated that the regulations were justified by "important health objectives."

The court also invalidated the Health Facilities Planning Act's certificate of need process, ruling that the possible state interest in preventing duplicate facilities in a geographic area did not supersede the constitutional rights involved, and that the public veto aspect of the act "exacerbated" its unconstitutionality.

Finally, the court rejected the state's plea to sever the unconstitutional parts of the statute and regulations, and to uphold the remaining provisions. Explaining that severance was improper because the Ambulatory Surgical Treatment Center Act's licensing requirement was an integral part of the statutory scheme, with one minor exception, the appellate court affirmed the lower court injunction. The court further justified its decision to invalidate the statute and regulations because, while they might have appeared to be cohesive when written, they were actually so "riddled with exceptions" as to be "unintelligible." Thus, the court's solution was to force the state to rewrite its abortion regulations so as to make them coherent and constitutional.

JUDGE JOHN COFFEY DISSENTS

In his dissent, Judge John Coffey, the third member of the panel, argued that the court should have severed the unconstitutional abortion-specific portions of the statute and confined itself to striking those while upholding the abortion-neutral medical practice provisions. Coffey criticized the majority for failing to utilize the statutes' severability clauses and overriding the legislature's legitimate attempt to regulate surgical centers to enhance public safety.

In Coffey's view, certain sections of the Medical Practice Act and the Ambulatory Surgical Treatment Act were enacted as abortion regulations and should be enjoined. He agreed with the majority that these provisions were invalid and that plaintiffs would likely succeed in a hearing on the merits. Basically, Coffey differed from the majority in the extent to which he believed abortion facilities should be regulated. His major concession to the plaintiffs was to agree that it was improper to treat all abortion facilities as ambulatory surgical treatment centers, regardless of the number of abortions performed there. However, in his view, it was permissible to regard abortion facilities that primarily perform abortions—such as Ragsdale's—as surgical centers. Abortions, Coffey argued, were surgical procedures and should be regulated like other surgical procedures performed in the centers. He insisted the courts should judge these "abortion neutral" regulations under a minimal scrutiny analysis common to economic and social legislation, not the strict scrutiny invoked by the majority.

Quoting extensively from the legislative history, Coffey denied that the Ambulatory Surgical Treatment Center Act was intended "primarily" to regulate abortions; rather, it was intended to regulate an increasingly popular vehicle for the delivery of medical services—the ambulatory treatment center. He proclaimed it as mere "historical coincidence" that the act was a companion bill to the 1973 Illinois Abortion Act. Further, he argued that the Health Facilities Planning Act was merely designed to rationalize and control the utilization of medical facilities in the state.

Coffey agreed that, by failing to distinguish between first and second trimester abortions, parts of the Ambulatory Surgical Treatment Center Act and the Medical Practice Act improperly prevented physicians from performing first trimester abortions in their offices. He disputed the majority's application of strict scrutiny to the remaining provisions of the statutes because, he insisted, they had been enacted primarily to regulate surgical facilities, not abortions. He said the U.S. Supreme Court never intended to allow *Roe* to deny a state the opportunity to regulate facilities in which minor surgery, including abortions, are performed. He charged that the majority's use of strict scrutiny effectively denied the state the power to enact reasonable legislation to protect the health and welfare of its citizens.

Coffey further urged that removing the offending sections of both acts rendered them constitutional, even under strict scrutiny, because they no longer burdened a woman's right to abortion. He rejected plaintiffs' claims that the Ambulatory Surgical Treatment Center Act and the Health Facilities Planning Act substantially impinged the right to abortion by limiting availability, increasing cost, and subjecting the delivery of abortion services in a geographic area to a public veto with a certificate of need requirement.

With physicians offices available, he argued, the Ambulatory Surgical Treatment Center Act did not limit the accessibility of abortions. Even if abortions were made more expensive (a point he did not concede), Coffey felt that the advance in public safety justified any potential increase in cost. Lastly, he defended the certificate of need process of the Health Facilities Planning Act as a valid exercise of the public's First Amendment right to freedom of speech. Similarly, Coffey noted that, with one exception, he would uphold the center act's regulations, including those relating to abortions. In his view, they conformed to *Roe* because they did not burden the woman's right to abortion.

Ending on a note of judicial restraint, Coffey said he was persuaded that a state's interest in protecting the health and safety of its citizens justifies its ability to license medical facilities where minor surgery, including first trimester abortions, are performed. He concluded by reiterating his belief that the courts should not override a state's decision to discharge these obligations.

THE PARTIES SETTLE THE CASE

Pending appeal by the state, the parties reached an accord on the issues of the case and asked to have the litigation dismissed by the Supreme Court. They returned to Judge Nordberg. The settlement agreement, entered by Nordberg on March 22, 1990, enjoined the enforcement of several of the challenged provisions of the statutes.

The consent decree created a scheme that divided surgical facilities into two types: those performing "abortions beyond 18 weeks gestational age, or with general, epidural, or spinal anesthesia, or with incisions exposing the patient to a risk of infection," and those performing "abortions within 18 weeks gestational age, under local anesthesia." The parties agreed that the facilities in the first category fell under the auspices of the Ambulatory Surgical Treatment Center Act and Health Facilities Planning Act; those in the second category, although still requiring licenses, would be governed by a new set of specially designed regulations.

Following a "fairness" hearing, Nordberg concluded that the parties' agreement to permanently enjoin a number of Ambulatory Surgical Treatment Center Act regulations was both reasonable and fair because it preserved the plaintiffs' constitutional right to have or perform abortions as well as the state's right to regulate clinics "devoted to abortions."[2]

Why the Settlement

The consent decree erected a legal barrier to the efforts of anti-abortion forces to drastically reduce, if not eliminate, the delivery of abortion services in Illinois. The *Ragsdale* settlement was greeted with sighs of relief by most pro-choice advocates and cries of "sell-out" by most anti-abortion forces who charged that it was contrary to the Supreme Court's 1989 decision in *Webster v. Reproductive Health Services*.

Nordberg, however, denied that *Webster* affected this litigation. He pointed out that a negotiated settlement required both parties to compromise, and dismissed as "mere speculation" the notion that the Supreme Court would uphold the challenged provisions of these laws. He noted that "settlement proceedings are not an appropriate occasion for resolving the merits of undecided legal issues."

The politics of the 1990 Illinois gubernatorial race played a major role in inducing the defendants to yield ground, while the 5-4 *Webster* (1989) vote in favor of the state was the primary impetus for the plaintiffs' decision to settle out of court. At the time of the settlement, Attorney General Neil Hartigan, with a reputation for being anti-abortion, was seeking the Democratic Party's nomination for governor. If successful, his likely Republican opponent would be Secretary of State Jim Edgar, who had a long-standing pro-choice position. The perception of Hartigan as anti-abortion was fortified by his role as attorney general in appealing the adverse lower court decisions in *Ragsdale* to the Supreme Court.

The immediate political fallout from *Webster* (1989), according to the media, was the narrow defeat of Republican gubernatorial candidates in Virginia and New Jersey; both losses were attributed in significant part to the candidates' anti-abortion stances. There seems little doubt that Hartigan had to try to diffuse this issue in light of the fact that the *Webster* (1989) ruling had galvanized pro-choice forces.

For its part, why would the plaintiffs (and their lawyers, the American Civil Liberties Union) choose to settle *Ragsdale* rather than seek a ruling from the high court? It is most likely that the plaintiffs feared the Supreme Court would reverse the lower courts, possibly adopting the dissent's interpretation of the statutes' legislative histories. Their apprehensions undoubtedly stemmed from the *Webster* decision.

Although the facts and legal issues in *Webster* (1989) were largely unrelated to the questions raised in *Ragsdale*, the *Webster* ruling was the first major case approving abortion restrictions since *Roe* legalized abortion in 1973. It signalled that the Supreme Court was adopting a more lenient posture toward state abortion laws, and essentially invited states to adopt more stringent abortion regulations, suggesting it would look favorably on further incursions into abortion rights.

The *Ragsdale* settlement was a victory for the plaintiffs because they secured the right to a legal clinic abortion during the first four-and-a-half months (18 weeks) of pregnancy. Few abortions take place after 18 weeks (Henshaw, Binkin, Blaine, and Smith, 1985, p. 92). The plaintiffs obtained more from the consent decree than did the state because it was less likely that the doors of the Illinois abortion clinics would have remained open had the Supreme Court decided the case.

Implications and Prospects

The Alan Guttmacher Institute monitors the availability and utilization of abortion services in the U.S. through its periodic national surveys of all known abortion providers. Over the years, the institute's data have shown a steady decline in hospital abortions and a rise in abortions delivered by providers other than hospitals, such as specialized abortion clinics, nonspecialized clinics, and physicians' offices.

In 1974, hospitals represented 73 percent of abortion providers and 46 percent of all reported abortions (Weinstock, Tietze, Jaffe, and Dryfoos, 1976, pp. 65-66). By 1988, only 40 percent of the providers were hospitals, and they accounted for only a tenth of reported abortions (Henshaw and Van Vort, 1990, p. 108). A 1988 institute survey found that "in 1977, there were 1,654 hospitals offering abortion services; the number fell to 1,405 in 1982, to 1,191 in 1985 and to 1,040 in 1988" (Henshaw and Van Vort, 1990, p. 107).

The loss of 614 hospital providers since 1977 has been nearly offset by an increase of 487 non-hospital providers. Non-hospital facilities also have larger abortion caseloads than hospitals. In 1988, 45 percent of hospital providers reported fewer than 30 abortions; only 30 hospitals performed 1,000 or more procedures; and 62 percent of all reported abortions were performed in specialized abortion clinics with annual caseloads of 1,000 or more (Henshaw and Van Vort, 1990, p. 108).

Of the few implementation studies by political scientists, most focus on abortion rates (Hansen, 1980; Johnson and Bond, 1980) or sample hospitals in limited geographical areas (Brady and Kemp, 1976; Kemp, Carp, and Brady, 1978; Bond and Johnson, 1982; Johnson and Bond, 1982). One study (Tatalovich and Daynes, 1989) that canvassed the entire population of U.S. hospitals using 1986 American Hospital Association records found that only 34.5 percent of the 3,752 hospitals nationwide with obstetrics capacity actually provided abortion services. In 38 states, the majority of hospitals with obstetrics capacity did not perform any abortions. It is likely that this figure even overstates the availability of hospital abortion services because many hospitals restrict themselves to therapeutic, or necessary, abortions rather than elective abortions. But the fact that two-thirds of all U.S. hospitals with the capability to offer abortions do not presents an especially bleak picture of the availability of abortion services in the United States.

With the diminished role of hospitals in the delivery of abortion services, specialized abortion clinics have undertaken the task of performing most U.S. abortions. The crucial role of these clinics in providing abortion services explains why the dispute represented by *Ragsdale v. Turnock* held the potential for an enormous impact on legalized abortion in the U.S. Because of the number of women affected, the *Ragsdale* suit could have led to the most significant judicial ruling since the landmark *Roe v. Wade* decision in 1973.

All the district and appellate court judges who were involved in the *Ragsdale* litigation were appointed by Republican presidents: Ronald Reagan appointed Nordberg, Eschbach, and Coffey; Gerald Ford appointed William Bauer. The fact that two of three Reagan appointees voted to invalidate key portions of the Illinois law (notwithstanding the intense effort by the Reagan administration to recruit judicial conservatives) signals how much they were guided by existing precedent—the *Roe* and *Akron* decisions specifically.

Looking back on these rulings, however, Coffey's dissent may prove to be prophetic. Webster was decided in 1989; three years later, in *Planned Parenthood of Southeastern Pennsylvania v. Casey* (1992), the Supreme Court further restricted abortion rights. In *Casey*, the federal district court struck most of the provisions of a Pennsylvania abortion control law. But all, save the provision for spousal notification, were upheld on a 2-1 vote by a panel of the Third Circuit Court of Appeals.

The majority opinion by Circuit Court Judge Walter Stapleton (a Nixon appointee) demonstrated *Webster*'s effect in eroding the "fundamental right" to abortion by denying the continued applicability of the strict scrutiny test to evaluate abortion restrictions. Stapleton believed that no more than four justices would uphold the *Roe* doctrine. Convinced that a majority of the high court would apply a lesser standard of review, he ruled that Associate Justice Sandra O'Connor's "undue burden standard" was the proper test.

In a sharply divided opinion, the U.S. Supreme Court upheld the circuit court. However, to the surprise of many, the joint opinion by O'Connor, Anthony Kennedy, and David Souter reaffirmed the "essential holding" of *Roe*. But the three, joined by Chief Justice William Rehnquist and Justices Antonin Scalia, Clarence Thomas, and Byron White, held that most of the provisions of the Pennsylvania law did not impose an "undue burden" on women seeking abortions. The latter four justices rejected the continuing validity of *Roe*, and would have upheld the entire law. Justices Harry Blackmun and John Paul Stevens, the only two justices left from the original *Roe* majority, concurred in the joint opinion's reaffirmation of *Roe*, and argued that the statute was wholly unconstitutional.

Although there are two new Democratic appointees on the high court (Ruth Bader Ginsburg and Stephen Breyer, replacing White and Blackmun respectively), pro-choice advocates cannot be sanguine about the future of abortion policy in the U.S. This is a very different court from the one that decided *Roe*. At a minimum, a more conservative majority now holds the decisive vote in abortion cases. Illinois-type regulations, if considered today, would certainly stand a much better chance of approval under an "undue burden standard," particularly if couched within the medical practice laws of a state. Given the importance of clinics in the delivery of abortion services, *Ragsdale* serves as a chilling reminder of the fragility of the right to choose guaranteed in *Roe*.

ENDNOTES

[1] Section 16(1) of the Medical Practice Act (MPA) and the Ambulatory Surgical Treatment Center Act (ASTCA) were Senate Bills 1050 and 1051. As conceded by the Ragsdale dissent, these two bills were "initially" companion bills to the Illinois Abortion Act (Senate Bill 1049) of 1973. The debate between the majority and the dissent was over the extent to which the ASTCA was enacted primarily to regulate abortions.

[2] The state was enjoined from enforcing the portion of this paragraph that defined any facility performing an abortion as an ambulatory surgical treatment center. It was also enjoined from enforcing Subpart 16(1) of the Medical Practice Act. Similarly, the injunction enjoined the enforcement of most of Subpart G of the Ambulatory Surgical Treatment Center Act (ASTCA) as well as the licensing requirements for facilities only providing abortion and gynecological services. Additionally, the state could not enforce several sections of the Health Facilities Planning Act and could not apply many of the equipment, personnel, and physical plant requirements of the ASTCA regulations to clinics performing abortions on women with pregnancies of less than 18 weeks with local anesthesia.

REFERENCES

Bond, J. R. and Johnson, C. A. (1982). Implementing a permissive policy: Hospital abortion services after *Roe v. Wade*. *American Journal of Political Science, 26*(2),1-21.

Brady, D. W. and Kemp, K. (1976). The Supreme Court's abortion rulings and social change. *Social Science Quarterly, 57*(4), 535-546.

City of Akron v. Akron Center for Reproductive Health, 462 U.S. 416 (1983).

Hansen, S. B. (1980). State implementation of Supreme Court decisions: Abortion rates since *Roe v. Wade*. *Journal of Politics, 42*(2), 372-395.

Henshaw, S. K. and Van Vort, J. (1990). Abortion services in the United States, 1987 and 1988. *Family Planning Perspectives,* 22(3),102-108, 142.

Henshaw, S. K., Binkin, N. J., Blaine, E., and Smith, J. C. (1985). A portrait of American women who obtain abortions. *Family Planning Perspectives, 17*(2), 90-96.

Jaffe, F. S., Lindheim, B. L., and Lee, P. R. (1981). *Abortion politics: Private morality and public policy*. New York: McGraw-Hill.

Johnson, C. A. and Bond, J. R. (1982). Policy implementation and responsiveness in nongovernmental institutions: Hospital abortion services after *Roe v. Wade*. *Western Political Quarterly, 35*(4), 385-405.

Johnson, C. A. and Bond, J. R. (1980). Coercive and noncoercive abortion deterrence policies. *Law & Policy Quarterly,* 2(1), 106-128.

Kemp, K. A., Carp, R. A., and Brady, D. W. (1978). The Supreme Court and social change: The case of abortion. *Western Political Quarterly, 31*(2), 19-31.

Planned Parenthood of Southeastern Pennsylvania v. Casey, 112 S.Ct. 2791 (1992).

Planned Parenthood of Southeastern Pennsylvania v. Casey, 947 F.2d 682 (3d Cir. 1991).

Planned Parenthood of Southeastern Pennsylvania v. Casey, 744 F.Supp. 1323 (E.D. Pa. 1990).

Ragsdale v. Turnock, 625 F.Supp. 1212 (N.D. Ill. 1985).

Ragsdale v. Turnock, 841 F.2d 1358 (7th Cir. 1988).

Roe v. Wade, 410 U.S. 113 (1973).

Rosenberg, G. (1991). *The hollow hope*. Chicago: University of Chicago Press.

Tatalovich, R. and Daynes, B. W. (1989). The geographical distribution of U.S. hospitals with abortion facilities. *Family Planning Perspectives, 21*(2), 81-84.

Turnock v. Ragsdale, 492 U.S. 916 (1989).

Turnock v. Ragsdale, 493 U.S. 987 (1989).

Turnock v. Ragsdale, 734 F.Supp. 1457 (N.D. Ill. 1990).
Webster v. Reproductive Health Services, 492 U.S. 490 (1989).
Weinstock, E., Tietze, C., Jaffe, F. S., and Dryfoos, J. G. (1976). Abortion need and services in the United States, 1974-1975. *Family Planning Perspectives, 8*(2), 58-69.

THE UNCONSTITUTIONAL CONDITIONS DOCTRINE AND MANDATING NORPLANT FOR WOMEN ON WELFARE DISCOURSE

LAURENCE C. NOLAN*

> [E]ven though a person has no "right" to a valuable governmental benefit and even though the government may deny him the benefit for any number of reasons, there are some reasons upon which the government may not rely. It may not deny a benefit to a person on a basis that infringes his constitutionally protected interests. . . .
>
> Perry v. Sinderman, 408 U.S. 593, 597 (1972).

The current movement for welfare reform[1] has caused both federal[2] and state legislatures[3] to seek quick solutions to revive a failed welfare system. The system is perceived to have failed mothers who receive welfare benefits for their dependent children. Three themes underlie many of the recommended solutions: 1) mothers receiving Aid to Families with Dependent Children (hereinafter AFDC) benefits for a long period of time become increasingly

---

* Associate Professor of Law, Howard University School of Law; B.S., Howard University, 1961; J.D., University of Michigan School of Law, 1974. I am grateful to Cynthia Dial, a third year student, for her research assistance, to George H. Nolan for his helpful comments, and to Professor Madelyn Squire for her encouragement.

1. *See* Lucy A. Williams, *The Ideology of Division: Behavior Modification Welfare Reform Proposals*, 102 YALE L.J. 719 (1992) (discussing the recent attempts to "reform" welfare by modifying the behavior of recipients).

2. *E.g.*, Family Support Act of 1988, Pub. L. No. 100-485, 102 Stat. 2343 (codified in scattered sections of 42 U.S.C.) (requiring some welfare mothers to enroll in training programs as a condition for welfare). One such training program is: The Job Opportunities and Basic Skills Training Program, Title II, 42 U.S.C. § 602 (1988). *See also* Martha L. Fineman, *Images of Mothers in Poverty Discourses*, 1991 DUKE L.J. 274, 277 (1991) (discussing the Family Support Act of 1988).

3. *See, e.g.*, Act of Jan. 21, 1992, ch. 526, 1991 N.J. Sess. Law Serv. 2782 (West) (denying any additional benefits for recipients who parent new children while on welfare); 1987 Wis. Laws 27 (conditioning the retention of welfare benefits on a child's school attendance, popularly known as "learnfare").

dependent on aid and have their sense of a work ethic undermined;[4] 2) mothers, especially minority, unwed, teenage mothers, receiving AFDC benefits continue having children in order to receive more benefits;[5] and 3) these mothers and their children are costly to taxpayers.[6] Given its potential for controlling women's reproduction, it is little wonder, then, that the Norplant contraceptive device[7] has been heralded as an important factor in welfare reform.

Since the 1960s, very few new birth control methods for females have been developed; however, this trend changed in 1990, when the Food and Drug Administration (hereinafter FDA) approved the use of Norplant in the United States.[8] The introduction of Norplant, a contraceptive designed to provide up to five years of continuous, but reversible, birth control,[9] immediately attracted the attention of trial courts and state legislatures. Courts sought to use it as a condition for probation in the criminal justice system,[10] and state legislatures

---

4. *See, e.g.*, Fineman, *supra* note 2, at 277 (noting that the Family Support Act of 1988 "focus[es] on reinforcing the work ethic and dominant individualistic norms of self-sufficiency through the imposition of 'workfare' provisions for mothers of young children"); Williams, *supra* note 1, at 719-20 (outlining the "rhetoric of the current 'welfare reform' debate [which includes] . . . dysfunctional mothers incapable of fitting into mainstream society . . . [who] are economically and emotionally atrophied because of their 'dependence' on welfare") (*comparing* Charles A. Reich, *The New Property*, 73 YALE L.J. 733, 785 (1964) *with* LAWRENCE N. MEAD, BEYOND ENTITLEMENT: THE SOCIAL OBLIGATIONS OF CITIZENSHIP 41 (1986)). *See also* GEORGE GILDER, WEALTH AND POVERTY 111-13 (1981) (noting that in 1979, the average income for a welfare family of four was greater than the average American family median income which encouraged those who could be self-supporting to remain on welfare); CHARLES MURRAY, LOSING GROUND 148-53 (1984) (describing the Negative Income Tax (NIT) experiment that conservative economists advocated instead of a welfare system since they felt that all welfare subsidies discouraged the poor from working).

5. *See* Fineman, *supra* note 2, at 280 n.11 (citing several statements made by a Congressional Representative in opposition to the Family Support Act, which hypothesized a welfare mother staying on welfare indefinitely by having a child every two years).

6. *Primetime Live* (ABC television broadcast, Sept. 9, 1993).

7. *See infra* part II (discussing the development, history, and use of Norplant as a tool to control contraception).

8. *Long-Term Contraceptive Approved*, [1990-91 Transfer Binder] Food Drug Cosm. L. Rep. (CCH) ¶ 43,066 (Dec. 26, 1990); Melissa Burke, Note, *The Constitutionality of the Use of the Norplant Contraceptive Device as a Condition of Probation*, 20 HASTINGS CONST. L.Q. 207, 207 (1992).

9. Burke, *supra* note 8, at 207 (describing how the implant can be easily removed).

10. The most notorious case was People v. Johnson, No. 29390 (Cal. Super. Ct., Tulare County 1991) (ordering a convicted child abuser to use Norplant for three years as a condition for probation). *See* Stacey L. Arthur, *The Norplant Prescription: Birth Control, Woman Control, or Crime Control?*, 40 UCLA L. REV. 1 (1992) (providing a comprehensive discussion of the *Johnson* case and the constitutionality of mandating Norplant as a condition for probation). *See also* Tracy Ballard, *The Norplant Condition: One Step Forward or Two Steps Back?*, 16 HARV. WOMEN'S L.J. 139 (1993) (describing the statutory problems with imposing Norplant on welfare recipients); Burke, *supra* note 8 (exploring the constitutional problems with conditioning probation on the use of Norplant); Janet F. Ginzberg, Note, *Compulsory Contraception as a Condition of Probation: The Use and Abuse of Norplant*, 58 BROOK. L. REV. 979 (1992) (arguing that the imposition of Norplant is unconstitutional and poor public policy); Madeline Henley, Comment, *The Creation and Perpetuation of the Mother/Body Myth: Judicial and Legislative Enlistment of Norplant*, 41 BUFF. L. REV. 703 (1993) (discussing the myth that women's conduct can be explained by and controlled

began to see it as part of welfare reform.[11] States have even made Norplant available for welfare recipients through Medicaid.[12] Some states proposed legislation to provide financial incentives to welfare recipients who use Norplant.[13] Two states proposed legislation to mandate its use for some welfare recipients.[14]

This article contends that the unconstitutional conditions doctrine must be part of the discourse regarding welfare reform and contraceptives, especially if the government begins to mandate the use of contraceptives such as Norplant. The simplest statement of the doctrine is that in conditioning the receipt of a government benefit, the government must not be allowed to do indirectly what it cannot do directly.[15] This article consists of four parts. Part I provides a historical background on welfare reform. Part II describes the Norplant contraceptive system. Part III discusses how linking the use of the Norplant system to welfare reform affects various individual rights. Part IV addresses the role of the unconstitutional conditions

---

through their reproductive capacity); Kristyn M. Walker, Note, *Judicial Control of Reproductive Freedom: The Use of Norplant as a Condition of Probation*, 78 IOWA L. REV. 779 (1993) (discussing the imposition of Norplant as the latest of several attempts by the government to control reproductive freedom).

11. Legislatively encouraging the use of Norplant typifies the dramatic turnaround in state contraceptive policy. Before Griswold v. Connecticut, 381 U.S. 479 (1965), and its progeny, the states' policies towards contraceptives were to restrict their availability. *See, e.g.*, Poe v. Ullman, 367 U.S. 497 (1961) (discussing CONN. GEN. STAT. §§ 53-32, 54-196 (1958), which, although later repealed, criminalized both using and providing medical advice on contraceptives).

12. *See, e.g*, TENN. CODE ANN. § 71-5-133 (Supp. 1993) (requiring the Tennessee Department of Human Services to provide information on the availability of Norplant through the Medicaid program); WASH. REV. CODE § 74.09.800 (Supp. 1994) (requiring the Washington Department of Health and Social Services to provide information about Norplant as part of its family planning services).

13. *See, e.g.*, H.B. 2089, Kan. 74th Leg., 2d Sess. (1991) (offering women receiving AFDC benefits a free Norplant implant and a $500 cash grant); David S. Coale, Note, *Norplant Bonuses and the Unconstitutional Conditions Doctrine*, 71 TEX. L. REV. 189, 195-96 (1992) (discussing proposed bills in Kansas and Louisiana to give cash grants to female welfare recipients who agree to use Norplant).

14. H.B. 3207, S.C. (1993); S.B. 2895, Miss. (1992). *See* John Robert Hand, Note, *Buying Fertility: The Constitutionality of Welfare Bonuses for Welfare Mothers Who Submit to Norplant Insertion*, 46 VAND. L. REV. 715, 718 (1993) (describing proposed legislation in Mississippi which would mandate Norplant use in some cases); Henley, *supra* note 10, at 749-51 (discussing a bill considered by the Mississippi Senate which would have "required women with four or more children to be implanted with Norplant in order to qualify for or continue to be eligible for public assistance[,]" as well as several other states' proposed programs offering incentives to encourage women to be voluntarily implanted); ALAN GUTTMACHER INSTITUTE, NORPLANT: OPPORTUNITIES AND PERILS FOR LOW-INCOME WOMEN (Dec. 1992) (Special Report #1).

15. *See infra* part IV.A (describing in greater detail the "unconstitutional conditions doctrine"). The conditions that this article considers are ones that influence conduct. The government provides a benefit on the condition that the recipient alters some type of conduct which is constitutionally protected from government interference. For a further discussion of this doctrine, *see generally* Lynn A. Baker, *The Prices of Rights: Toward a Positive Theory of Unconstitutional Conditions*, 75 CORNELL L. REV. 1185, 1189 (1990); Albert J. Rosenthal, *Conditional Federal Spending and the Constitution*, 39 STAN. L. REV. 1103, 1114 (1987); Kathleen M. Sullivan, *Unconstitutional Conditions*, 102 HARV. L. REV. 1413, 1421-22 (1989).

doctrine in the discourse and is divided into three sections. Section A describes the doctrine itself. Section B explores the reasons why the doctrine must be part of the discourse. Section C discusses how the courts have applied the doctrine and concludes that the fairest way to apply it in welfare benefits cases is to treat the condition as though it were direct.

PART I. WELFARE REFORM BACKGROUND

Historically, the policy underlying assistance to the poor contained the idea that only the "deserving poor," such as the blind, deaf, insane, and orphaned, should receive assistance.[16] Strands of this type of reasoning continue to appear today in discussions concerning public assistance policy.[17] Prior to 1935, there was essentially no national policy for public assistance to the poor;[18] instead, private groups or local cities and counties provided welfare relief.[19]

Because of the depression in the 1930s, the number of persons in need of public assistance dramatically increased.[20] States were overwhelmed by the costs of such assistance.[21] In an effort to provide some relief to the states, Congress enacted the Federal Emergency Relief Act of 1933,[22] thus beginning a national policy for public assistance.[23] Two years later, Congress enacted the Social Security Act of 1935.[24] This Act created three programs: the unemployment insurance program, the old-age insurance program, and a federal aid program to those states providing cash relief to "unemployables," such as the old, the blind, and the orphaned.[25]

The AFDC program[26] provided assistance for needy children

---

16. J.M. Wedemeyer & Percy Moore, *The American Welfare System*, 54 CAL. L. REV. 326, 327-28 (1966); Williams, *supra* note 1, at 721; Fineman, *supra* note 2, at 279-80. *See also* King v. Smith, 392 U.S. 309, 320 (1968).

17. Fineman, *supra* note 2, at 279-80 (indicating that the purpose of the Family Support Act of 1988 was to promote a new work ethic, thereby implying that the poor lacked such an ethic).

18. *See generally* Wedemeyer & Moore, *supra* note 16 (discussing the patchwork of public assistance programs that existed prior to the passage of the Social Security Act of 1935).

19. Wedemeyer & Moore, *supra* note 16, at 326.

20. Williams, *supra* note 1, at 722.

21. Williams, *supra* note 1, at 722.

22. Pub. L. No. 73-15, ch. 30, § 4(a), 48 Stat. 57 (1933) (codified at 15 U.S.C. § 724 (1988)).

23. Williams, *supra* note 1, at 722.

24. Pub. L. No. 74-271, ch. 531, 49 Stat. 620 (1935) (codified at 42 U.S.C. §§ 301-1397f (1988)).

25. Williams, *supra* note 1, at 722 (discussing the target populations of the Social Security Act of 1935); *see* Wedemeyer & Moore, *supra* note 16, at 329-34 (providing a comprehensive discussion of groups targeted for federal assistance).

26. *See* Social Security Act of 1935, ch. 531, 49 Stat. 620, 627 (codified as amended at 42 U.S.C. §§ 301-1304 (1988)) (stating that this program was initially named Grants to States for Aid to Dependent Children (ADC)). *See also* Public Welfare Amendments of 1962, Pub. L. No.

whose parent was deceased, absent, or incapacitated.[27] The program was targeted to provide relief for children living with their widowed mothers; however, the states were allowed to impose eligibility requirements, such as conditioning benefits upon the sexual behavior of women, which often made children ineligible for aid.[28] Through subsequent legislation in 1950, the program was extended to cover the caretakers of dependent children[29] and in 1962, to cover families with an unemployed parent and dependent children.[30] During the 1960s and 1970s, the program continued to expand[31] as part of the "War on Poverty" initiatives[32] and because of court decisions that overruled state regulations which prevented eligibility.[33] During that period, the government focused on the eradication of poverty and did not blame the poor for their economic plight.[34] Overall, the concept of welfare as an entitlement emerged.[35]

By the 1980s, many people concluded that the AFDC program had to be reformed.[36] Books by scholars George Gilder[37] and Charles

87-543, § 104(a)(3), 76 Stat. 172, 185 (codified as amended at 42 U.S.C. §§ 301-1394 (1988)) (noting that the name was changed to Aid and Services to Families with Dependent Children (AFDC) in 1962).

27. Social Security Act of 1935, ch. 531, 49 Stat. 620, 629 (codified as amended at 42 U.S.C. §§ 301-1304 (1988)).

28. *See* Williams, *supra* note 1, at 723 (contending that the program targeted white, widowed mothers and their needy children, but intentionally excluded single black women); King v. Smith, 392 U.S. 309, 311-22 (1968) (citing a judicial summary of welfare for the "worthy poor").

29. Social Security Act Amendments of 1950, Pub. L. No. 81-734, § 323, 64 Stat. 477, 551 (codified at 42 U.S.C. § 606(b) (1988)).

30. Social Security Act Amendments of 1962, Pub. L. No. 87-543, §§ 108(a), 109, 152, 76 Stat. 185, 189-90, 206-07 (codified at 42 U.S.C. § 606(b) (1988)); *see* Wedemeyer & Moore, *supra* note 16, at 333 (noting the extension of benefits to groups other than children).

31. *See* Williams, *supra* note 1, at 724-25 (citing that the welfare movement began to include groups previously excluded).

32. *See generally* Thomas Ross, *The Rhetoric of Poverty: Their Immorality, Our Helplessness*, 79 GEO. L.J. 1499, 1506 (1991) (describing the initiatives, which were developed during a time of economic growth, to help the poor without stigmatizing them as immoral).

33. *See, e.g.*, New Jersey Welfare Rights Org. v. Cahill, 411 U.S. 619 (1973) (overturning a regulation that denied AFDC benefits to families with children born out of wedlock); Lewis v. Martin, 397 U.S. 552 (1970) (striking down a regulation which included for AFDC qualification the income of the man who shared the mother's home and had no obligation to support the child); King v. Smith, 392 U.S. 309 (1968) (invalidating a regulation which disqualified from AFDC any mother cohabiting with a man who was not obligated to support the child).

34. *See* Ross, *supra* note 32, at 1507 (referring to notions that welfare recipients are undeserving and not truly needy based on their supposed immorality); Williams, *supra* note 1, at 724-25 (describing welfare recipients as moral deviants who manipulate public assistance funds for personal economic gains).

35. Williams, *supra* note 1, at 724 nn.34, 37 & 39. *See generally* Charles Reich, *The New Property*, 73 YALE L.J. 733 (1964) (introducing the term 'New Property' and extending the idea of property rights in government largess in the form of public assistance).

36. *See* GILDER, *supra* note 4, at 111 (emphasizing that AFDC presents a moral dilemma because AFDC increases fatherless, dependent families); MURRAY, *supra* note 4, at 159-66 (noting that the receipt of AFDC funds creates a disincentive to marry and acquire employment).

37. *See generally* GILDER, *supra* note 4 (using the tools of sociology to describe the limitations of contemporary economics in analyzing the disparities between the rich and the poor).

Murray[38] greatly influenced the welfare reform debate. They concluded that public assistance increased poverty, created dependence and immorality, and undermined traditional values.[39] Murray also concluded that poor people were motivated by economic incentives.[40] Many current welfare reform initiatives are based on these conclusions.[41] One of the numerous ways legislatures have proposed to remedy welfare's ills has included predicating a welfare mother's receipt of benefits on her use of Norplant.[42] Before discussing the constitutionality of such a requirement, it is helpful to gain first a better understanding of Norplant itself. With this goal in mind, the following section describes the Norplant contraceptive system in greater detail.

PART II. THE NORPLANT CONTRACEPTIVE SYSTEM.

The FDA approved the Norplant contraceptive system in December 1990 for general use by women.[43] The system consists of six small silastic rods, each containing the synthetic hormone levonorgestrel.[44] This device is implanted surgically under the skin of the woman's arm near the elbow joint.[45] The surgical removal of the system is a longer, more difficult procedure than the implantation.[46] The hormone slowly filters through the rods into the woman's blood-

38. *See generally* MURRAY, *supra* note 4 (discussing the change in character of the poor and the growth of welfare expenditure in the United States from 1950 to 1980).

39. *See* GILDER, *supra* note 4, at 111-12, 114-27 (stating how the prospect of receiving welfare benefits eliminates the breadwinner role of the father, dissolves the traditional two parent family, encourages public assistance fraud, increases government spending for income programs well over GNP earnings, and overall makes welfare funds more attractive than a minimum wage job); *see also* MURRAY, *supra* note 4, at 147-66, 178-91 (discussing the disincentives to work, marry, and maintain a certain socio-economic status due to the increase of welfare expenditures by the United States Government); Williams, *supra* note 1, at 725 (referring to works by authors who identified the differences between the truly needy and unworthy poor).

40. MURRAY, *supra* note 4, at 156-62.

41. *See* Williams, *supra* note 1, at 725 (contending that current welfare reform stems from the idea that there are unworthy recipients); *see also id.* at 726-46 (discussing the Learnfare and Family Cap welfare reform programs in Wisconsin which stemmed from the belief that some welfare recipients are unworthy).

42. *See supra* notes 13-14 (referring to states that have proposed conditioning the receipt of welfare benefits on a woman's use of Norplant).

43. *Long-Term Contraceptive Approved*, *supra* note 8; *see* Coale, *supra* note 13, at 189 n.3 (noting that Wyeth-Ayerst Laboratories is the United States' distributor of Norplant).

44. *See* WYETH-AYERST LABORATORIES, WOULD YOU LIKE UP TO 5 YEARS OF CONTINUOUS BIRTH CONTROL THAT IS REVERSIBLE? (1992) (describing levonorgestrel as a progestin that has been used in the pill since the 1960s).

45. *See Long Term Contraceptive Approved*, *supra* note 8, at 43,066 (describing the 10-15 minute procedure performed under a local anesthetic and in a doctor's office).

46. *See* WYETH-AYERST LABORATORIES, *supra* note 44 (estimating that removal of Norplant is a 15-20 minute procedure).

stream.[47] The rods contain enough levonorgestrel to prevent pregnancy for five years.[48]

Norplant is considered ninety-nine percent effective as a method of birth control.[49] Once removed, the subsequent conception rate for women formerly implanted with Norplant is the same as that of comparable women who were not implanted.[50] Common side effects may include menstrual cycle irregularity, headache, nausea, nervousness, and dizziness.[51] Norplant, however, does not protect against the transmission of Human Immunodeficiency Virus (HIV) and other sexually transmitted diseases (STD).[52]

The effectiveness of the contraceptive device, once implanted, does not depend upon the user or the user's sexual partner.[53] Furthermore, the user cannot remove it without medical assistance.[54] Thus, the woman is powerless to decide that she wants to try to conceive and then act on that decision. For these reasons, some proponents of welfare reform find Norplant an attractive contraceptive for women receiving AFDC benefits, even though individual rights may be affected.

PART III. LINKING NORPLANT TO WELFARE REFORM AFFECTS VARIOUS INDIVIDUAL RIGHTS[55]

After the FDA's approval of Norplant, the federal and state governments rapidly added it to their welfare programs by making it available to welfare recipients.[56] For instance, some state legislatures proposed "bonus" programs awarding cash bonuses to women

47. Irving Sivin, *Norplant Clinical Trials*, *in* DIMENSIONS OF NEW CONTRACEPTIVES: NORPLANT AND POOR WOMEN 1 (Sarah E. Samuels & Mark Smith eds., 1992) [hereinafter NORPLANT AND POOR WOMEN].

48. *Id.*

49. Coale, *supra* note 13, at 189.

50. WYETH-AYERST LABORATORIES, *supra* note 44.

51. WYETH-AYERST LABORATORIES, *supra* note 44.

52. NORPLANT AND POOR WOMEN, *supra* note 47, at xiv-xv.

53. NORPLANT AND POOR WOMEN, *supra* note 47, at xii-xiii, 1.

54. NORPLANT AND POOR WOMEN, *supra* note 47, at xi.

55. Norplant legislation also raises equality issues. For instance, Norplant legislation uniquely affects females. Although contraceptives are *not* uniquely female, the history of contraceptive technology focused upon the female. Further, this legislation overtly discriminates against lower socio-economic classes. The legislation, facially neutral as to race, also disproportionately affects minorities. For example, in the 1991 fiscal year, 40.1% of the children receiving AFDC funds were black and 18.5% were Hispanic. U.S. Department of Health and Human Services, 1991 Aid to Families with Dependent Children Recipient Characteristics Study 6.

56. *See* Coale, *supra* note 13, at 189 (identifying the various governments that include Norplant as an option in their welfare programs).

receiving AFDC benefits as long as they agreed to use Norplant.[57] Others, going further, proposed legislation mandating that some women use Norplant as a condition of receiving AFDC benefits.[58]

The focus of the debate for the proponents of these bills centers upon the following: the scarcity of the state's resources; the allegation that the welfare system creates intergenerational dependency; the opinion that welfare undermines the work ethic; and, the opinion that only mothers who can afford to care for their children, independent of public resources, should have children.[59] The debate's focus, however, fails to discuss the serious constitutional issues raised by these bills.

Legislation that links Norplant to welfare benefits infringes upon the protected fundamental rights of mothers.[60] The Constitution protects such rights from government interference unless the state can show a compelling need.[61] Since 1965, the right of privacy umbrella has protected the fundamental rights affected by Norplant legislation.[62]

Two branches of the right to privacy exist, and Norplant legislation would affect both. One branch involves the broad right to privacy in sexual matters,[63] including the fundamental rights to procreate,[64] to use contraceptives,[65] to parent,[66] and to define the family.[67]

---

57. H.B. 2089, Kan. 74th Leg., 2d Sess. (1991); H.B. 1584, La. 17th R.S. (1991). *See generally* Coale, *supra* note 13, at 189-94 (stating that Kansas and Louisiana introduced bills which established bonus programs).

58. S.B. 2895, Miss. (1992); H.B. 3207, S.C. (1993). In contrast, a proposed California bill provided that no person would be required to use contraceptives as a condition for eligibility for public assistance. Cal. Assembly 3593, Reg. Sess. (1993) .

59. *See Primetime Live, supra* note 6 (citing statements made by Roland Corning, a South Carolina state legislator, that mandating Norplant for women who want welfare would benefit states financially).

60. *See* Arthur, *supra* note 10, at 61-63 (noting that court-ordered Norplant for criminal cases in connection with probation also impinges on fundamental rights).

61. Griswold v. Connecticut, 381 U.S. 479, 485 (1965).

62. *Id.* at 485-86 (addressing, more generally, the interrelation of a contraceptive mandate and constitutional rights such as personal security and liberty).

63. *See Griswold*, 381 U.S. at 485 (identifying the right of privacy as a constitutionally protected right even though this right is not explicitly found in the Constitution). The majority and concurring opinions in *Griswold* relied on the First, Third, Fifth, Ninth, and Fourteenth Amendments to infer a general right of privacy applicable to the federal government and to the states. *Id.* Since that time, such fundamental rights as procreation, contraceptive use, and parenting have been described as being under the privacy umbrella. *Id.* at 485-86; *see also* Skinner v. Oklahoma, 316 U.S. 535, 541 (1942) (extending the concept of fundamental rights to procreation); Meyer v. Nebraska, 262 U.S. 390, 400 (1923) (extending fundamental rights to parenting).

64. *See Skinner*, 316 U.S. at 541 (recognizing the right to procreate as a fundamental right). The Court identified the right to marry and procreate as basic civil rights "fundamental to the very survival of the race." *Id.*

65. *See* Eisenstadt v. Baird, 405 U.S. 438, 453 (1972) (recognizing the right of unmarried people to use contraceptives); *Griswold*, 381 U.S. at 485-86 (1965) (recognizing the right of

The other affected branch involves the privacy right in personal or individual autonomy.[68] For example, one has a constitutionally protected right to bodily integrity regarding nonconsensual medical treatment, known as the principle of informed consent.[69] The principle of informed consent involves two elements: being informed about the treatment and giving one's consent.[70] Mandating the use of Norplant violates this principle because consent must be given freely, without any form of coercion. The principle of informed consent is central to medical ethics and legal doctrine.[71] It cannot be ignored with impunity. Further, the First Amendment Free Exercise Clause regarding religious objections to medical treatment may also protect an individual's right to bodily integrity.[72] Mandating the use of Norplant violates the Free Exercise Clause of the First Amendment when such contraceptive practices oppose religious

---

married couples to use contraceptives). *Accord* Carey v. Population Servs. Int'l, 431 U.S. 678, 685-88 (1977) (stating that access to contraceptives is essential to exercising the protected right of deciding whether to conceive a child).

66. Mandatory contraception laws preventing mothers on welfare from conceiving would interfere with their fundamental right to parent. The Constitution protects this right under the principle of family or parental autonomy. *See* Santosky v. Kramer, 455 U.S. 745, 753 (1982) (stating that the natural parents' fundamental liberty interest in the care, custody, and management of their child is protected by the Fourteenth Amendment); Meyer v. Nebraska, 262 U.S. 390, 400 (1923) (recognizing the fundamental right of parents to determine a child's education); Laurence C. Nolan, *Honor Thy Father and Mother: But Grandparent Visitation in the Intact Family?*, 8 B.Y.U. J. PUB. L. 51, 53-54 (1993) (discussing the principle of parental autonomy as part of the right to privacy interest).

Norplant legislation suggests that mothers on welfare are unfit to become parents. The Supreme Court held that a conclusive presumption that all unwed fathers were unfit parents violated the due process clause of the Fourteenth Amendment. Stanley v. Illinois, 405 U.S. 645, 649 (1972). Via a similar analysis, it would appear that such a presumption about mothers on welfare would violate due process as well. *Accord* Charles W. Murdock, *Sterilization of the Retarded: A Problem or a Solution?*, 62 CAL. L. REV. 917, 928-32 (1974) (discussing issues surrounding mentally retarded persons' fitness for parenthood as a basis for sterilization).

67. The Supreme Court has held, in cases of blood and marital relationships, that the Constitution protects the right of family definition. *See* Moore v. City of East Cleveland, 431 U.S. 494, 499 (1977) (allowing a grandparent to define her family to include her children and grandchildren). Norplant legislation would interfere with the welfare mother's right to define her family to include additional children.

68. The branch of the right of privacy—expressed in terms of the autonomy of the individual as a liberty interest—is protected by the Due Process Clause of the Fifth and Fourteenth Amendments. U.S. CONST. amend. V (stating that no person shall be "deprived of life, liberty, or property without due process of law"); U.S. CONST. amend. XIV (restating the Due Process Clause of the Fifth Amendment as applicable to the states).

69. *See, e.g.*, Washington v. Harper, 494 U.S. 210, 221-22 (1990) (recognizing a liberty interest in avoiding unwanted administration of antipsychotic drugs); Parham v. J.R., 442 U.S. 584, 597 (1979) (recognizing a child's liberty interests when involuntarily committed to a mental hospital).

70. *See* Arthur, *supra* note 10, at 93-94 (providing a general discourse on informed consent).

71. Arthur, *supra* note 10, at 93-96.

72. *See* Coale, *supra* note 13, at 206-08 (explaining the Free Exercise Clause and various religious objections to contraceptives).

beliefs.[73] Clearly, Norplant legislation significantly affects individual rights.

PART IV. THE ROLE OF THE UNCONSTITUTIONAL CONDITIONS DOCTRINE IN THE DISCOURSE

Unless considered in the light of the unconstitutional conditions doctrine,[74] it is likely that Norplant legislation will be judged merely as social and economic legislation, rather than legislation adversely impacting on the aforementioned constitutionally protected rights. Courts use a rational standard to determine the constitutionality of social and economic legislation, which requires the court to give deference to the legislation.[75] However, if Norplant legislation directly affects constitutionally protected rights, the Court would determine its constitutionality under a higher standard of review.[76] Thus, bringing the unconstitutional conditions doctrine into the discourse may result in such legislation being reviewed under a higher standard.[77]

SECTION A. THE UNCONSTITUTIONAL CONDITIONS DOCTRINE

Legal scholars recognize the classical theory that the government does not interfere in areas of personal decision-making.[78] With the growth of the modern regulatory state, however, the government disburses benefits in all areas of American life.[79] Consequently,

73. *See generally* JOHN T. NOONAN, JR., CONTRACEPTION: A HISTORY OF ITS TREATMENT BY THE CATHOLIC THEOLOGIANS AND CANONISTS (enlarged ed. 1986) (identifying the religious debates behind contraceptive use).

74. The doctrine states that although the government may choose not to provide a benefit, if it does so, then the government may not place a condition on the granting of that benefit. TRIBE, AMERICAN CONSTITUTIONAL LAW § 10-8, at 681 (2d ed. 1988). *See also* sources cited, *supra* note 15.

75. *See* Dandridge v. Williams, 397 U.S. 471, 485 (1970) (stating that in the area of economic and social welfare, a state needs only a reasonable basis to satisfy the Equal Protection Clause).

76. *See generally* Griswold v. Connecticut, 381 U.S. 479 (1965) (establishing that a violation of fundamental rights receives strict scrutiny by courts rather than rational review). *See also infra* notes 146-49 and accompanying text (defining the two standards).

77. *Carey v. Population Servs. Int'l* indicated that the Supreme Court would not only apply strict scrutiny to those statutes entirely prohibiting abortion, but would also apply this standard to those statutes limiting a woman's access to an abortion. Carey v. Population Servs. Int'l, 431 U.S. 678, 688 (1977).

78. Thornburgh v. American College of Obstetricians & Gynecologists, 476 U.S. 747, 772 (1986) ("Our cases long have recognized that the Constitution embodies a promise that a certain private sphere of individual liberty will be kept largely beyond the reach of government."); Seth F. Kreimer, *Allocational Sanctions: The Problem of Negative Rights in a Positive State*, 132 U. PA. L. REV. 1293, 1324-26 (1984); Gary A. Winters, Note, *Unconstitutional Conditions as "Nonsubsidies": When Is Deference Inappropriate?*, 80 GEO. L.J. 131, 135 (1991).

79. *See* Cass R. Sunstein, *Why the Unconstitutional Conditions Doctrine is an Anachronism (With Particular Reference to Religion, Speech, and Abortion)*, 70 B.U. L. REV. 593, 593 (1990) (identifying

government regulations increasingly affect constitutionally protected rights indirectly.[80] Such governmental activity as it relates to constitutional issues is problematic. For instance, citizens cannot claim an affirmative right to government aid.[81] Since the government is under no obligation to provide benefits, should the government be able to attach conditions to achieve indirectly that which it cannot do directly? The unconstitutional conditions doctrine answers this inquiry.

The essence of the doctrine is that a government may not grant a benefit with the condition that the recipient forego a constitutionally protected right, even if the government has no duty, in the first place, to provide the benefit.[82] The doctrine, then, protects individuals from a government's indirect actions which compromise constitutionally protected rights.

Having its genesis in the nineteenth century,[83] the doctrine was molded during the "Lochner era" to protect economic liberties which, at that time, were protected under substantive due process.[84] This judicially created doctrine is still applied post-"Lochner."[85] Opponents, such as Justice Holmes, argue that this doctrine does not exist since the greater power of the state not to create a benefit includes the lesser power to impose the condition.[86] The next section

spending, licensing, and employment as areas of increased government regulation).

80. Sunstein, *supra* note 79, at 598-99; Kreimer, *supra* note 78, at 1296-97.

81. The government generally has no affirmative duties, only negative ones. *See* DeShaney v. Winnebago County Dep't of Social Servs., 489 U.S. 189, 196 (1989) ("[O]ur cases have recognized that the Due Process Clauses generally confer no affirmative right to governmental aid, even when such aid may be necessary to secure life, liberty, or property interests of which the government itself may not deprive the individual.").

82. Sullivan, *supra* note 15, at 1415. *Accord* Perry v. Sinderman, 408 U.S. 593, 597 (1972) (explaining the doctrine in terms of the protected interest in freedom of speech).

83. Richard A. Epstein, *The Supreme Court 1987 Term-Foreword: Unconstitutional Conditions, State Power, and the Limits of Consent*, 102 HARV. L. REV. 5, 10 (1988).

84. Sullivan, *supra* note 15, at 1416 (commenting that traces of the statement of the doctrine appeared in earlier cases where courts have held that states cannot condition privileges on the surrender of constitutional rights). *See, e.g.*, Doyle v. Continental Insurance Co., 94 U.S. 535, 543 (1876) (Bradley, J., dissenting) ("Though a State may have the power, if it sees fit to subject its citizens to the inconvenience, of prohibiting all foreign corporations from transacting business within its jurisdiction, it has no power to impose unconstitutional conditions upon their doing so."); Lafayette Insurance Co. v. French, 59 U.S. (18 How.) 404, 407 (1856) (maintaining that Ohio could condition a foreign corporation's consent to do business in the state "provided they are not repugnant to the [C]onstitution or laws of the United States").

85. Epstein, *supra* note 83, at 5-7; *see also* Sullivan, *supra* note 15, at 1428-42 (describing how the doctrine developed in three areas of the law: state regulations against corporate rights, federal encroachment on state autonomy, and federal and state regulations against individual rights).

86. *See, e.g.*, Western Union Telegraph Co. v. Kansas, 216 U.S. 1, 53 (1909) (Holmes, J., dissenting) ("Even in the law the whole generally includes its parts. If the State may prohibit, it may prohibit with the privilege of avoiding the prohibition in a certain way.").

contends that the doctrine is still necessary because of the effect the growing regulatory state has on welfare benefits.

SECTION B. THE ROLE OF THE DOCTRINE IN THE WELFARE REFORM DISCOURSE

Today, the government has very broad powers to dispense welfare benefits with conditions that affect constitutionally protected rights, though indirectly.[87] Such conditions make it easier for the government to be less cognizant of these protected rights. The doctrine fulfills the need "to mediate the boundary between constitutional rights and government prerogatives in the area of spending."[88] Making the doctrine part of this discourse on Norplant and welfare reform, therefore, provides legislatures and courts with the proper framework against which to view the constitutionality of such conditions.

Foremost, the doctrine serves to safeguard the individual's protected rights from government manipulation.[89] Since the New Deal, the doctrine's protector role has gained significance in light of the government's expanded regulatory powers over social and economic programs.[90] Before the 1930s, poor people relied primarily upon private sources for welfare benefits.[91] The government has now assumed the central role in providing these benefits.[92] As a result, private funding sources have become scarce or have disappeared altogether, and the government's role today is basically

---

87. *See* Hand, *supra* note 14, at 743, 753 (commenting that the Supreme Court can examine the right involved, determine that the right is not unduly burdened by the condition, and uphold the condition). The Court typically does not overturn conditional laws when the benefit involved is welfare. *Id.*

88. Sunstein, *supra* note 79, at 593.

89. *See* Epstein, *supra* note 83, at 28 (commenting that the doctrine is a "second best approach" to controlling government discretion which is used to "take back" some of the power which was originally conferred upon government).

90. *See* Joel F. Handler, *Controlling Official Behavior in Welfare Administration*, 54 CAL. L. REV. 479, 492-500 (1966) (discussing the governmental interference and manipulation that occurs in welfare administration and the limited judicial procedures established to protect welfare recipients' rights against such intrusion).

91. *See* Kreimer, *supra* note 78, at 1296 (stating that poor individuals previously relied on the church and family for support); *see also* Michael M. Maddigan, *The Establishment Clause, Civil Religion, and the Public Church*, 81 CAL. L. REV. 293, 329 (1993) (commenting that after the New Deal, the government created numerous welfare programs which were traditionally functions exercised by the private sector).

92. Kreimer, *supra* note 78, at 1296 (observing that since the government has inherited the educational and welfare functions from the church and family, there are opportunities for government intrusion through the allocation of benefits).

monopolistic.[93] The doctrine is useful in this discourse to provide a check "against the political perils of monopoly."[94]

With the government in this "monopolistic" role as welfare provider, it can easily abuse its power. Norplant legislation may be an illustration of such abuse. A law making Norplant mandatory for a woman on welfare leaves two unsatisfactory choices for the woman: either accept welfare benefits conditionally or not at all. Such choices allow a government to foster its goals[95] at the expense of the mother's protected reproductive rights. Similarly, if bonuses are provided for women on welfare who agree to use Norplant, the government gains increased leverage over a welfare mother's constitutionally protected reproductive rights.[96] In this light, the government's offer of a bonus is not benign legislation. It is a cunning incentive to poor mothers to forego their constitutionally protected rights in exchange for their most immediate need: money.[97] It encourages poor women to forego their reproductive rights in choosing whether to use a contraceptive as well as what type of contraceptive to use.

---

93. *See* Maddigan, *supra* note 91, at 329-30 (stating that the government's acquisition of welfare functions, which had traditionally been the responsibility of the private sector, limited the private sector's ability to provide opportunities for its citizens). *But cf.* WALTER I. TRATTNER, FROM POOR LAW TO WELFARE STATE: A HISTORY OF SOCIAL WELFARE IN AMERICA 257-61 (3d ed. 1984) (stating that the inability of the private sector to provide the financial resources necessary was a factor in the growth of federal welfare).

94. Epstein, *supra* note 83, at 15. *See also* Kathleen M. Sullivan, *Unconstitutional Conditions and the Distribution of Liberty*, 26 SAN DIEGO L. REV. 327, 330 (1989) (describing this function as "state checking" since the power of the state is checked by maintaining the boundary between public and private spheres).

95. For example, Roland Corning, the author of a South Carolina bill, said on ABC Primetime that taxpayers are tired of making welfare payments to women who continue to have children: "They can have all the children they want. They just have to pay for them." *Primetime Live*, *supra* note 6.

96. *See* Sullivan, *supra* note 15, at 1492-93 (stating that the government overreaches when it forbids actions protected by individual rights of reproductive privacy); *see also* Charles A. Reich, *Beyond the New Property: An Ecological View of the Due Process*, 56 BROOK. L. REV. 731, 738 (1990) ("If benefits necessary to the survival of the individual are the property of the government, then these benefits become an instrument of social control. The government can impose conditions, supervise the behavior of the recipients, or deny them the control over their lives that most other citizens take for granted.").

97. Commenting on the proposed Kansas bill to give a bonus to welfare mothers who choose to use Norplant, Kansas Representative Kathleen Sibelius stated, "The idea of choice, I think, is removed, if you have a starving person and you offer them food if they will do something. I don't think that's a very realistic choice." *60 Minutes: Norplant* (CBS television broadcast, Nov. 10, 1991, cited in Henley, *supra* note 10, at 769). *See also* Charlotte Rutherford, *Reproductive Freedoms and African American Women*, 4 YALE J.L. & FEMINISM 255, 262 (1992) (commenting that some argue that offering cash bonuses should not be allowed because it emphasizes the wrong issue for the cause of poverty). By limiting poor women's procreational rights, issues of racism, sexism, and classism will not be eliminated. These issues are the root cause of poverty. *Id.*

The government may argue that since welfare benefits are gratuities, it may attach conditions. These conditions become of great concern when the state is acting as a "monopoly." Citizens, who do not have real options,[98] are placed in the unfair position of having to give up constitutionally protected rights in order to accept the government's gratuities. On the other hand, the government may argue that when it provides welfare benefits, the government and its citizens are acting more along "contract principles." Citizens are free either to accept or to refuse the government's offer. However, the government may become coercive in its demands when welfare mothers look to it as their only source of income.

Even if the government is not acting monopolistically, the doctrine preserves "private ordering"[99] in decision-making in which the government should have no power to control. Decisions regarding reproductive rights are in the private realm.[100] Norplant legislation allows the government to be in a position to take advantage of a welfare mother's plight and to influence her constitutionally protected reproductive decisions. The fact that Norplant does not require the woman's cooperation to be effective, once implanted, allows the government to assert control over her fertility with relative ease.[101] There is but one male contraceptive—the condom—and it is not comparable to Norplant.[102] Whenever the government can assert such control over a woman's decisions about her reproductive rights, these decisions are removed from the private realm. If the government could not enact legislation which directly asserted this control without showing a compelling state interest, neither should it be able to do so indirectly.

As a direct consequence of legislation allowing the government to influence a woman's reproductive decisions, the government is able to engage in the social engineering of family structures by gender, class, and race.[103] Norplant legislation affects only poor women

98. There is always the argument that there is an alternative: "Get a job."

99. *See* Sullivan, *supra* note 15, at 1492-96 (discussing the occurrence of "private ordering" when the government shifts the boundary between private and public ordering through its allocation of benefits). The unconstitutional conditions doctrine preserves this boundary.

100. Sullivan, *supra* note 15, at 1492.

101. *See supra* part II (discussing the fact that Norplant does not depend upon the cooperation of the user to be effective since it cannot be removed without medical assistance).

102. *See* Ballard, *supra* note 10, at 161 (commenting that establishing Norplant as a condition for probation violates equal protection because there is no approved long-term male contraceptive). Indeed, other than sterilization or the use of condoms, there are no controls over male fertility.

103. *See generally* Rutherford, *supra* note 97 (conveying that Norplant, like cash bonuses, is linked to issues of race, gender, and class).

receiving aid for dependent children.[104] Many proponents advocate Norplant legislation to reduce this particular class of welfare recipients.[105] These proponents believe this class should be reduced because it continues intergenerational dependency on welfare.[106] They say that only mothers who can afford additional children without welfare benefits should procreate: no more children should be born to a family that depends on welfare benefits.[107] By manipulating reproductive decisions, the government would be directly engineering the family structure of these welfare recipients. Such decisions on family structure are, however, in the private realm of decision-making, not in the government's realm.[108]

The unconstitutional conditions doctrine would encourage "government evenhandedness" or neutrality among its citizens in their constitutionally protected decision-making.[109] The Supreme Court has held that courts should defer to democratic decision-making in ordinary distributive matters.[110] Professor Stephen Loffredo notes that the Supreme Court bases its deference to social and economic legislation upon both the theory that decision-making in the legislature is democratic, and the assumption that poor people have fair access to this process.[111] He argues, however, that poor people do not have fair access.[112]

If poor people do not have fair access to the democratic process, the unconstitutional conditions doctrine provides a greater likelihood of the government's evenhandedness in its enactment of Norplant

---

104. *See supra* note 55 (illustrating how Norplant legislation affects a large number of minority women).

105. *E.g.*, Donald Kimelman, *Poverty and Norplant: Can Contraception Reduce the Underclass?*, PHILADELPHIA INQUIRER, Dec. 12, 1990, at A18 [hereinafter PHILADELPHIA INQUIRER] (recommending that Norplant be made available to African-American women who are members of the underclass).

106. *See supra* note 4 and accompanying text (explaining the proponents' fear of the lack of work ethic instilled in recipients of AFDC benefits).

107. *See Primetime Live*, *supra* note 6 (explaining the argument that taxpayers do not want to support women who continue to have more children).

108. Carl R. Schneider, *Moral Discourse and the Transformation of American Family Law*, 83 MICH. L. REV. 1803, 1833-42 (1985).

109. See Sullivan, *supra* note 15, at 1496-97 and Sullivan, *supra* note 94, at 331 which state that this doctrine imposes on the government an obligation of neutrality, such as that found in speech or religion, to not use benefits to shift viewpoints.

110. *See, e.g.*, Bowen v. Gilliard, 483 U.S. 587, 596-97 (1987) (holding that "unless the Legislative Branch's decisions run afoul of some constitutional edict, any inequities created by such decisions must be remedied by the democratic process").

111. Stephen Loffredo, *Poverty, Democracy and Constitutional Law*, 141 U. PA. L. REV. 1277, 1277-85 (1993).

112. *See generally id.* (offering the argument that because poor persons do not have access to the democratic process, their constitutional rights are not adequately protected).

legislation.[113] Norplant legislation allows the government to favor the reproductive rights of non-welfare mothers, as they are free to exercise their reproductive rights while welfare mothers must pay a higher price to do so.[114] The unconstitutional conditions doctrine affords a method to ensure that the government is evenhanded in its dealings between these two classes of women.[115] If the doctrine is not part of the discourse, Norplant legislation may be judged merely as social and economic legislation under the ruse that it does not directly affect reproductive rights.

The unconstitutional conditions doctrine also deters the government from making hasty decisions in times of crises regarding Norplant.[116] Norplant legislation first arose during a climate of severe budgetary constraints.[117] Its proponents argued that it would save money and reduce the budget,[118] ignoring that other consequences would also follow. This legislation will shape the child-bearing rights of poor women and will allow the government to regulate an intimate aspect of the lives of one group of its citizens.[119] It will allow the government to engage in social engineering of family structures based on gender, race, and class. Ironically, it may encourage promiscuous sexual behavior because of the spontaneity that Norplant allows, increasing the risks of AIDS and other sexually transmitted diseases.[120] Moreover, legislation mandating Norplant for welfare recipients will deny the state's most vulnerable citizens, dependent children, a subsistence income if their mothers refuse to give up their constitutional rights. Legislatures must confront these consequences of Norplant legislation. Without including the unconstitutional conditions doctrine in this discourse,

---

113. *See* Skinner v. Oklahoma, 316 U.S. 536, 546 (1942) (Jackson, J., concurring) (stating that "there are limits to the extent to which a legislatively represented majority may conduct biological experiments at the expense of the dignity and personality and natural powers of a minority").

114. *See generally* Coale, *supra* note 13, at 208-10 (stating that Norplant legislation places surcharges on poor women).

115. Sullivan, *supra* note 15, at 1496-97.

116. *See, e.g., supra* notes 2-3 (describing Congress and states' quick solutions to the welfare problem).

117. *See Primetime Live, supra* note 6 (describing the pro-Norplant argument that AFDC recipients are costly to taxpayers).

118. *Primetime Live, supra* note 6; *see also* Kerry Patrick, *Poor Women and Society Benefit by Linking Norplant, Welfare Aid*, WICHITA EAGLE, Mar. 4, 1991, at A11 (advocating the passage of his Norplant bonus bill because it would eventually save thousands of dollars in welfare payments).

119. *See supra* note 66 and accompanying text (indicating that Norplant legislation would interfere with the fundamental right to parent).

120. *See* NORPLANT AND POOR WOMEN, *supra* note 47, at xiv-xv (explaining that Norplant prevents pregnancy but does not protect against transmission of AIDS or other sexually transmitted diseases).

the government may not stop and re-evaluate the conditions attached to the welfare benefits.[121]

In times of budgetary crisis, legislatures may be more likely to follow popular opinions and prejudices in their haste to meet the crisis and silence critics. Norplant legislation promotes the theme that the government should only help the deserving poor.[122] Norplant legislation proponents argue that the undeserving poor, now termed the underclass, should be reduced.[123] The stereotype of this class includes welfare mothers who stay on welfare in order to avoid working and welfare mothers who continue to have children in order to continue receiving welfare.[124] This stereotype is unsupported in fact and contrary to recent studies.[125] As part of the discourse, the unconstitutional conditions doctrine would moderate legislative vulnerability to biased, popular beliefs and would encourage more careful analysis of the constitutional issues involved.[126]

The unconstitutional conditions doctrine serves to prevent the government from creating what Professor Kathleen Sullivan terms "a caste system among right-holders."[127] Some constitutional rights are too fundamental to be distributed in creating a hierarchy among

---

121. Coale, *supra* note 13, at 203 n.105.

122. *See supra* notes 16-17 and accompanying text (noting that the proponents of this legislation assert that AFDC creates a class of undeserving poor: those who do not work, those who continue to be dependent upon government largess, and those who continue to have children out of wedlock).

123. Ross, *supra* note 32, at 1507-08; *see* PHILADELPHIA INQUIRER, *supra* note 105 (describing how society would be better off if Norplant were used to reduce the number of children born into poverty).

124. Henley, *supra* note 10, at 753 n.315.

125. Henley, *supra* note 10, at 753 nn.316-17; *see also* Barbara Vobejda, *Gauging Welfare's Role in Motherhood: Sociologists Question Whether 'Family Caps' Are a Legitimate Solution*, WASH. POST, June 2, 1994, at A1 (reporting that social scientists doubt the existence of a causal link between welfare benefits and birth rates).

126. Popular opinions and prejudices are similar to myths: partly true and partly false. *See* Greg J. Duncan & Saul D. Hoffman, *Teenage Welfare Receipt and Subsequent Dependence Among Black Adolescent Mothers*, 22 FAM. PLAN. PERSP. 16, 16 (1990) (showing how the argument that welfare creates dependency and is systemic to future generations can be supported by studies that show black teenage mothers receiving AFDC are more likely to continue to be dependent on AFDC at age 26 than those who do not have children); Table prepared by the U.S. Department of Health and Human Services, Administration for Children and Families, on file with DHS (providing statistics showing that welfare recipients have additional children). *But see Unplanned Pregnancy is Main Cause of Welfare Reliance, Survey Finds*, 13 FAM. PLAN. PERSP. 186 (1981) (noting that studies show an unplanned pregnancy is the main cause of welfare dependency); Henley, *supra* note 10, at 753 nn.316-17 (noting that numerous studies do not support the argument that welfare mothers have additional children to increase their payments and arguing that Norplant legislation was inspired by the myth that women's conduct can be explained by and controlled through their reproductive capacity). *See also* Dorothy E. Roberts, *Punishing Drug Addicts Who Have Babies: Women of Color, Equality, and the Right of Privacy*, 104 HARV. L. REV. 1419, 1438, 1444 (1991) (discussing the myth of the promiscuous black woman which was prevalent from the time of slavery).

127. Sullivan, *supra* note 15, at 1497-99.

classes who would make the same choice except for the government's intervention.[128] Dependency on welfare benefits defines the class. Conditioning the exercise of procreational rights on the receipt of welfare benefits creates a caste constitutionally inferior to those not receiving welfare benefits.[129] This class is deemed unfit to procreate.[130] Additionally, such legislation punishes poor women and their children who do not choose to accept the condition. The unconstitutional conditions doctrine would, however, require the same standard for judging the constitutionality of laws affecting procreational rights, regardless of whether such laws are direct or indirect.[131]

Norplant legislation resembles legislation from the eugenics movement in the first quarter of this century, the premise being that only the biologically fit should procreate.[132] Mandatory Norplant legislation would also deny procreation rights to certain people, i.e., mothers who are AFDC recipients. The development of surgical techniques for sterilization, such as vasectomy and tubal ligation, increased the call for the passage of eugenic sterilization laws in some states[133] because castration could be supplanted by these more aesthetic and humane procedures.[134] Similarly, the attractiveness of Norplant as a contraceptive increased the call for Norplant legislation.[135] Once Norplant is implanted, the state does not depend on the user to control her fertility: the contraceptive stays implanted until surgically removed; it is easy to monitor; it is nearly ninety-nine percent effective; and its effectiveness can last up to five years.[136]

Finally, the doctrine serves as a warning that the government may be unconstitutionally affecting a protected right.[137] As previously discussed, Norplant legislation would affect the fundamental right to procreate.[138] The Supreme Court stated in *Meyer v. Nebraska*[139]

---

128. Sullivan, *supra* note 15, at 1498.

129. Sullivan, *supra* note 15, at 1498.

130. *See* Murdock, *supra* note 66, at 928-32 (discussing how one might perceive the foundation of sterilization laws to be unfitness for parenthood).

131. Sullivan, *supra* note 15, at 1498.

132. Walker, *supra* note 10, at 780 (explaining that the biological "fit" is defined according to the definer). *See* Robert J. Cynkar, Buck v. Bell: *"Felt Necessities" v. Fundamental Values?*, 81 COLUM. L. REV. 1418, 1431-32 (1981) (stating that sterilization laws were also premised on punishment and therapeutic motives).

133. Cynkar, *supra* note 132, at 1433.

134. Cynkar, *supra* note 132, at 1433.

135. *See supra* notes 45-54 and accompanying text (describing the Norplant contraceptive and its appeal to legislators).

136. NORPLANT AND POOR WOMEN, *supra* note 47, at xi.

137. *See generally* Sunstein, *supra* note 79 (noting the government's ability to affect lives through means other than criminal sanctions).

138. *See supra* part III (providing a description of other rights which are affected).

139. 262 U.S. 390 (1923).

that the right to start a family and raise children is constitutionally protected.[140] In *Skinner v. Oklahoma*,[141] the Court stated that marriage and procreation are fundamental to the very existence and survival of the human race.[142] More recently, the Court, in *Carey v. Population Services International*,[143] stated that the choice of whether to beget or bear a child is protected.[144] The right to procreate and its component parts are thus basic personal rights. The doctrine, therefore, allows for an analysis of the type of right involved and the consequences of attaching the condition.[145]

The unconstitutional conditions doctrine has an important role in this discourse. It provides the proper backdrop against which to view Norplant legislation when weighing the extent to which the condition will affect a welfare mother's constitutional rights.

SECTION C. THE APPLICATION OF THE DOCTRINE

Including the doctrine in this discourse helps one see why the standard for determining the constitutionality of Norplant legislation should be the same as if the legislation directly affected a constitutionally protected right. Since the 1930s, the Supreme Court has, as previously discussed, reviewed social and economic legislation under a rational basis standard.[146] This standard provides that the government must show that legislation has a rational relationship to a permissible state interest.[147] Legislation directly affecting constitutionally protected rights is reviewed under a strict scrutiny standard.[148] This standard provides that the state must show a compelling interest for the legislation which must also be narrowly drawn to achieve only that legitimate interest.[149]

---

140. *Id.*

141. 316 U.S. 535 (1942).

142. *Id.* at 541.

143. 431 U.S. 678 (1977).

144. *Id.* at 685.

145. *See infra* note 161 (suggesting that the standard of review under the doctrine should take an equal protection approach). Although equality issues almost always arise with unconstitutional conditions, the doctrine centers on the right that is affected. *Id.*

146. *See* Dandridge v. Williams, 397 U.S. 471 (1970) (upholding a rational basis standard for evaluating welfare classifications); Lyng v. Castillo, 477 U.S. 635 (1986) (upholding a provision of the Federal Food Stamp Act under rational review and rejecting the argument that this classification is subject to heightened scrutiny).

147. *Dandridge*, 397 U.S. at 485-86.

148. *See* Santosky II v. Kramer, 455 U.S. 745 (1982) (affirming the need for a compelling state interest when deciding whether to terminate parental rights).

149. *See* Consolidated Edison Co. v. Public Serv. Comm'n, 447 U.S. 530 (1980) (holding that a regulation suppressing controversial inserts in utility bills is not justified by a compelling state interest).

The Supreme Court has never held that there is a constitutionally protected right to welfare benefits.[150] Since the government has no obligation to provide welfare benefits, welfare legislation is usually reviewed under a rational basis standard.[151] As previously discussed, among the fundamental rights that Norplant legislation affects is the right to procreate.[152] Norplant legislation, therefore, may be reviewed under this minimum level of scrutiny since it does not directly affect a constitutionally protected right.

The unconstitutional conditions doctrine provides that the government should not be able to do indirectly what it cannot do directly.[153] The Supreme Court has not, however, applied the doctrine as it is literally stated.[154] That is, any condition indirectly affecting a protected right will be reviewed as if its impact were direct.[155] Legislation directly affecting a protected right, such as the right to procreate, would be reviewed under a strict scrutiny standard. Instead, the Court has sought to determine which conditions affecting constitutional rights should come under the doctrine.

Furthermore, the Court's approach in determining which conditions trigger the doctrine has been inconsistent. The Court uses

---

150. *See* Dandridge v. Williams, 397 U.S. 471 (1970) (upholding a Maryland AFDC program provision imposing a maximum monthly grant per family as constitutionally valid).

151. The Court has applied different standards of scrutiny based on the legislation involved. *See* Mississippi Univ. for Women v. Hogan, 458 U.S. 718 (1982) (explaining that although the legislation discriminates on the basis of gender, current law provides for review under a standard lower than strict scrutiny); *see also* Yick Wo v. Hopkins, 118 U.S. 351, 373-74 (1886) (holding that discrimination which is based on race is subject to a strict scrutiny standard of review); *but see* Washington v. Davis, 426 U.S. 229 (1976) (holding that legislation which has a disparate impact on race does not violate the Equal Protection Clause unless objective evidence shows that the law was purposefully designed to discriminate invidiously on the basis of race). The Supreme Court has been reluctant to find a violation of the Equal Protection Clause unless the legislation facially discriminates on the basis of race. *Id.* *See also* Henley, *supra* note 10, at 148 n.271 (explaining that Norplant legislation is facially neutral as to race, but disproportionately affects minorities since there is a high percentage of minority women who have families below the poverty line); David Robert Baron, *The Racially Disparate Impact of Restrictions on the Public Funding of Abortion: An Analysis of Current Equal Protection Doctrine*, 13 B.C. THIRD WORLD L.J. 1 (1993) (describing statutory restrictions on abortion as a form of government discrimination against minorities and a violation of equal protection laws).

152. *See supra* notes 64-66 and accompanying text (discussing Norplant's relationship to fundamental rights, such as procreation, parenting, contraceptive use, and other personal decisions which should remain free from governmental interference).

153. *See* Charles R. Bogle, *"Unconscionable" Conditions: A Contractual Analysis of Conditions on Public Assistance Benefits*, 94 COLUM. L. REV. 193 (1994) (discussing government methods for dealing with undesirable behavior in society).

154. *See supra* text accompanying note 111 (noting the Supreme Court's historical position on applying the doctrine to social and economic legislation).

155. *See* Sullivan, *supra* note 15 and accompanying text (explaining that the doctrine must be interpreted in a way that prevents government from indirectly impinging on protected rights).

theories of germaneness,[156] inalienability,[157] and coercion[158] to determine whether a particular condition triggers the doctrine. The Court has most frequently used various reformulated versions of the coercion theory.[159] Presently, the Court appears to be approaching the doctrine from the standpoint of government's presumptive power to allocate resources according to its selections. There is no coercion if the allocation results as a nonsubsidy instead of a penalty.[160]

Commentators have tried to reconcile the various approaches of the Supreme Court and have concluded that there is no reconciliation.[161] One is not able, therefore, to predict with any certainty whether the Supreme Court would determine that the unconstitutional conditions doctrine applies to the Norplant legislation.[162]

---

156. Robert L. Hale, *Unconstitutional Conditions and Constitutional Rights*, 35 COLUM. L. REV. 321 (1935) (noting that the less germane the purpose of the condition is to the underlying benefit, the more it looks like an unconstitutional condition). *See* Nollan v. California Coastal Comm'n, 483 U.S. 825 (1987) (holding that the Commission cannot condition the issuance of a building permit on reasons which are not germane, even to an outright denial of the permit).

157. Sullivan, *supra* note 15, at 1477-89 (noting that some rights, such as procreational ones which are inherent and biological, should not be traded because they are central attributes of personal identity).

158. Sullivan, *supra* note 15, at 1428-56 (noting that the coercion theory identifies unconstitutional conditions as those that have a coercive effect on the individual's exercise of his/her right).

159. *See* Sullivan, *supra* note 15, at 1428-56 (describing the Court's inconsistencies in determining what constitutes coercion). The Supreme Court has not, however, developed a paradigm for determining coercion. It has focused at times on whether the condition acts as a penalty, at other times on whether the condition acts as a deterrent, and at other times on the coercive effect of making the choice. *Id.*

160. Winters, *supra* note 78, at 132-33 (emphasizing the Court's change from a penalty focus to one of encouraging or discouraging alternative choices). The government can choose to subsidize or not to subsidize. *Id. See also* Rust v. Sullivan, 500 U.S. 173 (1991) (noting that funding to subsidize non-abortion family planning instead of abortion-related family planning is a legitimate choice); Harris v. McRae, 448 U.S. 297 (1980) (stating that funding to subsidize childbirth expenses instead of abortion expenses is a legitimate choice).

161. *See, e.g.*, Baker, *supra* note 15, at 1196-97 (proposing some theories on when to apply the doctrine); Epstein, *supra* note 83, at 6-14 (referring to the unconstitutional conditions doctrine as "mysterious" since it has different meanings in different contexts); Sullivan, *supra* note 15, at 1415-16 ("[R]ecent Supreme Court decisions on challenges to unconstitutional conditions seem a minefield to be traversed gingerly. Just when the doctrine appears secure, new decisions arise to explode it."). *See also* Sullivan, *supra* note 15, at 1491-99 (focusing on the effect of the doctrine being raised when a condition affects any of the following: private ordering, government evenhandedness, and constitutional caste). Professor Sullivan has rejected the theories of coercion, germaneness, and inalienability as unhelpful. *Id. See* Kreimer, *supra* note 78, at 1359-74 (suggesting that an unconstitutional condition arises if predictive, historical, and equality baselines show the condition penalizes the exercise of a constitutional right); Patricia M. Wald, *Government Benefits: A New Look at an Old Gifthorse*, 65 N.Y.U. L. REV. 247, 256 (1990) (suggesting an equal protection type of analysis, applying a heightened standard of review, where the state would have to show that the condition was substantially related to important purposes since a condition that limits a constitutional right creates a suspect classification); Gary Feinerman, *Unconstitutional Conditions: The Crossroads of Substantive Rights and Equal Protection*, 43 STAN. L. REV. 1369 (1991) (expanding on Judge Wald's approach).

162. *See* Baker, *supra* note 15, at 1187 (writing that the Supreme Court articulates a positive rule for unconstitutional conditions cases that involve public assistance programs). According

This article posits that the unconstitutional conditions doctrine should be applied to Norplant welfare legislation as if the legislation were directly affecting a woman's procreational rights.[163] The purpose of the doctrine is to prevent the government from doing indirectly what it cannot do directly regarding constitutional rights.[164] An analysis of whether the doctrine applies ends by determining if a condition is affecting a constitutionally protected right. Further analysis beyond this determination defeats the purpose of the doctrine and creates the possibility that identical affected rights will be judged by different standards.[165]

There is risk to individual freedom, as this article argues, in allowing the government to use conditions to do indirectly what it cannot do directly. To minimize this risk, it is crucial that one standard be applied to legislation affecting identical constitutionally protected rights. This approach is the fair way to determine the constitutionality of Norplant legislation. It protects the rights of the individual from government's indirect regulations.

It is only when the appropriate standard has been identified, and is applied, that the constitutionality of the condition should be judged. A welfare reform statute should stand or fall under the same standard as if its effect were direct. This approach, however, does not mean that the application of the same standard to direct and indirect legislation would yield the same result.[166] It only means that

---

to Baker's theory, if the Court determines that a constitutional right is involved, it then determines whether the condition requires those unable to earn a subsistence income — who are otherwise eligible for AFDC benefits — to pay a higher price to exercise the right than those earning an income. *Id.* *See also* Coale, *supra* note 13, at 193, 208 (concluding that Norplant incentive bonuses do require poor women to pay a higher price to exercise their constitutional rights).

163. Sunstein, *supra* note 79, at 595 (writing that the doctrine is an anachronism and should be abolished). Instead, the constitutionality of conditions should be "an approach that asks whether, under the provision at issue, the government has constitutionally sufficient justifications for affecting constitutionally protected interests." *Id.* It is not clear, however, whether "constitutionally sufficient justifications" are the equivalent to this article's approach: namely, that the standard for evaluating the constitutionality of the condition should be the same standard that would be used if the legislation directly affected the protected interest.

164. *See* Sullivan, *supra* note 15 and accompanying text (asserting that an individual should not change his/her constitutionally protected behavior or choices because of governmental interference or conditions).

165. For example, if the legislation required women to use Norplant as a condition to receiving AFDC, and if the legislation were determined to be merely social and economic legislation, the constitutionality of the condition would be judged under the rational basis standard. *See* Dandridge v. Williams, 397 U.S. 471 (1970). If the legislation required all women to use Norplant, it would be judged under the strict scrutiny standard. Both laws would affect procreational rights. *See supra* note 77 and accompanying text (discussing the application of strict scrutiny to the parallel issue of abortion rights).

166. It is during the application of the standard that the court weighs such interests of the state in Norplant legislation against the curtailment of the individual's right. For example, the state's interests in Norplant legislation may vary from increasing the health of poor women, to

legislation affecting the same right would be reviewed under a common standard. In this way, protected rights would be judged and treated alike. The constitutionality of Norplant legislation should, therefore, be judged by the same standard as if the legislation directly affected constitutionally protected rights.

CONCLUSION

As the government has become more powerful in its regulation of benefits in American life, the conditions it imposes in exchange for benefits become extremely significant. This is especially true regarding AFDC benefits. These benefits provide subsistence income for poor mothers and their children. The options for such families to obtain subsistence income elsewhere are negligible. Under such circumstances, the power of the government to impose conditions on AFDC benefits increases while the recipients' other options, if they choose to forego the benefit, are reduced. Moreover, the pervasive potential of the government's use of Norplant to influence reproductive rights critically increases the significance of conditioning AFDC benefits. The unconstitutional conditions doctrine serves to prevent possible overreaching when constitutional rights are affected. The doctrine should be part of the discourse regarding welfare benefits and Norplant legislation, for it will serve as a constant reminder that the government should not be able to do indirectly that which it cannot do directly.

---

ecological concerns of population control, to the welfare of unborn children, to the judicial recognition that "states have considerable latitude in allocating their AFDC resources." Dandridge v. Williams, 397 U.S. 471, 478 (1970). These interests may or may not be compelling when the standard is applied to the legislation, whether direct or indirect. And, applying the standard to Norplant bonus legislation may give a different result than when applying the standard to Norplant legislation mandating its use to all recipients. What is crucial, however, is that the same standard is being applied to legislation affecting identical constitutionally protected rights.

COMMUNITARIANISM AND FEMINISM: THE CASE AGAINST THE PREFERENCE FOR THE TWO-PARENT FAMILY

*Susan B. Apel**

I. Introduction

The conversation has become deafening. Political scientist Charles Murray wants to bring back the social stigma that used to attach to unwed mothers and place their babies in orphanages.[1] Academics write scholarly articles that portray rational[2] and economic[3]

* Professor, Vermont Law School; J.D., Northeastern University School of Law, 1977. In addition to my career as an academic, I practiced law for approximately twelve years. The subject matter of my practice varied, but always contained a substantial family law component. I wish to thank Claudia Comstock and Melissa Crawford for their assistance in the production of this article. I am also indebted to readers of various drafts and to many colleagues and friends who forbore my conversations on this subject. These include Richard O. Brooks, Sheldon Novick, and Celia Campbell-Mohn who issued the invitation to join the Vermont Law School faculty discussion group on the meaning of community. Finally, I wish to thank Cecil Reniche and the Vermont Law School chapter of the National Lawyers' Guild for inviting me to present an earlier draft of this work.

1. While Charles Murray's idea first appeared in his book, Losing Ground (1984), the most recent flurry of media attention appeared to result from an op-ed piece he authored for the Wall Street Journal. Charles Murray, *The Coming White Underclass*, Wall St. J., Oct. 29, 1993, at A14. Many columnists and others have responded, often intertwined with commentary about the current welfare reform debate. *See, e.g.*, Mary McGrory, *Reducing Reform to Baby Talk*, Wash. Post, July 19, 1994, at A2; Barbara Vobejda, *Conservative Welfare Idea Criticized; Bills Would Cut Funds for Unwed Mothers*, Wash. Post, June 24, 1994, at A18. Murray is not alone; on this point, he travels in the company of, among others, former Secretary of Education William J. Bennett. *See* William J. Bennett, *The Best Welfare Reform: End It*, Wash. Post, Mar. 30, 1994, at A19.

2. Elizabeth S. Scott, *Rational Decisionmaking About Marriage And Divorce*, 76 Va. L. Rev. 9 (1990).

reasons for marriage over divorce, while Christopher Lasch argues for a constitutional amendment to ban divorce in those marriages with children.[4] New Democrat Bill Clinton ponders restructuring welfare so as to remove what he believes to be incentives for single mothers to have more babies. Even liberal Health and Human Services Secretary Donna Shalala pontificates: "I don't like to put this in moral terms, but I do believe that having children out of wedlock is just wrong."[5]

Never, it seems, have the mass media, politicians, academics, leaders of task forces and think tanks vied so energetically with one another to decry the present state of the American family. And never, despite the continuing evolution of the family from the traditional patriarchy of the 1950's to more diverse forms of intimacy, have unmarried, single, and/or homosexual parents been found so absolutely wanting in the proper means of raising children. And finally, never have those who decry been so adamant or so abundant in their agreement that the solution is clear—point the society back to its traditional patriarchal, nuclear family structure by whatever modes of carrots and sticks necessary, and the nation, including children, will be the better for it.

During the past three decades, feminists have been gathering in pockets on the edges of this conversation and have voiced their own concerns about the status and problems of families in our present social and legal culture. Particularly in the areas of divorce reform and family leave, feminist academics and legal practitioners have spoken loudly, and sometimes effectively, about the need to take family matters seriously. In doing so, they have succeeded in reframing family issues as part of our public rather than just our private selves and have thereby caused issues such as domestic violence, child and elder care, and the economic inequities of divorce to be seen as candidates for legal reform.

At least one strand of feminist legal and sociological culture has developed another, and simultaneous, focus equal in importance to its concern with families. The work of Carol Gilligan[6] and others has engaged the feminist mind with an inquiry about the benefits and burdens of individual autonomy versus connectedness.[7] In feminist

3. Martin Zelder, *The Economic Analysis of the Effect of No-Fault Divorce Law on the Divorce Rate,* 16 HARV. J.L. & PUB. POL'Y 241 (1991).

4. Christopher Lasch, *Who Owes What to Whom?,* HARPERS, Feb. 1991, at 47-48.

5. Joe Klein, *The Out-of-Wedlock Question,* NEWSWEEK, Dec. 13, 1993, at 37.

6. CAROL GILLIGAN, IN A DIFFERENT VOICE: PSYCHOLOGICAL THEORY AND WOMEN'S DEVELOPMENT (1982).

7. The following are illustrative, but certainly not exhaustive, of articles on this theme: Leslie Bender, *A Lawyer's Primer on Feminist Theory and Tort,* 38 J. LEGAL ED. 3 (1988); Susan G. Kupfer, *Autonomy and Community in Feminist Legal Thought,* 22 GOLDEN GATE U. L. REV. 583 (1992); Linda C. McClain, *"Atomistic Man" Revisited: Liberalism, Connection and Feminist Jurisprudence,* 65 S. CAL. L. REV. 1171 (1992); Carrie Menkel-Meadow, *Portia in a Different Voice: Speculations on a Woman's Lawyering Process,* 1 BERKELEY WOMEN'S L.J. 39 (1985); Robin L. West, *Jurisprudence and Gender,* 55 U.

jurisprudential thought, one ponders the law's relationship to the dichotomy of liberalism with its emphasis on individual rights and what is sometimes described as an ethic of care, based on the concepts of responsibilities and relationships with others. While the debate shows no sign of waning, it is fair to say that many feminist academics and others have offered potent criticism of the law's focus on individual rights and have urged that the law reframe disputes in ways that recognize, and perhaps strengthen, the ties that bind us as members of a community.

This article explores what happens when feminism's concerns about families and its quandary about the balance between individuals and communities meet the deafening conversation that hopes to persuade us to define families in more, rather than less, restrictive ways. The conversation is not only loud, it is ubiquitous, emanating from the mouths of people in a variety of fields and, perhaps more interestingly, from liberals and conservatives alike.

I have chosen not to focus on the extreme, right-wing rhetoric that feminists might find so easy to denounce. Rather, I have opted for a feminist critique of one aspect of communitarianism. I have done so for several reasons. First, while communitarianism styles itself as drawing from all points on the political spectrum, it is probably seen as more liberal than conservative (perhaps best described as progressive) and not what most feminists find, on the whole, politically repulsive. Second, while the term "communitarianism" has yet to reach the status of the proverbial household word, it shows no evidence of extinction. Robert Bellah's books, such as *Habits of the Heart*, continue to sell; Amitai Etzioni has found his way to the popular media, e.g., The MacNeil-Lehrer News Hour, to promote his new book, *The Spirit of Community: Rights, Responsibilities, and the Communitarian Agenda.* I was introduced to communitarianism by *The Responsive Community*, a scholarly journal devoted to the communitarian viewpoint, which continues to publish. In addition, this is a movement that has made its way into the White House in the person of William Galston, self-styled communitarian and adviser to the President.

A third reason for selecting communitarianism as representative of the many components of the conversation about families is that it has defined itself and its views on this subject in a clear and systematic fashion. It has developed a written platform[8] with signatories, as well

---

CHI. L. REV. 1 (1988); Joan C. Williams, *Deconstructing Gender*, 87 MICH. L. REV. 797 (1989); Anne C. Dailey, *Feminism's Return to Liberalism*, 102 YALE L.J. 1265 (1993) (book review). *See also* ELIZABETH FOX-GENOVESE, FEMINISM WITHOUT ILLUSIONS (1991).

8. Communitarian Network, *The Communitarian Platform: Rights and Responsibilities*, 2 RESPONSIVE COMMUNITY: RTS. AND RESPS. 4 (1991-92).

as a position paper on the family.[9] The platform stands for and fuses together the views of dozens of its signatories; thus, it speaks for a movement, rather than for any particular individual. Platform signatories and luminaries, including Robert Bellah, Jean Bethke Elshtain, and Mary Ann Glendon, have written individually in the communitarian vein as well. For purposes of this article, however, I have chosen to focus on the clear and unified expression represented by the movement's platform.

However, the main reason for focusing on communitarianism is a fondness for paradox. The communitarian movement shares these two interests—the importance of families, and the focusing on our collective, rather than our individual, selves—with those of at least some feminists. Therefore, one might expect the communitarian position on the family to be compatible with feminism. The two seem a natural alliance, sharing ground in a developing feminist view of families and the law. Unfortunately, once the initial fascination is over, the communitarian view of family strikes me as the fish that has worked its way into feminism's kitchen—after the proverbial three days and with due consideration, it has started to produce an offensive odor.

Various aspects of the communitarian platform may be either more or less appealing to feminist sensibilities. The most dangerous to women, however, and the focus of this paper is the platform's unfortunate insistence on the primacy of the two-parent family. The drafters of the Communitarian platform appear blind to the difference that gender makes to one's role in nuclear families. In addition, its certainty that two-parent families are necessary for rearing children in a positive way is misplaced.

II. Why the Preference for the Two-Parent Family is Harmful to Women

The drafters of the platform are clear and direct in declaring their preference for the two-parent family, claiming that both children and society are better served when there are two parents in the household. The drafters of the position paper, perhaps in anticipation of criticism that theirs is a retrospective or nostalgic view of the family of the 1950's, add that they "do not seek a return to traditional families, to discrimination against women, or the notion that parenting is only a mother's job. We strongly support, and recognize as morally correct, the fundamental equality of fathers and mothers: both command the same basic rights and responsibilities."[10] How then, in

9. Communitarian Network, The Communitarian Network and The Responsive Community: Rights and Responsibilities, A Communitarian Position Paper on the Family (1993).

10. *Id.*

light of such explicit and enlightened sentiment, can a stated preference for two-parent families harm women?

Nothing in the specific language of the communitarian platform speaks of marital status; one might presume, then, that the unmarried two-parent family would suffice as well as the married version. A fair reading of the rest of the platform and position paper on the family, however, yields a different result. Reminiscent of the early and continuing sentiments of Daniel Patrick Moynihan,[11] the paper's references to *unwed* motherhood, juxtaposed with the notions of stability and healthy functioning, clearly remove the unmarried parent from the preferred two-parent category.

> The most important single indicator of childhood problems—from poor health to poverty to behavioral problems to school drop-out to criminality—is whether or not a child grows up in a stable, functioning family. *With out-of-wedlock birth rates approaching 80 percent in some areas,* we can anticipate a further rise in all these negative consequences unless something is done soon.[12]

Other parts of the platform and position paper speak to the need for the parents to remain *married*; additionally, the focus on *divorce reform* indicates that the 1990's communitarian two-parent family shares at least one characteristic of the 1950's traditional family: that marriage, not merely cohabitation, is an implicit prerequisite for children.

This presumption, both stated and implied, forms a definition of family (otherwise conspicuously absent from the platform) that excludes parents who are *unable* to marry—most notably, same-sex partnerships.[13] Despite its protestations to the contrary, this exclusion by communitarians of lesbian and gay partners is anything but subtle. In a letter to the author reviewing an earlier draft of this manuscript for publication, the editors of *The Responsive Community*, the journal of the communitarian movement, responded to the claim that they were biased against homosexual parenting by stating that "nowhere in either document [the communitarian platform and the position paper on the family] do we disapprove of homosexual marriages or relationships."[14] It is at best unfortunate that the editors do not view their repeated references to "marriage" (which, absent another specific def-

11. Martha Albertson Fineman, *Legal Stories, Change, and Incentives—Reinforcing the Law of the Father*, 37 N.Y.L. Sch. L. Rev. 227, 243 (1992).

12. Communitarian Network, *supra* note 9, at (emphasis added).

13. As of the writing of this article, same-sex marriages are banned in every state. In Baehr v. Lewin, 852 P.2d 44 (1993), the Supreme Court of Hawaii held that sex is a suspect category for purposes of equal protection analysis under the Hawaii Constitution and that the statutory prohibition of same-sex marriages must be subject to the more rigorous level of judicial review, or strict scrutiny. Whether the Hawaii statute will ultimately be upheld or struck down is unknown, nor can one predict all of the consequences if same-sex marriages become legal in one state of fifty.

14. Letter from Sarah Horton, Assistant Managing Editor, *The Responsive Community* (Apr. 19, 1993) (on file with author).

inition, must be taken in its traditional legal sense) and to the terms "mother and father" as excluding homosexual persons, for whom marriage is legally impossible and for whom the pairing of "mother and father" is not descriptive. Surely the drafters of the position paper do not need to be reminded of the power, and the politics, of language. Sandra Pollack may be too kind and understated when she questions the inclusivity/exclusivity of the term "mother," in saying that "for lesbians, the whole question of naming and the terms we use to identify ourselves is more complex than it might be for straight women."[15] Additionally, in his book *The Spirit of Community*, Amitai Etzioni, the founder of communitarianism, indulges in a curious use of quotation marks when he says "Radical Individualists use these dismal statistics [showing a decline in the number of nuclear families] to bolster their argument that the nuclear family cannot be resurrected because it has been replaced by a wide variety of other 'families,' from single-parent households to gay couples."[16] What can such a use of punctuation mean if not that while radical individualists may consider gay couples to be "families," they are not; to call them so is an abuse of the term. In any event, one questions why, if communitarians "do not disapprove" of homosexual relationships, they do not just say so. Why not explicitly and affirmatively state that sexual orientation does not disqualify gay and lesbian partners from inclusion in families that are acceptable for healthy child-rearing?

More to the point, it is already long past the time to consider whether homosexual or lesbian parents are "as good as" their heterosexual counterparts; perhaps they are better, or at least capable of contributions to a child's development in unique ways. Pollack notes the existence (although paucity) of research on the possible benefits of being a child of a lesbian mother and quotes poet Audre Lorde, who states:

> There are certain basic requirements of any child—food, clothing, shelter, love. So what makes our children different? We do. Gays and Lesbians of Color are different because we are embattled by reason of our sexuality and our Color, and if there is any lesson we must teach our children, it is that difference is a creative force for change, that survival and struggle for the future is not a theoretical issue. It is the very texture of our lives . . . .[17]

While this exclusion of same-sex partnerships alone should give women pause, the preference for the married, two-parent family may be even more dangerous to heterosexual women who become mothers. The legal availability of marriage makes unmarried motherhood seem even more suspect. Lesbians and gays are currently pre-

15. Sandra Pollack, *Lesbian Parents: Claiming Our Visibility*, *in* FEMINIST FRONTIERS III 263, 263 (Laurel Richardson & Verta Taylor eds., 1993).

16. AMITAI ETZIONI, THE SPIRIT OF COMMUNITY: RIGHTS, RESPONSIBILITIES, AND THE COMMUNITARIAN AGENDA 72 (1993).

17. Pollack, *supra* note 15, at 267.

vented by law from marrying their chosen mates.[18] Heterosexual women who do not marry may be guilty of the sin of choice.

While the vast majority of parents (perhaps even a greater majority of single mothers) would not dispute the communitarian logic that the availability of more hands makes child rearing easier, to require, or even to prefer, marriage as the purchase price for the extra hands may serve the interests of some, but not all and maybe not even most, women.

To dispense with the easiest example, there are the women/mothers represented by the overly-maligned, fictional television character, Murphy Brown.[19] It does not matter at this point that most single mothers do not have the resources depicted in this fictional situation. Some women do, and their rarity[20] is not just cause for dismissing them or their concerns. Amitai Etzioni, in response to criticism that his model family could best be described as "yuppies," responded that "[w]ell, first of all, they too have children,"[21] a point I do not dispute. But some yuppies are single, and "they too have children" or want them. There are a significant number of women, many approaching the end of their childbearing years, who are unmarried and desire to have a child, either biologically or by adoption. Factors that tend to characterize this group of women could, but will not always, make up for the missing pair of hands that marriage is reputed

18. This does not imply that legal prohibitions of same-sex marriages are the only, or even the primary reasons, why same-sex couples do not marry. Within the lesbian and gay communities, views of marriage are as divergent as in groups of heterosexual persons. For a classic debate on this issue from the gay and lesbian viewpoint, *see* Paula L. Ettelbrick, *Since When Is Marriage a Path to Liberation?*, OUT/LOOK, Fall 1989, at 9; Thomas B. Stoddard, *Why Gay People Should Seek The Right to Marry*, OUT/LOOK, Fall 1989, at 9. The text of their arguments has been incorporated in several textbooks and anthologies, including LESBIANS, GAY MEN, AND THE LAW 398-405 (William Rubenstein ed., 1993).

19. Murphy Brown is the title character in a CBS television comedy. She is portrayed as a single woman who is employed as a reporter for a nationally-acclaimed news magazine program. She is highly educated, economically independent, presumably well-paid, and is able to afford a private nanny to care for the child she decided to have out of wedlock. This fictional character became news when then Vice President Dan Quayle, in a speech delivered in San Francisco, sought to explain and condemn the riots that erupted in Los Angeles following the acquittal of police tried for the beating of Rodney King. Claiming that the civil unrest resulted from "a poverty of values," Quayle said, "[i]t doesn't help matters when prime time TV has Murphy Brown—a character who supposedly epitomizes today's intelligent, highly paid, professional woman—mocking the importance of fathers by bearing a child alone and calling it just another 'lifestyle choice.' " Christopher Connell, *Quayle Blames the Rioters for the Los Angeles Riots*, CLEV. PLAIN DEALER, May 20, 1992, at A1.

20. One estimate is that births to Murphy Brown-type mothers, defined as white women with college degrees, account for four percent of all births to unmarried women. Stephanie Mencimer, *The Welfare Mommy Trap; Want the Poor to Have Fewer Kids? Fund Abortions*, WASH. POST, Jan. 9, 1994, at C1 (quoting U.S. Census Bureau statistics). *See also* William Raspberry, *That Disturbing Charles Murray*, WASH. POST, Dec. 1, 1993, at A25.

21. ETZIONI, *supra* note 16, at 67.

to provide: maturity and experience; greater financial resources; better planning for parenthood; availability of other child care arrangements; and, not the least, the conscious, deliberate choice in favor of parenthood. What do we tell these women? If we think we can dissuade or prevent them from becoming mothers simply because there is no man in the house, recent census figures have proven us wrong.

The Murphy Browns are the easiest category, however, and admittedly, one could make the case that their numbers, though significant, are dwarfed by what many on all points of the political continuum view as the real problem: unwed teenage (read: poor) mothers. My guess is that the reference in the position paper to "eighty percent out-of-wedlock births in some areas" really means the fourteen-year-old unmarried mother in the inner city.[22] The specter of such teenaged children attempting, and often failing, to provide even the rudiments of basic child care to their infants is a horrible one. Virtually everyone recognizes the problems generated by the infliction of parenthood on young and poor women. We ache for the babies that are ill-cared for and often severely neglected and abused. The question is not, then, whether motherhood is a good idea for poor teenagers and their offspring; with rare exception, it isn't.

The question is this: how will the marriage of parents in these cases better a bad situation? The mostly-teenaged fathers of these babies are no better equipped to parent than the mothers; most can offer little, if any, financial support and sadly (like some fathers everywhere) many have no interest in parenthood at all. Can we honestly suppose that "marrying the young woman off" to the father of her baby will solve her parenting problems? Does not the encouragement of marriage, in addition to unprepared-for parenthood, exacerbate rather than help solve the problem? Will marriage take what is usually

22. A reviewer of an earlier draft of this article made the point that "race is often part of the equation for unwed teenage (read: poor) mothers." I have purposefully decided not to tease out all of the ways in which race might make a difference to my argument for several reasons. First, my research indicates that the issue of births to unwed women of various racial backgrounds is a complex one. Within the African-American community alone, for example, there are many viewpoints on this subject, some of them conflicting. *See* Regina Austin, *Sapphire Bound!*, WISC. L. REV. 539 (1989) (casting teenage pregnancy in a somewhat more positive light and arguing that the African-American community has been historically tolerant of single teenage motherhood). *But see* Faye Wattleton, *Teen-Age Pregnancy: The Case for National Action*, THE NATION, July 24, 1989, at 138-139 (stating that teenage motherhood is "both cause and consequence of a host of social ills," and is "more a product of class than race"). Moreover, race cannot always be divided from ethnicity or nationality or culture. For an excellent discussion of what the author calls "Afrocentric and Eurocentric traditions" in family structure, *see* Marilyn Bowens, *The Cleaver Paradigm as an Anachronism in a Multi-Cultural Society*, 7 WISC. WOMEN'S L.J. 25 (1992-93). *See also* Dorothy E. Roberts, *Racism and Patriarchy in the Meaning of Motherhood*, 1 AM. U. J. GENDER & L. 1 (1993). The high rate of teenage pregnancy is often touted as a peculiarly American (independent of race) phenomenon, at least among industrialized nations.

a miserable situation and somehow transform it into a healthy family life for all concerned?[23]

The problem with these out-of-wedlock births is not that the home lacks a father, but that the mother is herself still a child, unprepared for parenthood and lacking in child rearing skills. Add to that the reality of poverty that provides for insufficient everything—food, heat, education, medical care, plumbing, simple sleeping space—and that extra pair of hands, attached as they may be to another teenager facing the same grinding poverty and who may, or may not, want to be there, has little power to strengthen the family.

This brings us to what may be the largest segment of women who are or who want to be mothers—adult women who are neither remarkably poor nor remarkably wealthy. Given that many of these women would themselves voice a preference for marriage and for a two-parent family, why should the platform's same preference cause concern to women?

This question may slam us up against the hardest truth, and one that many of us, women included, find difficult, if not impossible, to face. There is something wildly and irrepressibly comforting about the picture of the happy, traditional family, either because many of us grew up in one, or perhaps because many of us did not. Regardless of which reality was ours, we have all been imprinted with the specter of family as meaning one kind of family, where each parent (one of each sex) loves the other and both love their biological children, where all problems (mostly minor and often portrayed as comic) are resolved quickly and to the satisfaction of all. It is an infectious image, and one from which we feel we have strayed and to which we long to return.

Communitarians, however immune they may try to hold themselves from such nostalgia, partake of a vision of the family that is essentially traditional, with a paper-thin 1990's veneer. Who, if not Murphy Brown, might serve as the ideal communitarian family? Why, the Huxtables, of course. Dr. Etzioni states: "I am sorry that Dan Quayle stirred up such a fuss about single mother Murphy Brown; it would have been more positive to point to the Huxtables. Here are two professionals, a lawyer and a doctor, both working full-time but very dedicated to their children and one another."[24]

Despite its success as entertainment, The Cosby Show suffered from a considerable amount of criticism that its portrayal of family life

23. It appears as if Amitai Etzioni has subsequently tried to distance himself from the position that would demand marriage of teen parents. He chides Charles Murray and others whom he calls "arch-conservatives" for insisting that marriage is the answer to teen pregnancy and welfare problems. He claims that Murray and others are really preaching abstinence, while he favors more education about contraception. Amitai Etzioni, *Marrying Off Teens Won't Solve the Welfare Problem*, PHILADELPHIA INQUIRER, July 26, 1994, at A11. Perhaps Dr. Etzioni should call for a revision of the communitarian platform to make this point clear.

24. ETZIONI, *supra* note 16, at 67.

was not particularly realistic, although it probably was not the worst offender in this regard. While a feminist rhetorical analysis of the show is beyond the scope of this article, my own view is that it was a sort of the "Father Knows Best" of the 1990's, an enlightened and often amusing patriarchy, but a patriarchy all the same—from the opening credits in which the father is the central figure, surrounded by his dancing wife and children (too trivial? picture the same scene with gender roles reversed), to the plot lines of one more marriage in which the wife polices her husband's diet. More to the point, the parents seldom *actually* worked; they were known to have careers that were peripheral to the setting of the show, neither of which made real demands on the time they had to devote to parenting. In short, they were professional parents, the kind that no family, married, gay or single, could possibly afford to be, but who can sell nostalgia as surely as June Lockhart in her kitchen in Lassie.

That such a traditional image of the family, rooted in the 1950's mind-set, was more fantasy than reality seldom intrudes upon our consciousness. In her book, *The Way We Never Were*, Stephanie Coontz finds it necessary to remind us that "contrary to popular opinion, 'Leave It to Beaver' was not a documentary."[25] That 1950's image of the family, however true or (probably) false in past decades, is now no longer true or attainable for significant numbers of people. While some women marry, have children, and exist in happy, healthy, nuclear families, just as many, if not more, do not.

As but one example of the problems women face in marriage, and a crucial one, is the obvious: in the past decade, statistics about severe physical abuse of women by their mates have been replicated in study after study. Forty-one percent of all women who are murdered are murdered by their husbands. Recently, columnist Anna Quindlen summarized the most recent and widely reported study from the Surgeon General of the United States: "The greatest public health threat for many American women is the men they live with. The Surgeon General has said that violence is the leading cause of injury to women between the ages of fifteen and forty-four, more common than car accidents and cancer deaths combined."[26]

Despite our preoccupation with public violent crime and

> the fact that Americans suffer from a murder rate that is four times higher than Europe's and a rape rate that is seven times higher, "the home is actually a more dangerous place for American women than the city streets." Each day, four women are killed by their male partners; in 1989, more women were abused by their husbands than got married in the same period.[27]

---

25. STEPHANIE COONTZ, THE WAY WE NEVER WERE: AMERICAN FAMILIES AND THE NOSTALGIA TRAP 29 (1992).

26. Anna Quindlen, *Give Abused Women a Sporting Chance on Super Sunday*, CHI. TRIB., Jan. 19, 1993, at 15.

27. COONTZ, *supra* note 25, at 3 (quoting, in part, the U.S. Surgeon General).

Another study, by the Senate Judiciary Committee, said that over one million women *report* abuse each year.[28] While the numbers contained in these and other reports never fail to stun the reader (one hopes), the real shock comes with the recognition that studies measure only those incidents of abuse that are reported. Most experts agree that the majority of domestic crimes against women are *un*reported. Given that the reported numbers have caused the incidence of domestic violence against women to be labeled (by Quindlen and others) as an "epidemic,"[29] what would we call this phenomenon if it included *all* severe physical abuse? What should we call it if, in addition, *less* severe forms of physical abuse were added? To what shameful number would we have to refer if *all* abuse, emotional and psychological as well as physical, were tallied? No one knows, but it's a big number, too big to allow us to think that the threat of abuse lurks only in a few, easily identifiable, bizarre marriage relationships.

While the law's response to this violence has in some ways evolved to protect women, the reality is that the law has failed in many respects. At its best, it can be responsive to incidents of domestic abuse, but because it is the nature of most law to act after the fact, it can do little to prevent the problem. In the domestic arena, sometimes it cannot even act after the fact. One most heinous aspect of the law is that incidents of violence that are sexual in nature may be altogether beyond the reach of the law. In many states, it is *still* legal for a man to rape a woman to whom he is legally married.[30]

---

28. *Hearings on S. 102 Before the Comm. on the Judiciary*, 101st Cong., 2d Sess. 117 (1990); S. REP. NO. 102, 102d Cong., 2d. Sess. 197 (1991) (report accompanying The Violence Against Women Act of 1991, Pub. Law No. 102-199).

29. Quindlen, *supra* note 26, at 25. *See also* Linda L. Ammons, *Discretionary Justice: A Legal and Policy Analysis of a Governor's Use of the Clemency Power in the Cases of Incarcerated Battered Women*, 3 J.L. & POL'Y 1 (1994) (summarizing studies and articles in the Journal of American Medical Association documenting the frequency of domestic violence).

30. Diana Russell's study of 930 women showed that 14% of married women had been the victims of attempted or completed rape by their husbands. Based on other studies in this area, Russell believes this estimate to be low. DIANA E.H. RUSSELL, RAPE IN MARRIAGE 57-58 (1990).

It is difficult to pin down the extent to which the marital rape exemption (permitting men to rape their wives without criminal responsibility) continues to exist across the nation. Scholars differ on whether all, or most, states have eliminated it. The confusion stems from the fact that the exemption was in some states statutory and, in others, by common law. In addition, while it is fair to say that the exemption has changed in recent times, the changes have been varied. Some states attach criminal responsibility to the act of rape, regardless of whether the victim is the perpetrator's wife. Others permit rape to be defined as a crime only when the parties are living separate and apart, and there are various definitions and requirements of proof concerning this factor. Some states prescribe the time in which the rape must be reported to the police. In some cases, marital rape is a crime separate from nonmarital rape and carries a lighter penalty. For an excellent description and examples of what the author calls "the marital allowance," *see* Jaye Sitton, *Old Wine in New Bottles: The "Marital" Rape Allowance*, 72 N.C. L. REV. 261 (1993).

Conservative estimates that physical abuse occurs in at least one out of four marriages (other studies claim the ratio is one out of two)[31] seem to underwhelm many of us. No matter what the statistics say about the prevalence of domestic violence, we manage somehow to pretend that it is an aberration. To be fair, William Galston, a signatory of the communitarian platform and author of *A Liberal-Democratic Case for the Two-Parent Family*, states that "to avoid misunderstanding, I want to make it clear that a general preference for the intact two-parent family does not mean that this is the best option in every case."[32] Presumably, he would not endorse the two-parent family in which abuse occurs. However, Galston's very choice of language sends the message that overall, abuse happens so infrequently that we might allow it as an exception, exceptions being defined as something not just apart from the norm, but far apart from the norm. Statistics, even the most conservative ones, simply do not bear out this view.

Moreover, while the act of leaving an abusive spouse has been known to precipitate additional violent acts, staying in the marriage is no less dangerous. The degree of violence usually escalates, and, therefore, the longer a woman remains in an abusive marriage, the greater her chances for serious injury.[33] The failure of the communitarian platform to address this issue head on permits the inference that preservation of the marriage is the primary goal, against which negative factors, such as physical abuse (or even the possibility of it) must be duly weighed.

Certainly, then, the greatest harm to women posed by marriage is the threat to life and health caused by abusive partners. In addition, many other, more pervasive, problems exist. From a legal perspective, marriage may protect women by securing theoretical rights to support and inheritance, but while the law may provide for women, it is often far from benign. Even a cursory view of the law reveals that it may explicitly or implicitly limit where a married woman may live and by

31. A study by Joan Zorza estimates that fifty percent of married women will be beaten on at least one occasion by their husbands. Joan Zorza, *The Criminal Law of Misdemeanor Domestic Violence, 1970-1990*, 83 J. CRIM. L. & CRIMINOLOGY 46 (1992).

32. William Galston, *A Liberal-Democratic Case for the Two-Parent Family*, 1 THE RESPONSIVE COMMUNITY: RTS. AND RESPS. 14, 19 (1990-91).

33. LENORE WALKER, TERRIFYING LOVE: WHY BATTERED WOMEN KILL AND HOW SOCIETY RESPONDS 42 (Harper Perennial 1990) (1989). While staying in an abusive relationship is dangerous, a woman taking steps to leave, or actually leaving, also provokes violence. Walker states:

> Battered women especially are in a double bind. Leaving the batterer is, in many cases, more dangerous than remaining; . . . separation is the time of greatest volatility and peril in battering relationships. But staying with the batterer brings with it inevitable physical abuse, an ever-lingering threat of death. Whatever the battered woman does, she simply is not safe.

Id. at 256. At another point in her book, Walker opines that there is very little hope for improvement in abusive relationships and counsels: "The best hope for the battered woman to stop the violence is to end the relationship altogether." *Id.* at 46.

what name she and her children are known. The law may redefine her property interests and may prescribe general behavior that it views as the proper wifely role, attaching legal penalties (most often upon divorce) to those who have chosen to behave differently.[34]

34. There are many instances in which a change in one's marital status has profound legal consequences. Among the more well-known (and one might add, not necessarily per se sex discriminatory) examples are one's filing status for tax purposes (and, hence, the amount of taxes one is required to pay), one's status as a dependent that gives rise to eligibility for certain Social Security and welfare benefits, obligations to support one's spouse, and continued right to receive alimony payments. Less commonly recognized instances include the law's treatment of property, which can be characterized as marital, community, or separate, each with attendant legal and practical consequences. While married couples may choose from among several forms of ownership of their marital home, the form known as tenancy by the entireties usually prevails by default. The legal effect includes the inability of either spouse to alienate his or her interest, but may also grant to the husband the right of possession and control during the life of the tenancy. D'Ercole v. D'Ercole, 407 F. Supp. 1377 (D. Mass. 1976). Depending on a single woman's financial situation, including her current sources of income; whether or not she has a career; her educational training; and what, if any, property she currently owns and the source of that property (purchase, gift, inheritance), marriage could prove to be a financially questionable choice.

But all of the above apply to men contemplating marriage as well. Other issues are not so (arguably) gender-neutral. Choice of name is one example. While it has become more common for married women to retain their surnames and to retake their surnames upon divorce, the United States Supreme Court has let stand opinions in at least two federal cases that held that states (Kentucky and Alabama) could require women to use their husbands' names when applying for a driver's license. Whitlow v. Hodges, 539 F.2d 582 (6th Cir.1976), *cert. denied*, 429 U.S. 1029 (1976); Forbush v. Wallace, 341 F. Supp. 217 (M.D. Ala. 1971), *aff'd per curiam*, 405 U.S. 970 (1972). The 1990's has not brought complete relief for those seeking drivers' licenses. *See* Jorgensen v. Larsen, 930 F.2d 922 (10th Cir. 1991) (holding once again that a petty bureaucracy's "important interest in regulation" trumps individual liberty). Some state statutes require a showing of "good cause" to effectuate a name change, and courts have construed personal and professional reasons to be insufficient in the case of four female plaintiffs. *See In re* Mohlman, 216 S.E.2d 147 (N.C. Ct. App. 1975). For an example of this issue at its most outrageous, consider the case of Attorney Wolvovitz, who was threatened with jail for contempt of court by U.S. Senior District Judge Hubert Teitelbaum, who insisted that she call herself "Mrs. Lobel" while in his courtroom. When her co-counsel objected, he was found in contempt and given a suspended sentence of thirty days. Judge Teitelbaum offered the following resolution of the matter: "What if I call you sweetie?" he asked. FAMILY LAW: CASES, COMMENTS, AND QUESTIONS 181 (Krause ed., 3rd ed. 1976) (quoting N.Y. TIMES, July 14, 1988, at 23). Finally, the Uniform Marriage and Divorce Act allows a woman to resume her former name upon divorce, but states that "the court may, and if there are no children of the parties shall, order her maiden name or a former name restored." UNIF. MARRIAGE AND DIVORCE ACT § 314(d), 9A U.L.A. 475 (1973). Thus, if there are children, a woman's desire to use her own name is not her right, but is at the discretion of the court.

A second example of law that operates to the detriment of women concerns the establishment of a marital domicile. Until recently, the choice of where a married couple made their home was solely the province of the husband. Aside from denying women equal authority within the marriage, the issue of who defines the domicile arises when one party raises the issue of desertion in a divorce proceeding. Two recent developments *seem* to have rendered moot the notion of who chooses the marital domicile: 1) most, if not all, states have dispensed with this overtly gender-based requirement and 2) the prevalence of no-fault grounds in divorce has eclipsed the use of the fault-based ground of desertion. On closer inspection, however, vestiges of male power in defining the domicile remain. In a recent New York case, the court lauded itself for acknowledging and then rejecting "the naked power of the husband to select the place of the marital abode," and in dicta called for equality in decision-making. It even held that the wife, who refused to leave Poland to join her husband in the United States, was not guilty of desertion. The rationale, however, centered on the woman's employment outside of the home, the fact that she held her teaching job for six years, and that it supported herself and her children. Szumanski v. Szumanska, 611 N.Y.S.2d 737, 738 (N.Y. Sup. Ct. 1994). One wonders: had defendant been unemployed outside the home, would the court revert to the old standby of male prerogative? A second case presents a more unusual fact pattern. In Defeo v. Defeo, 605 N.Y.S.2d 202 (N.Y. Sup. Ct. 1993), the court granted plaintiff's request for a divorce when his wife ceased to visit him in prison. It held that such conduct constituted abandonment from the marital domicile, which the court found to be the penitentiary.

Moreover, while fault is becoming less utilized in determining *grounds* for divorce, it is still a viable part of the equation when it comes to financial matters. Thus, a marital domicile is defined and a woman's departure from it may redound to her detriment in support and/or property settlements. In Reid v. Reid, 375 S.E.2d 533 (Va. Ct. App. 1989), Ms. Reid was denied a right to support from her husband because she legally deserted the marriage when she moved out of the marital domicile. A reading of the court's opinion is chilling. While finding Ms. Reid's account of her marriage to be credible, it felt that the totality of the circumstances did not establish sufficient excuse for her to leave. Among other things, she claimed that her husband's work schedule left the raising of the children almost exclusively to her; that he did not seem interested in family life or a sexual relationship; that there was a total lack of intimacy in the marriage; and that despite her attempts to tell him of her distress, her husband persisted in believing that their marriage was a happy one. In furtherance of her marital commitment, Ms. Reid sought and obtained marital counseling for both of them, but this was unsuccessful because Mr. Reid still did not understand the extent of the problem. Finally, in an effort to get through to her husband that a problem existed and that she "couldn't go on with it without doing something about it," she moved into an apartment for a trial separation. *Id.* at 537. Claiming that this was unlike marriages with clear-cut cases of physical abuse, the court held that Ms. Reid's decision to "desert" the marital home was not justified. Consequently, she was not entitled to support from her husband. In Skannal v. Skannal, 631 So. 2d 558 (La. Ct. App. 1994), Ms. Skannal's claim for support was denied when she left the marital home in fear for her safety. She testified that her husband threatened to kill her if she left, that he had a violent temper, that he was an alcoholic and a recluse, and that he fired a gun at trespassers on the property. The court noted that Ms. Skannal failed to produce any corroborating witnesses, while (reminiscent of the Hill-Thomas hearings) Mr. Skannal produced witnesses (one of them Mr. Skannal's business partner) who, despite their acquaintance with Ms. Skannal, had no idea that Ms. Skannal believed there to be marital problems.

A third area centers on the traditional view that marriage can provide a form of economic security for women due to the husband's obligation to support her during the marriage. (While it is true that the duty to provide financial support is now fairly gender-neutral, the difference between most men's and women's incomes shows that men's greater incomes would make them, as a group, more likely to be cast in the provider role.) The right to be supported within the marriage is largely illusory, however, since courts routinely refuse to involve themselves in the dynamics of intact families. *See, e.g.*, Commonwealth *ex rel* Goldstein v. Goldstein, 413 A.2d 721 (Pa. Super.

The law, however, is only one dimension of the marriage relationship. Putting it aside and removing the issue of domestic violence for the moment, there is the day-to-day impoverishment of the quality of some married women's lives that occurs in thousands of minute examples that are difficult to quantify. Pepper Schwartz, in her response to William Galston's argument in favor of the two-parent family, tries to describe it as follows:

> On what grounds could I possibly dissent? How about *gender*? How about man's inhumanity to woman? How about a thousand years of female sacrifice unnoticed, almost unmentioned? And how about the .amily as the primary institution of women's subordination and oppression? Put another way, while there are kind and equitable marriages, there are many more which exist because women do not—or cannot—demand a fair deal, either in the domestic division of labor or in decision-making.[35]

Schwartz's concern about the division of domestic labor is echoed in the work of others. Even Robert Bellah, writing in 1985, recognizes that "working wives and mothers come home to families where the men still expect them to do the preponderance of the housework and child care. There have been considerable changes in expectation in

---

Ct. 1979); *see also* Posner v. Posner, 233 So. 2d 381 (Fla. 1970); McGuire v. McGuire, 59 N.W.2d 336 (Neb. 1953).

Child custody is an area of further concern. Should divorce occur, the married woman who wants custody of her children has a husband to contend with. Certainly, cases exist in which the parties agree or the court determines that the father would be the better custodial parent. The "best interest" test used by the vast majority of courts in deciding custody issues can be a nightmare for all concerned, due to its indeterminacy and almost total judicial discretion. For women, however, the stakes are worse. Sexist assumptions about the proper role of women (it has been argued) may favor the traditional, stay-at-home mother. By the same token, they punish other mothers, such as those who work outside the home. *E.g.*, Linda R. v. Richard E., 561 N.Y.S.2d 29 (N.Y. App. Div. 1990), in which the trial court denied custody to the mother, who worked three to four days per week as a nurse investigator for a law firm, and whose work hours were considerably fewer than her pediatrician husband. The trial court found that she was "a remote control" mother. The appellate court reversed. In Lane v. Schenck, 614 A.2d 786 (Vt. 1992), a mother who had been awarded custody was denied the right to move out of state to attend law school because it would interfere with the father's visitation. Despite some who believe that it is less risky for married mothers to simply stay home with their children (assuming they have the choice), this may actually increase these mother's chances of impoverishment upon divorce. LENORE WEITZMAN, THE DIVORCE REVOLUTION: THE UNEXPECTED SOCIAL AND ECONOMIC CONSEQUENCES FOR WOMEN AND CHILDREN IN AMERICA (1985). *See also* Joan C. Williams, *Deconstructing Gender*, 87 MICH. L. REV. 797 (1989).

Finally, that the *combination* of legal and social factors operates to disadvantage women in marriage is not a novel suggestion. Two excellent expositions, one historical, the other more political, are JOAN HOFF-WILSON, LAW, GENDER, AND INJUSTICE: A LEGAL HISTORY OF U.S. WOMEN (1991), and SUSAN MOLLER OKIN, JUSTICE, GENDER, AND THE FAMILY (1989).

35. Pepper Schwartz, *Gender and the Liberal Family*, 1 RESPONSIVE COMMUNITY: RTS. AND RESPS. 87 (1990-91).

this area but not much change in actual behavior."[36] A five-year study by the Families and Work Institute shows that in ninety-one percent of households, men perform household repairs, while in eighty-one percent of households, women take greater responsibility for cooking; in seventy-eight percent for cleaning, in eighty-seven percent for shopping; and in sixty-three percent for paying bills.[37] A more extensive treatment of this subject is Arlie Hochschild's *The Second Shift*, in which she finds that still, in these supposedly egalitarian times, wives are performing the bulk of household work, even when they work for pay outside of the home. Hochschild rightly sees this as no small matter, finding that the *extra month's worth* of work performed by women results in their fatigue, sickness and emotional exhaustion. Nor is the unequal distribution of work uncommon. Hochschild states:

> the women whose husbands fully share [household work] still consider themselves "unusual," while the women whose husbands don't, consider themselves "normal" . . . . One woman, a bank clerk and mother of two young children, who did nearly everything at home, ended her interview as many women did, talking about how lucky she felt. She woke at 5:00 a.m., crammed in housework before she set off for work, and after she got back, asked her husband for help here and there, getting very little. She didn't seem lucky to me. Did she feel lucky because her husband was doing more than the "going rate" for men she knew? As I gradually discovered, husbands almost never talked of feeling "lucky" that their wives worked, or that they "did a lot" or "shared" the work of the home. They didn't talk about luck at all . . . .[38]

Hochschild goes on to say that

> if women who have an equal deal feel "lucky" because it is so rare and precious and unusual and precarious an arrangement to have—if all of us who have some small shard of help are feeling "lucky"—maybe something is fundamentally wrong with the usual male outlook on the home, and with the cultural world of work that helps create and reinforce it.[39]

Whatever the source of the disparity, Hochschild's findings dispute the notion that a husband and father in the home will necessarily provide help with the housework. Thus, the rationale for preferring the two-parent family—the extra pair of hands—is faulty.

It would be reassuring to be able to say that none of this is new, that over the past ten years we have achieved a collective enlighten-

---

36. ROBERT BELLAH ET AL., HABITS OF THE HEART: INDIVIDUALISM AND COMMITMENT IN AMERICAN LIFE 111 (1985).

37. *Survey: Workers Value Autonomy, Communication*, VALLEY NEWS (White River Junction, Vermont), Sep. 7, 1993, at 15 (citing Robin Schatz, *It's Satisfaction as Much as Cash; Workers Express Their Druthers*, NEWSDAY, Sep. 3, 1993, at 47.

38. ARLIE HOCHSCHILD, THE SECOND SHIFT: WORKING PARENTS AND THE REVOLUTION AT HOME at xii (1989).

39. *Id.*

ment on, if not the cure for, the prevalence of physical abuse and other forms of oppression that marriage means to many women. But we have not. Christian fundamentalists (no small group) continue to preach on the propriety of woman's subordinate role in marriage; other religions do so as well, albeit somewhat less stridently.[40] The cultural pressure for women to marry is still there, projecting the same rosy picture it always has, while ignorance of the reality remains. Students in my family law class, formally educated and more versed in the law than most women, are astonished when I ask them if they know what the legal ramifications of marriage may be and even more astonished when they find out. Those same women students authenticate Robert Bellah's view that expectations have changed concerning household work when they voice that their generation is committed to shared housework and child care in marriage; yet when a female faculty member schedules an annual presentation on how to combine career with family, *year after year, only women show up.*

What will it take for us to stop *en*couraging marriage for women and to *dis*courage the notion that marriage is a sort of haven, and the only legitimate one, for women and their children? Rather than selling the idea of marriage to women, a task the popular culture still does quite well, we should be simultaneously painting the truth of marriage as it is while working to make it an arrangement that may well operate to women's advantage in the future. While we cannot unequivocally cast marriage aside as an option for women, and even if we respect and admit to the existence of good marriages, is it not irresponsible to suggest that marriage, as it is currently conceived, ought to be the *preferred* arrangement for women and their children? Women who opt for a life outside of marriage, with children or without, have valid reasons for their choices. Their decisions must not be dismissed or trivialized, as William Galston would have us do, as "the easy relativism of the proposition that different family structures represent nothing more than 'alternative life-styles.' "[41]

The communitarian platform's adherence to traditional marriage is echoed in its position on divorce. Communitarians do not favor it, bringing as it does a host of problems for children, including undisputed economic catastrophe. As with communitarians, many feminists would not only approve of, but have been laboring for the past two decades to bring about, reforms that would address the documented inequality in the economic status of men and women (and

40. As an aside, it is interesting to note that the communitarian position paper "congratulate[s] . . . religious organizations that provide systematic preparation before marriage takes place and counseling when divorce threatens," without inquiring into the substance of that counseling. In my experience, many of my women clients who suffered years of physical abuse by their husbands were consistently advised by their clergy to "bear it" and to rectify the abuse by becoming "better" wives. COMMUNITARIAN NETWORK, *supra* note 9, at 13.

41. Galston, *supra* note 32, at 20.

hence children) following divorce. Even Dr. Etzioni acknowledges Lenore Weitzman's pioneer study showing that men experience a forty-two percent increase in their standard of living following divorce, while women (and children, since ninety percent of children reside in the custody of their mothers) experience a seventy-three percent decrease.[42] The reality is even worse, since the figures assume that the parties will act in compliance with the financial obligations imposed upon them by the divorce, e.g., child support; in fact, only half of custodial mothers receive the child support due them; twenty-five percent receive nothing at all.[43]

There is a difference in advocating fairer treatment upon divorce and suggesting that the problem is divorce itself. The communitarian platform states:

> It follows that *widespread divorce,* when there are children involved, especially when they are in their formative years, is indicative of a serious social problem. Though divorces are necessary in some situations, many are avoidable and are *not in the interest of the children,* the community, and probably not of most adults either. Divorce laws should be modified, not to prevent divorce, but to signal society's concern.[44]

Pepper Schwartz, in her response to William Galston, sees this as "an exhortation to the couple"[45] to stay married, but I see it as something more. It is a statement of belief that "widespread divorce" exists because too many people enter into and leave marriages at whim, without regard to the consequences to themselves or their children. That is what the drafters of the platform see as "a serious social problem." Such a belief evinces blindness to the more likely "serious social problem": traditional marriage, as it has been defined and as it has operated, has not worked for half of the couples who have tried it. A fifty-percent failure rate is a respectably sufficient ratio that should stop us from insisting that people simply keep on trying. That the odds are not with them speaks as much, if not more, of the problems with the institution of marriage as it does about the character or lack thereof of the individuals.

To posit that divorce is widespread because people are irresponsible or otherwise flawed is appealing for all of the reasons that blaming victims makes us feel better. It stops us from having to look at the real problem and allows us to feel superior. Marriage has been referred to as the cornerstone of our society for so long, that even with a fifty-percent failure rate, we continue to adhere to our belief that it is the preferable way to live. Consequently, if those involved cannot (or will not) hang in there, there must be something wrong with them.

42. ETZIONI, *supra* note 16, at 82.
43. WEITZMAN, *supra* note 34, at 283.
44. Communitarian Network, *supra* note 8, at 8.
45. Schwartz, *supra* note 35, at 89.

My own experience as a divorce lawyer speaks to this point. I am old enough to have practiced law before the no-fault divorce laws allowed people to divorce simply by agreeing that their marriage was irretrievably broken. At that time, the person seeking the divorce had to prove that there was a reason, or grounds, for ending the marriage. I represented hundreds of divorce plaintiffs, the vast majority of whom were women. Establishing grounds was not always difficult in a legal sense. In a real world sense, the vast majority of these people had what just about anyone would consider more than valid reasons for ending their marriages. Probably nine out of ten women had experienced physical abuse in their marriage; for at least half of those, physical abuse was the major reason for the divorce. For some, the husband's adultery was the final straw. But for many the reason was an amalgamation of many factors: physical abuse, absence of the husband from the home, his failure to support the family even when he was able, his refusal to assist in rearing the children, his drug or alcohol abuse, and his mistreatment of her family and friends.

My professional experience in representing hundreds of women and dozens of men in domestic matters allowed me a glimpse into the married lives of more people than the average non-lawyer. But it certainly is not at odds with my own personal observations nor with the experiences of others. Most people know at least several couples who have divorced. If asked, probably most of us could name one or two whom we might believe to have ended their marriages irresponsibly, or without the gravity we would want to see them exhibit. There may be another case or two in which the marriages were so outrageously bad that we would probably admit to wondering how the marriage lasted as long as it did. In the vast majority of these situations, however, the decisions appear to have been wrenching ones, last resorts, the reluctant best choice of several bad options. It dishonors these women who have sought to end their marriages to suggest that the inability to sustain a viable two-parent union stems from a whimsical rejection of responsibility and commitment. Such castigation is a superficial and overly simple response to a much more complex situation.

III. Why the Preference for the Two-Parent Family May Not Operate to the Advantage of Children

William Galston is correct when he states that children living in single-parent families tend to be economically poorer than those living in two-parent families.[46] He concludes that putting and keeping a father in the home is to the economic advantage of children.

Ending sex discrimination in the workplace would also help to eliminate the poverty faced by one-parent households. Most single-parent families are headed by women, who still earn approximately

46. Galston, *supra* note 32, at 16-17.

sixty cents for every dollar earned by men. Thus, economic hardship to single-parent families is not caused by the absence of one of the parents, but by the one parent whose earning power is greater. Moreover, child care responsibilities cripple women in the workplace, regardless of marital status. The single mother has difficulty in finding suitable child care that will enable her to work. Are we to presume that married women, because of the presence of their husbands in the home, do not experience the same difficulty?

Financial concerns aside, the supposed advantage of the two-parent home appears to be more sociological and psychological. The argument is that children who have two parents in the home are more secure, less prone to problems in school and on the street, and less likely to engage in use of drugs and inappropriate sexual activity. Most of the studies amounting to what Galston describes as "a mountain of scientific evidence"[47] are studies of "broken families," that is, families after divorce, and sometimes non-divorced, or "intact" families, used as a sort of control group.[48]

As a professor of family law, I am aware that the studies done on this issue could indeed form something approximating a mountain. I would also concede that most of them do in fact make the case that children of divorce, as a group, suffer ills that many children from non-divorced families do not. I also admit that the number of these studies and the similarity of conclusions are convincing and deserving of thoughtful attention. I do not, however, think that these studies may be cited as unequivocal proof[49] that the two-parent family, partic-

---

47. *Id.*

48. Among the best known of these studies is reported in JUDITH S. WALLERSTEIN & SANDRA BLAKESLEE, SECOND CHANCES: MEN, WOMEN, AND CHILDREN A DECADE AFTER DIVORCE (1989). Katha Pollitt notes that while the media gleefully trumpeted the Wallerstein/Blakeslee study (which Pollitt refers to as a "travesty of social science"), that study did not use a control group. She goes on to say:

> As it happens, a 1991 study in *Science* did use a control group. By following 17,000 children for four years, and comparing those whose parents split with those whose parents stayed in troubled marriages, the researchers found the "divorce effect" disappeared entirely for boys and was very small for girls. Not surprisingly, this study attracted absolutely no attention.

KATHA POLLITT, REASONABLE CREATURES: ESSAYS ON WOMEN AND FEMINISM 38 (1994).

49. Studies do exist which challenge the notion that divorce is the culprit responsible for social and psychological harm to children. *See, e.g.*, David M. Fergusson et al., *Parental Separation, Adolescent Psychopathology, and Problem Behaviors*, 33 J. AM. ACAD. OF CHILD AND ADOLESCENT PSYCHIATRY 1122 (Oct. 1994) (Data collected during a 15-year longitudinal study of 935 children of divorce showed a small, but detectable, elevated risk of a range of problems in adolescence including substance abuse, mood and anxiety disorders, and early sexual activity. However, the authors also concluded that much of the correlation between such problems and the parents' divorce "appeared to be spurious and arose from confounding social and contextual factors that were present in the child's family before parental separation."); Frank F. Furstenberg & Julien O. Teitler, *Reconsidering the Effects of Marital Disruption—What Happens to Children of Divorce in Early Adulthood*, 5 J. FAM. ISSUES 173 (Jun. 1994) (identifying predivorce factors which explain a portion of the effect of divorce on children and sug-

ularly the on-the-verge-of-divorce, two-parent family, is always better for children's development.

A minor, but important, reason is that many studies measure the effect of divorce upon already existing families as opposed to families that began with a single parent and remain so. Thus, it is difficult to determine to what extent the actual breakup of family results in problems for children and to what extent the problems are caused by the resulting single-parent status of the post-divorce family structure. Given that the number of children of divorce is substantial, however, shouldn't we say that the studies are convincing enough to confidently point the finger at divorce as the single biggest annihilator of children's well-being? My answer is a tentative maybe, which would not allow me to make the kind of assertions found in the communitarian platform.

If we accept that children suffer after divorce, it does not follow that those same children, without divorce, will prosper. No study has, or can, establish that. The better studies often include a control group of children from non-divorced families that almost inevitably score higher on whatever means are used to measure happiness, adjustment, and potential for success. But those are not the same children nor the same families, and that is a crucial distinction. Children who suffer after divorce may have suffered during the marriage and may well have suffered (perhaps more acutely) had the family remained together. Most divorces occur after periods of serious conflict within the family, which can cause problems for children that continue after the family breakup. No one knows, or can know, the effect upon children who live in intact families where parents want to divorce and do not. Anecdotal evidence exists, often from persons who are now adults, who say that their parents remaining together for the sake of the children was a horrendous mistake, one that subjected them to a childhood spent in unresolved and continual conflict. They describe the effects of such an arrangement as more deleterious to their own welfare than divorce could ever have been. In households in which family violence exists (whether directed at women only or at children as well), the research suggests that children who are a part of these violent households will replicate the role of abuser or victim in subsequent relationships and that "violence begets violence."[50] Di-

gesting that researchers in this area pay closer attention to the processes that lead up to divorce when evaluating the experience of children exposed to parental separation); Jennifer Clark & Bonnie L. Barber, *Adolescents in Postdivorce and Always-Married Families—Self-Esteem and Perceptions of Fathers Interest*, 56 J. MARRIAGE & FAM. 608 (Aug. 1994) (study measuring connection between adolescents' self-esteem and perceived interest by their fathers in post-divorce, mother-headed families and two-parent, always-married families found no difference in levels of self-esteem according to family structure).

50. ANGELA BROWNE, WHEN BATTERED WOMEN KILL 28 (1987).

vorce in these cases might then alleviate rather than cause suffering for the children involved as well as for their future families.

In short, I would not argue that divorce does not affect children nor that, most of the time, there are some harmful effects. I would say that the jury is still out on how best to fix that. Exhorting parents to stay married because it is best for the children overstates what we know about the subject, and the worst harm to children may result from our superficial acceptance of the available evidence.

Finally, despite the explicit disavowal of nostalgia, to suggest that by discouraging divorce one will preserve the family in a way that will nurture children is to assume there is an overabundance of nurturing going on in the two-parent household. The drafters of the position paper on the family themselves admit that while they believe in equality between fathers and mothers: "If there is any systematic difference between fathers and mothers it is that many more fathers neglect their children than do mothers."[51]

They then exhort parents to "rededicate themselves to their children."[52] (Do mothers need such an exhortation?) The statistics of fathers neglecting their children following divorce are shameful, but most fathers do not become neglectful as a result of a divorce decree. As the drafters acknowledge, the lesser known, but even more shocking statistics, speak to the failure of fathers *in intact families* to assume responsibilities toward their children.[53] Given that in one study, middle-class fathers who were living with their children were found to spend fifteen to twenty minutes per day (another study clocked the interaction at 37.3 *seconds* per day)[54] with their children, what is the point of pretending that divorce is the only, or even the major, culprit? As Pepper Schwartz states in her response to William Galston, "[h]ow, may I ask, are the problems of the family going to be ameliorated? . . . Is there a policy or plan that will help make children important to fathers, or make it easier or more rewarding to be a parent?"[55]

As in Hochschild's study of housework, child care is still the province of women, whether they be single or married mothers. The "extra pair of hands" appears to be neither mopping floors nor changing diapers.

---

51. COMMUNITARIAN NETWORK, *supra* note 9, at 15-16.

52. *Id.* at 16.

53. Karen Czapanskiy, *Volunteers and Draftees: The Struggle for Parental Equality*, 38 UCLA L. REV. 1415 (1991).

54. JUDITH AREEN, FAMILY LAW: CASES AND MATERIALS 195 (3d ed. 1992). A more recent study by the International Association for the Evaluation of Educational Achievement showed that in the United States, mothers average 10.7 hours each weekday caring for pre-school age children, while fathers spent approximately 45 minutes. Kelly Owen, *US Dads Lag in Child Care Duties, Global Study Finds*, L.A. TIMES, Feb. 20, 1995, at A5.

55. Schwartz, *supra* note 35, at 89.

IV. Contemplating Feminist Principles

The complexities of modern feminist theory probably prevent a unified response to the communitarian preference for the married, heterosexual parent. One must acknowledge that many women have signed the communitarian platform. Among the signatories are Jean Bethke Elshtain and Mary Ann Glendon, who have taken women's concerns as their own and have contributed to feminist thinking. Therefore, at the risk of anticipatorily appearing to protect myself from naysayers by admitting that I have no unified theory, I would like to offer the following as contributions to a conversation that I believe has strayed too far in the wrong direction. My thoughts have developed as I have pondered what I consider to be the simplest of feminist principles.

First, feminism is grounded in the concrete and the real, everyday lives of women. And while the number of never-married women continues to rise, most women still marry and raise children within heterosexual, nuclear families. One might even conjecture that they prefer to do so. To that end, feminism must continue to examine the inequities, legal and other, that women suffer in the marriage relationship and must not abandon its efforts to reform this cultural institution. Pepper Schwartz and Arlie Hochschild posit that a major problem in the current marriage relationship is that women, through cultural and other forces, are not able to bargain on an equal footing with their partners. Feminism needs to continue to identify what needs to change in order to place women in a fairer bargaining posture, even while we ponder whether contractualism is the ultimate solution to the problems within marriage.[56]

Equally important is feminism's commitment to inclusion. Women, who understand first-hand what it means to be an outsider, should not be in the business of privileging some above others. The words of Jean Bethke Elsthain deserve examination here:

> A second radical criticism holds . . . that in defending the family and intergenerational ties, one privileges a restrictive ideal of sexual and intimate relations. There are within contemporary American life those who believe that a society can and should stay equally open to all alternative arrangements, treating "life-styles" as so many identical peas in a pod. To be sure, families in modernity coexist with those who live another way, whether heterosexual and homosexual unions that are by choice or by definition childless; communalists who diminish individual parental authority in favor of the preeminence of the group; and so on.

But the recognition and acceptance of plural possibilities does not mean each alternative is equal to every other with reference to specific social needs . . . . Those excluded by, or who exclude themselves

56. HOFF-WILSON, *supra* note 34, at 280; Clare Dalton, *Deconstructing Contract Doctrine*, 94 YALE L.J. 1096-1113 (1985).

from, this authoritative norm should not be denied social space and tolerance for their own practices . . . .[57]

This passage both scares and appalls me, for many reasons. It bespeaks of an "us" and a "them," holding the traditional family as the norm and attaching the status of "other" to any and all other arrangements, denigrated ever so subtlety once again by the use of quotation marks around the description "life-styles." It appears to denote homosexual unions as "by definition childless," when that is clearly not the case. Elshtain may be right when she claims that recognizing plural possibilities is not the same as holding all of the those possibilities in equal regard. She does not, however, offer any insight into how to select the appropriate one(s) for addressing "specific social needs," that is, rearing children. The unstated assumption is that the traditional family structure is best, which induces one to ask, "best for whom?" Marilyn Bowens has an answer:

> The long-standing policy of favoring the (so-called) "traditional" family over others has operated to perpetuate a status quo in which white, middle- to upper- class heterosexual men (and, by association, the white women who "appropriately" attach themselves to them, and their white children) are validated and extended privilege in this society at the expense of everyone else. It cannot be mere coincidence that those most negatively impacted by the pro-Cleaver [so named after the Cleaver family of 1950's television] bias are women, African-Americans, and lesbians and gays . . . .[58]

Most of all, the problem with Elshtain's position is that it extends not full-bodied sisterhood, but a weak and condescending "tolerance" for the aberrant, establishing a hierarchy in which some choose to hand out "social space" to the almost worthy.

In addition, feminism, with its incessant questioning, prodding, and restructuring of everything, must be held to a higher standard of imagination than communitarianism provides. It is not just that after all of the searching, the same old tired notion of the patriarchal family is dusted off and resurrected as the solution once again, though that is certainly part of it. Martha Fineman posits that we reject the notion derived from patriarchal ideology that sexual connection stands as the determinative factor in defining family, stating that "single motherhood . . . could be considered deviant and threatening simply because it is a rejection of the primacy of the sexual connection as the core organizing familial concept . . . ."[59] More so, however, is the hope that feminism dare to dream no small dreams. Are we to believe that the reformed version of the family, even assuming that we could somehow cure the current ills it visits upon women, is our only choice if we

57. JEAN BETHKE ELSHTAIN, POWER TRIPS AND OTHER JOURNEYS: ESSAYS IN FEMINISM AS CIVIC DISCOURSE 59 (1990).

58. Bowens, *supra* note 22, at 43 (footnotes omitted).

59. Martha L. Fineman, *Images of Mothers in Poverty Discourse*, 1991 DUKE L.J. 274, 291 (1991).

desire to raise children? Frances Olsen writes that "[our] historical experience . . . makes us assume, often without realizing it, that the only alternatives are patriarchy on the one hand, and atomistic individualism on the other."[60] What of other possibilities: single parenthood, mothers sharing households with other mothers, women forming relationships with other women for the purpose of cooperatively raising children,[61] a reinvestigation of extended families. Instead of narrowing choices, we should not only be widening them, but conceiving and putting into place support structures that will make them possible. These support structures should include the obvious such as child and elder care, education, and job training for women. We should also envision and implement new housing arrangements that are safe and functional for women and children. We should question whether legal recognition of relationships is necessary and, if so, recognize different kinds of relationships with all of the attendant benefits that are now reserved for marriage. Finally, a true revolution needs to redefine all of those cultural elements that contribute to supporting the myth of the one-size-fits-all patriarchal, nuclear lifestyle. We should be working toward building a society in which women may raise children happily and healthily, without requiring a man and a marriage license as the price for admittance to the parenting club.

Tautologically, feminism must empower women. My critique of the communitarian preference for the traditional, two-parent family is not just a theoretical one. In the same way that we must realize that marriage will remain an option, perhaps even a preferred option for many women, we must also realize that our efforts at reform of marriage and parenthood are constrained by the willingness of men, in significant numbers, to change their own behaviors. As of this date, it is men, rather than women, who (generally speaking) have separated the notion of children from the notion of family. Women may respond to this reality in three ways. They can attempt to force men (through use of the law or other cultural forces) to live up to their responsibilities as fathers. Or, women can educate, plead, and cajole men to live with them in family relationships, hoping to change their hearts as well as their behaviors. While both of these may be worth doing, they have not proven highly effective. Regardless of whether one is optimistic or pessimistic concerning the willingness of men to partner and parent equally and the rate at which such changes will or will not occur, feminism must answer the question: "What do we do in the meantime?" Thus, the third alternative is needed. Feminism needs to empower women who cannot, or will not, wait for men to join them in the parenting enterprise.

---

60. Frances Olsen, *The Politics of Family Law*, 2 LAW & INEQ. J. 1, 11 (1984).

61. *See generally* BARBARA EHRENREICH, THE HEARTS OF MEN: AMERICAN DREAMS AND THE FLIGHT FROM COMMITMENT 180-82 (1983).

Finally, a word about cultural versus radical feminism, legal liberalism and autonomy versus communitarianism and connectedness. This paper does not resolve the question of which approach is better. I do not think it needs to, nor does my critique of certain aspects of communitarianism mean that it and radical feminism must or should forever part ways. The real issue is whether the building of community by definition must annihilate the rights of individuals, or more specifically, whether building community will require that same old kind of building block—and *only* that kind of building block—known as the traditional nuclear family. If so, and this particular form of communitarianism *does* appear willing to build itself on the backs of women, feminism must denounce it. Having done so, it ought then turn its attention to the more rewarding enterprise of leading all of us out of the false dichotomy between acceptance of human beings as individuals and connection among ourselves.

V. Conclusion

The communitarian platform on the family evinces a sensitivity to many issues that deserve attention and conversation. It appears to advocate not just any old two-parent family, but one in which women are not oppressed and in which fathers recognize and assume their familial responsibilities. However, the exclusivity of the preference for two-parent families does not accord sufficient respect for the diversity of arrangements, kin and otherwise, that can and do provide for the rearing of children. More to the point, it does not insist, loudly and without equivocation, that the changes it advocates must be brought about in the two-parent family as a precondition to its preferred status. Nor does it seem to comprehend the enormity of the consciousness change required. It is well to be optimistic, to hope for change and to find ways to create it, but one must not at the same time keep trying to sell the status quo, as though it will do until something better comes along. It will *not* do, and perhaps the best way to make change is to help women understand why it will not do. If we support women who just say no to marriage, rather than encourage those who just say no to divorce, that may be the greatest of all catalysts for change.

THE NORPLANT SOLUTION: NORPLANT AND THE CONTROL OF AFRICAN-AMERICAN MOTHERHOOD

Darci Elaine Burrell*

INTRODUCTION

> *Yet all through the darkest period of the colored women's oppression in this country her yet unwritten history is full of heroic struggle, a struggle against fearful and overwhelming odds, that often ended in horrible death, to maintain and protect that which woman holds dearer than life. The painful, patient, and silent toil of mothers to gain a fee simple title to the bodies of their daughters, the despairing fight, as of an entrapped tigress, to keep hallowed their own persons, would furnish material for epics.*[1]

On December 10, 1990, the United States Food and Drug Administration (FDA) approved the contraceptive Norplant for general use in the United States.[2] Norplant was immediately hailed as a revolutionary advance in reproductive technology.[3] The contraceptive device offers women a convenient and reliable

---

* J.D. Candidate, 1995 UCLA School of Law; B.A. 1991, University of California, San Diego. The author would like to thank her family for their continuing support, Professor Kimberlé Crenshaw for providing a framework in which this paper could be written, and the board of the UCLA Women's Law Journal for their true commitment to exploring the diversity of women's experiences.

1. Black Women in Nineteenth-Century American Life: Their Words, Their Thoughts, Their Feelings 329 (Bert J. Lowenberg & Ruth Bogin eds., 1976), *quoted in* Dorothy E. Roberts, *The Future of Reproductive Choice for Poor Women and Women of Color*, 12 Women's Rts. L. Rep. 59, 59 (1990).

2. *Long-Term Contraceptive Approved*, [1990-1991 Developments Transfer Binder] Food Drug Cosm. L. Rep. (CCH) ¶ 42,180 (Dec. 10, 1990).

3. *See, e.g.*, Judy Mann, *New Contraceptive Advances Freedom, Responsibility*, Wash. Post, Dec. 12, 1990, at B3.

alternative to traditional birth control devices.[4] Norplant consists of six, match-sized, silicon tubes which release a steady stream of the synthetic hormone levnorgestill into the bloodstream to prevent pregnancy. The tubes are surgically inserted under the skin of a woman's arm and prevent conception for up to five years.[5]

Not long after the FDA approved the use of Norplant, several controversial proposals for the use of the contraceptive emerged: (1) on December 12, 1990, the Philadelphia Inquirer published an editorial many perceived as racist and classist,[6] entitled, "Poverty and Norplant: Can Contraception Reduce The Underclass?";[7] (2) states quickly moved to incorporate Norplant in their welfare systems, providing either reimbursement for the cost of Norplant to women on AFDC or a cash bonus for those women who agreed to be implanted with the device;[8] (3) high schools across the country considered offering Norplant to teenage girls in order to prevent teenage pregnancy;[9] (4) on January 2, 1991, a California judge ordered Darlene Johnson, an African-American[10] woman convicted of child abuse, to receive Norplant as a condition of her probation.[11]

Proposals to coerce poor women of color[12] to use Norplant took many by surprise, including the contraceptive device's crea-

---

4. " 'In some ways [Norplant] is the dream method that people who work on birth control have looked for for decades,' said Dr. Daniel Callahan, director of the Hastings Center, a research institution in Briarcliffe Manor, N.Y., that studies issues of medical ethics. 'It is long-acting and would make a woman infertile until she takes an active decision not to be.' " Philip J. Hilts, *U.S. Approves Contraceptives Planted in Skin*, N.Y. TIMES, Dec. 11, 1990, at A1.

5. Janet F. Ginzberg, *Compulsory Contraception as a Condition of Probation: The Use and Abuse of Norplant*, 58 BROOK. L. REV. 979, 979 (1992).

6. *See, e.g.*, Claude Lewis, *Norplant Editorial was Offensive: The Thrust of the Editorial was Aimed at the Black Underclass. Unjustly So.*, PHILA. INQUIRER, Dec. 21, 1990, at A19; *see also infra* discussion part V.B.

7. Donald Kimmelman, *Poverty and Norplant: Can Contraception Reduce the Underclass?*, PHILA. INQUIRER, Dec. 12, 1990, at A18.

8. David S. Coale, *Norplant Bonuses and the Unconstitutional Conditions Doctrine*, 71 TEX. L. REV. 189, 189 (1992).

9. *See, e.g.*, Tim Larimer, *High School Offers Birth Control Implant, Blacks Disagree on Merits of Program*, DALLAS MORNING NEWS, Mar. 17, 1993, at A37.

10. I use the terms "African-American" and "Black" interchangeably in this article.

11. Ginzberg, *supra* note 5, at 979–80.

12. In this essay, I often discuss the impact of the abuse of reproductive technology on "poor women of color." I emphasize class, race, and gender in this manner because African-American women commonly fall at the intersection of these three categories. None of these characteristics can be neatly isolated, particularly in the area of reproductive rights. Just as there are appreciable differences between wo-

tor, Dr. Sheldon Segal.[13] "The team that worked on Norplant had been concerned that a government would misuse the device to enforce birth control. But, they were worrying about China, not California."[14]

In spite of the fact that the nationwide debate over the use and abuse of Norplant has focused[15] on poor African-American women, the legal community has largely abstained from addressing this issue. Although some law review articles have discussed the constitutionality of the coercive use of the Norplant contraceptive device,[16] at the time of this writing, none have specifically addressed the effect of the proposed uses of Norplant on poor women and women of color or have questioned the underlying sentiments behind these proposals.[17] Medical

---

men of different races, there are differences between women of different classes. For example, the abuse of reproductive technology, such as sterilization abuse, is far more likely to impact a poor African-American woman than a middle- or upper-class African-American woman. While a poor African-American woman may have no choice but to attend a particular physician or hospital, a wealthier African-American woman may have more health care options, and thus, more control over her reproductive capacities.

13. *See* Ellen Goodman, *Norplant is Changing Politics of Birth Control*, ATLANTA J. & CONST., Feb. 26, 1991, at A13 ("It took 24 years to develop, test and approve an implantable device that can prevent pregnancy for as long as five years. It took less than two weeks for Norplant to be billed as a new method of coercion.").

14. *Id.* Dr. Segal wrote an outraged letter to the editor in response to the *Philadelphia Inquirer* editorial, but Johnson's case was decided before the letter could even be published. *Id.*

15. By the word "focused," I mean that Norplant proposals both *impact* poor, African-American women disproportionately and that these proposals are, in a sense, *aimed* at poor, African-American women. For example, although Norplant bonus programs are facially aimed at all welfare recipients, most people erroneously believe that the majority of women on welfare are African-American. Thus, these so-called bonus programs are aimed at poor, African-American women because many Americans view these women as the problem. *See infra* part V.A. On the other hand, the use of Norplant as a condition of probation for women who use drugs during pregnancy may disproportionately impact poor African-American women because: (1) they are more likely to be reported for drug use during pregnancy; (2) the drugs poor Black women use are more likely to receive attention than the drugs used by white women; and (3) poor Black women may already be viewed as bad mothers. *See infra* part V.B.

16. *See, e.g.*, Kristyn M. Walker, *Judicial Control of Reproductive Freedom: The Use of Norplant as a Condition of Probation*, 78 IOWA L. REV. 779 (1993); Thomas E. Bartrum, *Birth Control as a Condition of Probation — A New Weapon in the War against Child Abuse*, 80 KY. L.J. 1037 (1991-1992); Coale, *supra* note 8, at 198.

17. While doctrinal analyses are certainly important and useful in this area, they may not go quite far enough. "[D]octrinally-focused inquiries fail to examine the historical antecedents of modern coercive interventions in pregnancy, particularly those directed against outsider women, and fail to explore the effect of race and class factors on the treatment women receive." Nancy Ehrenreich, *The Colonization of the Womb*, 43 DUKE L.J. 492, 498 (1993). It is one thing to say, for instance, that a

ethicists[18] and some members of the African-American community[19] have expressed concerns about potential Norplant abuses on poor and/or young African-American women. Some legal theorists, however, have dismissed these concerns as unrealistic or facile.[20]

Although feminists and feminist legal theorists have analyzed the patriarchal control of reproduction and the use of reproductive technology as social policy,[21] very few legal scholars have used this groundwork to explore the proposed uses of Norplant. Feminist legal scholars like Dorothy Roberts, Lisa Ikemoto, and Nancy Ehrenreich have demonstrated that, in the past, reproductive technology has been used to control the reproduction of those women whom society, for one reason or another, has labeled as deviant.[22] The abuse of reproductive technology has been based in almost every instance on the belief that these women were not "good women" or "good mothers" and therefore were less deserving of motherhood than other women.

In this Essay, I argue that the Norplant proposals aimed at poor, African-American women are based on the notion that poor, Black women are "deviant" and thus less deserving of motherhood than white women.[23] In Part I, I discuss the failure

---

particular reproductive regulation is or is not constitutionally permissible. It is another to look at the reasons behind the regulation and to explore how and why certain regulations impact one group disproportionately.

18. *See infra* note 55 and accompanying text.

19. *See* Larimer, *supra* note 9, at A37 (" 'The first solution proposed for all our social ills is to control the fertility of poor people,' said [Julia Scott, public education director for the National Black Women's Health Project]."); Dorothy E. Roberts, *Norplant's Threat to Civil Liberties and Racial Justice*, N.J. L.J., July 26, 1993, at 20.

20. *See infra* note 55.

21. *See* ANGELA Y. DAVIS, *Racism, Birth Control, and Reproductive Rights*, *in* WOMEN, RACE, & CLASS 202 (1981); Ehrenreich, *supra* note 17; Lisa C. Ikemoto, *The Code of Perfect Pregnancy: At the Intersection of the Ideology of Motherhood, the Practice of Defaulting to Science, and the Interventionist Mindset of Law*, 53 OHIO ST. L.J. 1205 (1992); Dorothy E. Roberts, *Punishing Drug Addicts Who Have Babies: Women of Color, Equality, and the Right of Privacy*, 104 HARV. L. REV. 1419 (1991).

22. *See infra* part III.

23. Although other women of color have also been subjected to the abusive use of reproductive technology, this paper will deal specifically with African-American women. I have no desire to perpetuate the myth that racism and sexism in this country are simply Black and white issues (so to speak), but I will concentrate on African-American women, in part because I am an African-American woman, and in part because, for reasons I will explore later in this paper, the Norplant proposals have been focused almost exclusively on poor, African-American women.

of feminists and legal theorists to examine the history of African-American women in the United States, and illustrate how this failure has led to a definition of "reproductive freedom" which fails to address the reproductive needs of Black women. In Part II, I explore the patriarchal ideal of motherhood and how poor women and women of color have been unable to meet that ideal. In Part III, I discuss the historical devaluation of African-American motherhood. In Part IV, I discuss how society's concern with the growth of the Black underclass and increasing welfare costs has led to the construction of the African-American welfare mother as the new deviant. I also analyze the Norplant proposals aimed at poor, African-American women in light of this country's growing obsession with the welfare mother. This Essay concludes that the proposed uses of Norplant are based on the belief that poor, African-American women, because of their deviant status, do not deserve to procreate.

I. The Reproductive Rights Movement: The Absence of Color and Critique

> Some problems we share as women, some we do not. You fear your children will grow up to join the patriarchy and testify against you, we fear our children will be dragged from a car and shot down in the street, and you will turn your backs upon the reasons they are dying.[24]

Most feminists recognize that reproductive freedom is essential to the liberation of women. As Dorothy Roberts states, "Without the ability to determine their reproductive destinies, women will never achieve an equal role in social, economic and political life and will continue to be politically subordinate to and economically dependent on men."[25] Throughout history, women have been subjected to the patriarchal control of sexuality, reproduction, and motherhood.[26] Although this control has taken many forms, the reproductive rights movement in this country has been very narrowly focused. The Supreme Court has recognized only two areas of reproductive rights — the freedom to decide whether to use contraceptives and whether to terminate a

---

This is a paper about the devaluation of African-American motherhood. This focus is not intended to lessen the seriousness of the abuse that other women or men of color have suffered. Nor is it intended to suggest that Black women are more oppressed than any other group.

24. Audre Lorde, Sister Outsider 119 (1984).

25. Roberts, *supra* note 1, at 60.

26. *See* Ikemoto, *supra* note 21.

pregnancy.[27] In addition, feminists have chosen to focus the reproductive rights movement almost exclusively on the right to abortion. Thus, in the United States "reproductive freedom" has largely been defined as the right to abortion.

Women of color have been noticeably absent from the reproductive rights movement. This is not necessarily because women of color are unaware of their oppression as women or of the importance of reproductive freedom. Rather, women of color have been dissatisfied with the reproductive rights movement, in part, because of the failure of white feminists to adequately address the needs of women of color. Access to abortion is certainly important to all women. What the leaders of the reproductive rights movement have failed to recognize, however, is that for women of color, and, in fact, for all women, reproductive freedom means far more than abortion rights.

"Reproduction encompasses a range of events and conditions from the ability to bear children, to conception, to carrying a fetus, to abortion, to delivering a baby, to caring for a child."[28] Therefore, to adequately ensure that all women in this country have true reproductive choice, the definition of "reproductive freedom" must be expanded to include, "for example, . . . decisions about sterilization and medical treatment; it must include access to fertilization technologies and to prenatal and perinatal care."[29] These issues are as necessary to reproductive freedom as access to safe and legal abortion. Unrestricted access to abortion will not equal true reproductive freedom to women of color, and thus, until traditional reproductive rights organizations expand their definition of reproductive freedom to include these concerns, "reproductive freedom" will continue to hold very little meaning.

By adhering to such a narrow definition, the reproductive rights movement has effectively ignored the reproductive freedom of women of color. Historically, poor women and women of color have been subjected to the patriarchal control of their reproduction. The patriarchy has abused reproductive technology and criminal sanctions to express its disapproval of these women and to label them as deviant. This abuse has

27. Roberts, *supra* note 1, at 62.
28. *Id.*
29. *Id.*

taken the form of coerced contraception,[30] sterilization abuse,[31] court ordered Cesarean sections,[32] and unequal prosecution for drug abuse during pregnancy[33] as well as for perinatal abuse.[34] Poor women and women of color have also been denied access to adequate prenatal and perinatal care. In addition, poor women and women of color have also been denied the right to raise their own children.[35] In spite of this, the reproductive rights movement has declined to address adequately any of these concerns. The leaders of the reproductive rights movement have largely failed to explore the history of patriarchal control over reproduction,[36] particularly where that control affects women of color, and therefore have been unable to define "reproductive freedom" so that it will have meaning for poor women and women of color.

Across the country, women of color are expressing their dissatisfaction with the reproductive rights movement, a movement "that historically has been almost lily-white."[37] These women perceive the abortion rights movement to be "white woman-led and white woman-defined."[38] Groups like the Women of Color Coalition of Planned Parenthood Southeastern Pennsylvania, the National Council of Negro Women, the National Black Women's Health Project, and the National Latina Health Organization are

---

30. For instance, a poor woman is denied federal funding for abortion and therefore is forced to carry an unwanted pregnancy to term.

31. Historically, African-American women, Latina women, and Native-American women have been subjected to forced or coerced sterilization. *See infra* part IV.C.

32. *See* Ehrenreich, *supra* note 17, at 510, 520–21 (discussing how poor women, single women, women of color, and other "outsider women" have been disproportionately subjected to court ordered sterilization).

33. *See* Roberts, *supra* note 21 (discussing the disparate impact on African-American women of laws against drug abuse during pregnancy).

34. *See* Ikemoto, *supra* note 21 (discussing the disproportionate prosecution of poor women, women of color, and older women for the death of their newborn infants).

35. Historically, these women have had their children taken from them and placed in foster care, orphanages, and other institutions. *See infra* part III. Interestingly, some of today's conservative political leaders are once again suggesting that the children of poor women (women on welfare) be placed in orphanages.

36. Some feminist legal theorists, like Nancy Ehrenreich, also suggest that white women have been far too accepting and uncritical of the patriarchal control over their own reproduction. Ehrenreich, *supra* note 17, at 493–95. Expanding the definition of reproductive freedom could very well benefit white women as well as women of color.

37. Fawn Vrazo, *Minorities are Joining Fight for Abortion Rights*, PHILA. INQUIRER, Mar. 10, 1993, at C1.

38. *Id.* at C3.

organizing to give women of color a stronger voice in the reproductive rights movement. They are questioning the quick surrender of feminists in the battle over Medicaid funding for abortions,[39] demanding greater representation for women of color on the National Abortion Rights Action League, and attempting to expand the definition of "reproductive freedom" to encompass the needs of women of color.[40] In addition, these organizations are trying to convince members of the majority-white reproductive rights movement to be more critical of new contraceptives like Norplant.[41]

Norplant, the first new contraceptive in twenty-five years, is undoubtedly a tremendous advance in reproductive technology. The device has many advantages over other forms of birth control. Norplant has been applauded as nearly infallible with a worldwide failure rate of 0.2 pregnancies per 100 women,[42] as compared with the pill (progestrogen and estrogen) and the minipill (progestrogen only) which have failure rates of 1 per 100, and the condom with a failure rate of 2 per 100.[43] In addition, because Norplant is inserted under the skin of the upper arm, releasing a steady stream of the progestrogen levonorgestrel into the bloodstream, women cannot forget to take it, and it does not interfere with regular sexual activity.[44]

Developers of Norplant also claim that the contraceptive device does not cause many of the same side effects as other contraceptives.[45] Unlike oral contraceptives, the hormone in Norplant enters the bloodstream directly and does not cause stomach upsets.[46] Moreover, Norplant ensures that the levonorgestill is released steadily, which prevents sudden bursts of a high level hormone which occurs when using injectable contraceptives.[47] Norplant, however, can cause irregular and prolonged menstrual

---

39. *Id.* at C1, C3. "Only 11 states, including New Jersey, now routinely allow Medicaid-funded abortions. In the other 39, including Pennsylvania, poor women are mostly left to scrape together the money for the procedure themselves." *Id.* at C3.

40. *Id.*

41. *Id.*

42. Jane Seymour, *When the Pill Gets Under Your Skin: Norplant*, New Scientist, Dec. 4, 1993, at 22.

43. *Id.*

44. *Id.*

45. Ginzberg, *supra* note 5, at 980.

46. *Id.*

47. Seymour, *supra* note 42, at 22.

bleeding, acne, and hair loss.[48] Nevertheless, studies indicate that most women are satisfied with Norplant and would recommend it to their friends.[49]

Ironically, despite the significant amount of control Norplant offers women over their reproductive lives, the contraceptive was immediately used to curtail rather than promote women's reproductive freedom. As Dr. Sheldon Segal, the contraceptive's developer, declared, "We created a method to enhance reproductive freedom and people keep finding ways to use it for the opposite purpose."[50] Similarly, journalist Ellen Goodman has noted that "[t]he simple, effective device offers women the liberating possibility of planning their families. But in America the first reactions are not about expanding possibilities. Norplant has been most publicly and ardently taken up by those who want to cap social problems by getting a lock on the womb."[51]

Many were not surprised that society quickly accepted Norplant as a means of social control. Shortly after Norplant's introduction, medical ethicists warned that it might be abused by those seeking to prevent certain groups of women, such as retarded women and women on public aid, from procreating.[52] " 'I told you so,' said Arthur Caplan, director of the Center for Biomedical Ethics at the University of Minnesota. 'I am not surprised by this, but I find it troubling when reproductive capacity is manipulated as a form of punishment. It sure didn't take long.' "[53]

In spite of the fact that Norplant proposals have been directed at curtailing the reproductive choices of poor women of color, particularly poor African-American women, the impact of these proposals on the reproductive freedom of poor women of color has received little attention. Norplant as a condition of probation or as part of a bonus program for women on public assistance has been debated to some extent on the pages of this country's law journals. Most articles, however, focus on the con-

48. *Id.*

49. Dick Kaukas, *Birth-Control Implants Work for Teens, Alliant Finds*, COURIER-J., Sept. 28, 1993, at 1C.

50. Ellen Goodman, *Norplant Dilemma: Choice vs. Coercion*, ST. LOUIS POST-DISPATCH, Feb. 20, 1991, at 3C.

51. *Id.*

52. Tamar Lewin, *Implanted Birth Control Device Renews Debate Over Forced Contraception*, N.Y. TIMES, Jan. 10, 1991, at A20.

53. *Id.*

stitutionality of the coercive use of the Norplant contraceptive device[54] without analyzing the motivation behind such proposals or the reproductive history of women of color in America. I have found none that explore or adequately address the concerns expressed by African-Americans and medical ethicists about the potential abuses of Norplant, although some legal theorists quickly dismiss these apprehensions.[55]

Whether one accepts or discounts the concerns of medical ethicists and African-Americans, such a decision should be based on a thorough examination of the history of the patriarchal control of reproduction and how that control has impacted women of color. It is unwise to reject these concerns in light of the historical abuse of reproductive technology and the reasons behind that abuse. I argue that the Norplant proposals which have been directed at poor African-American women are based on the same motives which have fueled the abuse of reproductive technologies in the past: the devaluation of African-American women as mothers and the attempt to control the reproduction of these women because of their deviant status.

II. Good Women/Bad Women: The Historical Patriarchal Construction of Motherhood

Motherhood has always been, and continues to be, a colonized concept — an event physically practiced and experienced by

54. *See supra* note 16.

55. *See, e.g.*, Stacey L. Arthur, *The Norplant Prescription: Birth Control, Woman Control, or Crime Control?*, 40 UCLA L. Rev. 1, 9 (1992) (footnote omitted) ("The argument that mandatory contraception for probationers amounts to unauthorized eugenic sterilization is . . . for the most part . . . a red herring."); John R. Kramer, *Introduction*, 67 Tul. L. Rev. 1725, 1733–34 (1993) ("Whites may or may not wish to discourage black reproductive freedom, but there is no evidence of a governmental program designed to accomplish this. Black women need not fear that their right to bear children is under serious attack through fetal abuse prosecutions nor do black birth rates suggest that they do so. This wolf is more imaginary than Peter's.").

I must emphasize that neither Professor Arthur nor Dean Kramer deal extensively with the abusive use of reproductive technology. Their written work does not reveal a thorough analysis of the reproductive history of African-American women in America which would warrant a conclusion that the present Norplant proposals bear no relation to the abuses of reproductive technology in the past. Dean Kramer's discussion of the prosecution of African-American women for drug abuse during pregnancy is only two paragraphs of a thirteen-page paper, and Professor Arthur dismisses charges of racism and sexism in Norplant proposals with a footnote.

> women, but occupied and defined, given content and value, by the core concepts of patriarchal ideology.[56]

Feminist legal scholar Lisa Ikemoto describes the "Code of Perfect Pregnancy" as "the idea and practice of controlling women with regard to conception, gestation, and childbirth in ways that express dominant cultural notions of motherhood."[57] The Code is based on the social definition of good motherhood, as established by the patriarchy, as well as the control of information and the disempowerment of women with regard to their own reproductive capacities.[58] The Code of Perfect Pregnancy is rooted in the ideology of motherhood developed during the Nineteenth and early Twentieth Centuries.

With the rise of manufacturing and the development of factories, the need for female employees began to increase.[59] Which women would fill that need, however, quickly became a question of class position, resulting in a differentiation between privileged motherhood and devalued motherhood.[60] "Distinctions were made between those women who could afford to cultivate domestic skills without pay . . . and those women . . . who needed income either for themselves or their families."[61] By the 1830s the distinction evolved into the "domestic code." The public and private spheres became increasingly separate, and women became the managers of the private sphere. The private sphere, and the mother as the manager of that sphere, became the source of virtue, morality, and tradition.[62] It became the role of the "good mother" to inhabit the private sphere and to "conceiv[e], giv[e] birth to, and rais[e] children who would grow up to contribute to the social order and not detract from it."[63] Those women who remained at home were defined as respectable, good women while those who worked outside of the home came to be viewed as deviant, selfish, and neglectful of their families.[64]

Due partly to plain economics, status as a "good mother" was reserved for privileged middle- and upper-class women. The

---

56. Martha L. Fineman, *Images of Mothers in Poverty Discourses*, 1991 Duke L.J. 274, 289–90.

57. Ikemoto, *supra* note 21, at 1207.

58. *Id.*

59. Joel F. Handler & Yeheskel Hasenfeld, The Moral Construction of Poverty 51 (1991).

60. Ikemoto, *supra* note 21, at 1210.

61. Handler & Hasenfeld, *supra* note 59, at 51–52.

62. Ikemoto, *supra* note 21, at 1211.

63. *Id.*

64. Handler & Hasenfeld, *supra* note 59, at 52.

middle-class and upper-class imposed their values on poor women, single women, and women of color regardless of their ability to conform. Although poor women, single women, and women of color were never morally excused from work, they were still expected to fulfill the good mother role.[65] In other words, at the same time these women were expected to work outside the home in order to resist becoming public charges, they were criticized for not staying home to fulfill their proper roles of wives and mothers. They were neither exempted from the work ethic that required them to work nor exempted from the domestic code that required them to stay at home.

From a moral standpoint, poor women, single women, and women of color were in a no-win situation. African-American women in particular suffered from this contradiction. They were seen as deviant for failing to fulfill the role of the good mother, although they were often expected to leave their own homes and children to be nannies and wet nurses for white women's children.[66]

In order to preserve the public/private dichotomy and maintain social order, the patriarchy promulgated laws restricting women to the private sphere[67] and imposed criminal laws that treated women who could not fit social norms, like race, class, religion, and sexual morality, "as deviants, as the cause of disorder."[68] Unsurprisingly, because poor women, single women, and women of color could not conform to the these norms, the use of state power to preserve the public/private distinction and to maintain social order fell most heavily on them.[69]

---

65. Ikemoto, *supra* note 21, at 1220.

66. *Id.*

67. *Id.* at 1212. For example, in Bradwell v. Illinois, 83 U.S. (16 Wall.) 130 (1872), the United States Supreme Court affirmed an Illinois Supreme Court decision denying a woman's application to practice law. "That God designed the sexes to occupy different spheres of action, and that it belonged to men to make, apply, and execute the laws, was regarded as an almost axiomatic truth." *Id.* at 132, *quoted in*, Ikemoto, *supra* note 21, at 1212.

68. Ikemoto, *supra* note 21, at 1215. The struggle to maintain the sanctity of the family and thereby save America from perceived social destruction included crusades against prostitution, alcohol, and pornography. *Id.*

69. *Id.* at 1220. Immigrant girls were disproportionately punished for juvenile delinquency and blamed for social problems. In addition, the stereotypical prostitute was often depicted as ethnic. *Id.*

The patriarchy apparently viewed these women's racial, class, and religious differences as extremely threatening to the social order. Perhaps this fear was based on the notion that these women held vastly different, even dangerous, moral beliefs,

The perpetuation of the private sphere and the use of state power to control poor women, single women, and women of color were closely related to society's desire to maintain social order. Working women threatened the traditional family, the vehicle for the promotion of traditional values.[70] Those opposed to working women attributed numerous social ills to the employment of women.[71]

Working women and poor women were also considered a threat to the social order by virtue of their roles as child rearers. Poor women, single women, and women of color, because of their failure to conform to the good mother standard and their deviant status as women, were seen as incapable of transmitting appropriate values, such as the work ethic, to their children.[72] "Throughout this period, the reformers were strongly concerned about deviant behavior, and especially the transmission of the wrong values and habits from the parent to the child."[73] The children of these women were considered to be "predelinquent — if not controlled, they were likely to become paupers and criminals."[74]

Social workers and so-called child protectors expanded the definition of cruelty to include all kinds of neglect.[75] Poor women and single women were over-represented in neglect cases.[76] Due to economic necessity, these women were forced to work outside the home and could not supervise their children like middle- and upper-class women could. This lack of supervision became one of the defining elements of neglect.[77] Social reformers

---

which, if not contained, would contaminate and destroy white, middle-class, Christian society.

70. HANDLER & HASENFELD, *supra* note 59, at 55. This criticism applied to all working women, regardless of race. It should not be surprising, however, that because women of color were often required to work to support their families, sanctions against working women affected them disproportionately.

71. Reformers feared that working women would become discontented, that their maternal functions would be impaired, and that they would produce deformed children. They also feared that working women would become prostitutes. *Id.*

72. Although African-American women often served as nannies for white women's children, they were only responsible for the physical needs of these children. They were not expected to pass on social values to their charges.

73. HANDLER & HASENFELD, *supra* note 59, at 65.

74. *Id.*

75. *Id.* at 60.

76. LINDA GORDON, HEROES OF THEIR OWN LIVES: THE POLITICS AND HISTORY OF FAMILY VIOLENCE — BOSTON, 1880-1960, 92, 94–95 (1988).

77. *Id.* at 96, 121.

were convinced that this type of neglect would lead to delinquency, and children were often removed from their homes.[78]

The fear that poor women, women of color, and other "bad women" would produce and rear children who would detract from the social order also led to the creation of juvenile courts. The juvenile courts were intended to intervene in "bad families" — those headed by poor women, single women, and women of color — and rescue delinquent, dependent, and neglected children.[79] The juvenile court judge could remove these "predelinquent" children from their homes and send them to reformatories and state schools for dependents, institutions intended to "help them."[80]

Because poor women, single women, and women of color were never morally excused from work, they found themselves caught between the proverbial rock and a hard place. If they stayed at home to devote the "proper" amount of care and attention to their children, they and their children would become public charges, and the women would be labeled as deviant because of their failure to conform to the work ethic and to take care of their families. If, however, they went into the workforce, they would be labeled as deviant for their failure to conform to the middle- and upper-class good mother standard. Thus, regardless of which path they chose (if one can say that they really had a choice if they wanted to feed their families), these women were viewed as deviant and as threats to the social order. In addition, they would pass on that deviance, one way or another, to children who would grow up to pose a continuing threat to the social order.

As discussed above, the patriarchy attempted to preserve social order by maintaining the private sphere, penalizing women who failed to conform to social norms, and removing children from atmospheres that would lead them to detract from the social order. Additionally, the patriarchy began to use reproductive technology to control deviant women. For example, during the eugenics period,[81] sterilization laws were used to prevent

78. *Id.* at 73.

79. *Id.* at 138–40. "*Delinquent* meant committing acts that would be criminal if done by an adult. *Neglected* meant abandoned, either in fact or by poor parental caretaking. *Dependent* meant poverty that was inflicted or caused by misconduct." *Id.* at 63.

80. *Id.* at 138–40.

81. *See infra* part IV.B.

prostitutes from genetically transmitting their deviant traits.[82] Other "bad" women, such as women who expressed their sexuality, women suffering from postpartum depression, and women of color, were placed in institutions, where they were more commonly forcibly sterilized.[83]

The use of reproductive technologies like sterilization to control the reproduction of deviant women did not end during this period. Reproductive technology increasingly is being used to control women who do not fit the ideal of womanhood. As Lisa Ikemoto points out, "over the past century, these laws [which speak to women's biological capacity to conceive and bear children] have begun to take the shape of an increasingly comprehensive regulatory scheme . . . . These regulations elaborate upon the ideology of motherhood by stating preferences about who should and who should not become pregnant and hence become mothers."[84] This regulatory scheme restricts a woman's ability to decide whether to conceive, how to conceive, whether to terminate the pregnancy, and how to behave during pregnancy.[85]

African-American women in particular have been subjected to the abuse of reproductive technology as a result of their deviant status. In the next Part, I will discuss various ways in which the reproductive capacities of African-American women have been governed as a means of social control.

III. The Historical Devaluation of African-American Motherhood

> It is not my child who tells me: I have no femaleness white women must affirm. Not my child who says: I have no rights black men must respect. It is not my child who has purged my face from history and herstory, and left mystory just that, a mystery; my child loves my face and would have it on every page, if she could, as I have loved my own parents' faces above all others . . . . We are together, my child and I. Mother and child, yes, but *sisters* really, against whatever denies us all that we are.[86]

---

82. Ikemoto, *supra* note 21, at 1230.

83. *Id.* at 1231–33.

84. *Id.* at 1221–22.

85. *Id.*

86. Alice Walker, *One Child of One's Own: A Challenging Personal Essay on Childbirth and Creativity*, Ms., Aug. 1979, at 47, 75.

Historically, African-American women have been seen as deviating from the ideal of true womanhood. "Real" women were expected to be pious, pure, submissive, and domestic, middle-class and white.[87] Black women, on the other hand, were presumed to conform to an entirely different set of characteristics — characteristics which precluded them from ever being seen as ideal women. Generally, four controlling images of African-American women have emerged, all of which deviate from the middle- and upper-class standard of womanhood: (1) "mammy," the faithful, obedient, nurturing, and caring domestic servant;[88] (2) the "matriarch," who is overly aggressive, unfeminine, and emasculating;[89] (3) the "welfare mother," who is irresponsible, lazy, and immoral;[90] and (4) the "Jezebel," who is sexually aggressive.[91] These images, linked both to the Black woman's sexuality and her maternity, combine to present African-American mothers in a negative light. Thus, the African-American woman is constructed as deviant, both in her role as mother and in her role as woman.[92] The traits associated with Black women stand in direct contrast to those associated with the ideal standard of motherhood. As a result, African-American women are seen as "somehow less female, perhaps even less human as well."[93] Thus, they are not maternal nor are they deserving of motherhood.[94]

The notion that African-American women are deviant, both in their roles as women and in their roles as mothers, has justified the historical abuse of Black women's reproductive liberty. Only through exploring this history can one truly understand and therefore evaluate the Norplant proposals aimed at poor, African-American women. Because dominant society has consistently used reproductive regulations and reproductive technology

---

87. Patricia Hill Collins, Black Feminist Thought: Knowledge, Consciousness, and the Politics of Empowerment 71 (1990).

88. *Id.*

89. *Id.* at 74.

90. *Id.* at 77.

91. *Id.*

92. In this context, the African-American woman's role as "mother" is almost indistinguishable from her role as "woman." Because she is a "bad" woman, the African-American woman is likely to pass her "bad" characteristics, such as laziness, irresponsibility, and immorality, on to her children. Therefore, she is a "bad" mother. The use of reproductive technology to curtail the reproductive freedom of women of color is based on this concept.

93. Ehrenreich, *supra* note 17, at 510.

94. Roberts, *supra* note 21, at 1436–44.

to curtail the reproductive liberty of so-called deviant women, any proposal to limit the reproductive freedom of African-American women should be evaluated in light of the historical denial of reproductive freedom to Black women. In this Part, I discuss various ways in which dominant society has used reproduction and reproductive technology to control deviant Black women.

A. *Slavery*

The experience of Black women during slavery typifies the control of African-American women's reproduction as an expression of their deviant status, as well as the devaluation of African-American motherhood. During slavery, Black women were valuable both for their labor and for their reproductive capacities.[95] The majority of slave women were field workers.[96] Like boys, when slave girls came of age, they were sent to the fields to "work the soil, pick the cotton, cut the cane, [and] harvest the tobacco."[97] Women, however, were also subjected to sexual abuse as a means of increasing the slaveowner's wealth. With the abolition of the international slave trade, slave women's reproductive capacities became essential to the perpetuation of the slave population.[98] For the young cotton-growing industry to flourish or for new slaveowners to increase their labor force, slave masters had to rely on Black women to reproduce.[99]

In order to reproduce the slave labor force, female slaves were encouraged and often forced to engage in sex frequently, both with other slaves and with their masters.[100] "Rape, in fact, was an uncamouflaged expression of the slaveholder's economic mastery and the overseer's control over Black women as workers."[101] As Barbara Omolade, an African-American feminist theorist, describes it, the Black woman became a fragmented

95. *See* DAVIS, *supra* note 21, at 202; Roberts, *supra* note 1, at 59.

96. DAVIS, *supra* note 21, at 5.

97. *Id.* at 6. A former slave described her introduction to slave labor: "I tended to the children when I was a little gal and tried to clean house just like Old Miss tells me to. Then as soon as I was ten years old, Old Master, he say, 'Git this here nigger to that cotton patch.' " TO BE A BLACK WOMAN: PORTRAITS IN FACT AND FICTION 16 (Mel Watkins & Jay David eds., 1970), *quoted in* DAVIS, *supra* note 21, at 6.

98. COLLINS, *supra* note 87, at 51.

99. *See* DAVIS, *supra* note 21, at 6–7.

100. Barbara Omolade, *Hearts of Darkness*, *in* POWERS OF DESIRE: THE POLITICS OF SEXUALITY 350, 354 (Ann Snitow et al. eds., 1983) ("My mother was young — just 15 or 16 years old. She had 14 chillen and you know that meant lots of wealth.").

101. DAVIS, *supra* note 21, at 7.

commodity, with every part of her body, including her womb, subject to the master's control:

> [H]er head and her heart were separated from her back and her hands and divided from her womb and vagina. Her back and muscle were pressed into field labor where she was forced to work with men and work like men. Her hands were demanded to nurse the white man and his family as domestic servant whether she was technically enslaved or legally free. Her vagina, used for his sexual pleasure, was the gateway to the womb, which was his place of capital investment — the capital investment being the sex act and the resulting child the accumulated surplus, worth money on the slave market.[102]

Thus, during slavery, Black women were not just seen as failing to conform to the good mother standard; they were never conceived of as mothers at all. They were breeders: incubators for the master's labor force, the tools of economic growth. This is the ultimate expression of the African-American woman's devalued status as woman and as mother. Her womb was seen as separate from her, like an empty box, unconnected to her humanity or any desire she might have to conceive, bear, and nurture a child.

The complete disregard of Black women's status as mothers was also codified in law. Around 1662 Virginia passed a law which ensured that "Children got by an Englishman upon a Negro woman shall be bond or free according to the condition of the mother."[103] Similarly, one year after the abolition of the international slave trade, a South Carolina court held that female slaves had no claims to their children.[104] As a result of such legislation, slave owners were encouraged to replenish their slave population through sexual exploitation of their female slaves, knowing that any children born as a result of such a union would legally be slaves. Thus, Black women lost all rights to their children. They not only lost control over their reproductive capacities, but were forced to watch as their children were taken away from them to be sold or pressed into work in the fields.

Society's refusal to view African women as human beings, much less mothers, justified both the cooption of their reproductive capacities and the loss of their children. Black women during this period were never given the chance to conform to the

---

102. Omolade, *supra* note 100, at 354.

103. Paula Giddings, When and Where I Enter: The Impact of Black Women on Race and Sex in America 37 (1984).

104. Davis, *supra* note 21, at 7.

good mother standard. They were never viewed as mothers. Black women were seen as having no more attachment or right to their children than animals. They were, in fact, seen as " 'natally dead,' with no obligations to their offspring . . . ."[105] Therefore, violation of their reproductive rights meant nothing.

B. *Margaret Sanger and the Eugenics Movement*

The control of African-American women's reproductive liberty during slavery was based on their devaluation as women and as mothers. In addition, their status as property precluded any consideration of their reproductive or maternal rights. However, even as Black women came to be viewed as people rather than property, they were still viewed as deviant. Although they had gained status as mothers, in the sense that they were allowed to nurture their own children, they were still far from the patriarchal notion of the ideal mother. Moreover, because of their deviant position, Black women began to be viewed more clearly as a threat to the social order as both women and mothers.[106] As a result, the eugenics movement, which sought to control the genetic transmission of deviant characteristics, attempted to control the reproduction of African-American women as a means of social control.

During the first decades of the Twentieth Century, the eugenics movement gained rising popularity in the United States. The movement embraced the notion that intelligence and other personality traits, both positive and negative, were hereditary.[107] "This hereditarian belief, coupled with the reform approach of the progressive era, fueled a campaign to remedy America's social problems by stemming biological degeneracy."[108] Eugenicists attempted to halt the reproduction of undesirable characteristics by preventing those believed to carry those characteristics from procreating:

> Society has been brought to a greater realization than ever of the evils that attend the presence of the growing number of the socially undesirable people in our population. The men-

105. Toni Morrison, Playing in the Dark: Whiteness and the Literary Imagination 21 (1992).

106. *See supra* part III.

107. Robert L. Burgdorf, Jr. & Marcia P. Burgdorf, *The Wicked Witch is Almost Dead:* Buck v. Bell *and the Sterilization of Handicapped Persons,* 50 Temp. L.Q. 995, 997–98 (1977); Dorothy E. Roberts, *Crime, Race and Reproduction*, 67 Tul. L. Rev. 1945, 1961 (1993).

108. Roberts, *supra* note 21, at 1961.

> tally diseased, the feeble-minded, the idiots, the morons, and the criminals are regarded as the Nemeses of our civilization and the prohibition of their propagation is considered the salvation of society and the race.[109]

Around the turn of the century, many states began to enact involuntary sterilization laws intended to preclude such undesirables from reproducing.[110]

Eugenic sterilization laws sought to control the reproduction of those who deviated from the social norm. Besides criminals, individuals suffering from mental instability or intellectual disability, and persons addicted to alcohol or drugs, this group included women who deviated from the ideal of true womanhood: prostitutes, single mothers, immigrant women, poor women, and women of color.[111] Those who supported eugenic sterilization saw the control of reproduction as a means of social control. By preventing individuals with undesirable characteristics from reproducing, society would be improved and problems like crime, poverty, and drug abuse could be prevented.

Some white reformers, increasingly concerned about the growth of the Black community, an inherently deviant group, and the decrease of the white race, began to pick up the eugenics banner as a means of controlling the procreation of African-Americans.[112] As early as 1903, white Americans, including President Theodore Roosevelt, began to fear "race suicide" or that "Yankee 'stock,' which displayed the lowest birth rates, would be overwhelmed, numerically and hence politically, by immigrants, nonwhites, and the poor."[113]

By 1919, the influence of the eugenics campaign on the birth control movement headed by Margaret Sanger was clear.[114]

---

109. *Id.* (quoting JACOB HENRY LANDMAN, HUMAN STERILIZATION: THE HISTORY OF THE SEXUAL STERILIZATION MOVEMENT vii (1932)).

110. *See generally* Elyce Z. Ferster, *Eliminating the Unfit — Is Sterilization the Answer?*, 27 OHIO ST. L.J. 591, 591–96 (1966) (describing various sterilization statutes around the turn of the century).

111. Ikemoto, *supra* note 21, at 1230. *See, e.g.*, Buck v. Bell, 274 U.S. 200 (1927). Although Carrie Buck, the plaintiff in *Buck*, was to be sterilized ostensibly to prevent the spread of mental illness, Ms. Buck was in fact sterilized because she was poor and had a child out of wedlock. Ms. Buck's daughter turned out to be an average child who even excelled in some areas. *See* Stephen J. Gould, *Carrie Buck's Daughter*, 2 CONST. COMMENTARY 331, 336 (1985); Paul A. Lombardo, *Three Generations, No Imbeciles: New Light on* Buck v. Bell, 60 N.Y.U. L. REV. 30, 51 (1985).

112. DAVIS, *supra* note 21, at 214.

113. *See* DANIEL J. KEVLES, IN THE NAME OF EUGENICS: GENETICS AND THE USES OF HUMAN HEREDITY 88 (1985).

114. DAVIS, *supra* note 21, at 213.

Sanger started her crusade for birth control based on a belief that working-class women, who could not afford multiple children, had a need for family planning and reproductive choice.[115] However, Sanger's vision of reproductive freedom did not extend to African-American women either as women or as mothers.[116] In fact, Sanger felt that the manipulation and domination of the reproduction of Black women was fully justified. Sanger stated in an article published in the American Birth Control League's journal that "the chief issue of birth control is more children from the fit, less from the unfit."[117] In a number of articles and books, Sanger promoted the use of sterilization, abortion, and contraception to eliminate "human weeds."[118] In addition, the American Birth Control League began to make room for other eugenicists in its ranks. The author of *The Rising Tide of Color Against White World Supremacy* was offered a seat on the board, and articles by the director of the American Eugenics Society began to appear in the *American Birth Control Journal*.[119]

Sanger and her associates on the Birth Control Federation of America began to directly undertake the control of the African-American population. In 1939 they started the "Negro Project" to regulate the "mass of Negroes, particularly in the South, [who] still breed carelessly and disastrously, with the result that the increase among Negroes, even more than among whites, is from the portion of the population less fit, and least able to rear children properly."[120] The Federation intended to enlist the aid of Black ministers to propagandize for control among African

115. *Id.* at 211. Sanger recounts the story of a twenty-eight-year-old woman who had attempted self-abortion. When the woman begged the doctor for information on birth prevention, he advised her to force her husband to sleep on the roof. When the woman died three months later after attempting another self-abortion, Sanger vowed to pledge herself to promoting birth control. "I went to bed, knowing that no matter what it might cost, I was finished with palliatives and superficial cures; I resolved to seek out the root of evil, to do something to change the destiny of mothers whose miseries were as vast as the sky." DAVIS, *supra* note 21, at 212 (quoting MARGARET SANGER, AN AUTOBIOGRAPHY 92 (1971)).

116. At this time, working women, as well as women of color, were considered to be deviant. *See infra* part III. I am not sure why Sanger relaxed her perception of womanhood to include working-class women but not Black women. I would guess, however, that her vision of motherhood was tainted by racism.

117. LINDA GORDON, WOMAN'S BODY, WOMAN'S RIGHT: BIRTH CONTROL IN AMERICA 281 (1976).

118. *See* GEORGE GRANT, GRAND ILLUSIONS: THE LEGACY OF PLANNED PARENTHOOD 63–65 (1992).

119. DAVIS, *supra* note 21, at 214.

120. GORDON, *supra* note 117, at 332.

Americans. Sanger wrote in a letter to a colleague, "We do not want word to get out that we want to exterminate the Negro population and the minister is the man who can straighten out that idea if it ever occurs to any of their more rebellious members."[121]

Thus, supporters of the eugenics movement saw African-Americans as an inherently deviant class, likely to produce individuals with undesirable characteristics. In addition, African-American women were not seen as proper mothers. Those in favor of eugenic sterilization and the use of birth control to restrict the reproduction of African-Americans feared that Black women would raise children destined to detract from the social order. Moreover, Margaret Sanger, a crusader for women's reproductive freedom, supported and promoted the use of reproductive technology to diminish the reproductive liberty of African-Americans. The devaluation of African-American women as mothers and the fear of their perceived inability to inculcate their children with proper values led to the denial of their reproductive freedom.

C. *Sterilization Abuse*

Sterilization as a method for controlling the reproductive capacities of African-American women did not occur only in the early Twentieth Century. As late as the 1970s, sterilization abuse of women of color in general and Black women in particular was rampant.[122] One study revealed that, in 1973, 43% of the women sterilized under a federally-funded program were African-American, although only 33% of the total number of patients were Black.[123] While the use of sterilization to control undesirable characteristics such as criminality fell into disfavor,[124] it was still acceptable to control the reproduction of women who failed to meet the ideal of true womanhood. As during the eugenics movement, the sterilization of African-American women during the 1970s served as a means of social control. The sterilization abuse of African-American women was closely linked to the notion that excessive childbearing by poor women of color was un-

121. DAVIS, *supra* note 21, at 215.

122. *Id.* at 215–19.

123. *See* Dick Grosboll, Note, *Sterilization Abuse: Current State of the Law and Remedies for Abuse*, 10 GOLDEN GATE U. L. REV. 1147, 1153 n.30 (1980).

124. *See, e.g.*, Skinner v. Oklahoma, 316 U.S. 535 (1942) (holding a state statute that provided for the sterilization of habitual criminals unconstitutional).

desirable.[125] This notion was grounded in the belief that poor women of color and their children placed a financial burden on the state,[126] as well, perhaps, as the fear that children raised in the neglectful atmosphere of poor homes would grow up to pose a threat to the social order.

Black women were often induced or coerced into submitting to sterilization. Some doctors required African-American women to consent to sterilization before they would deliver babies or perform abortions.[127] In 1974 in *Relf v. Weinberger*,[128] a federal district court found that 100,000 to 150,000 poor women had been sterilized under federally funded programs.[129] An unidentified number of women chose sterilization under the threat that their welfare benefits would be withdrawn unless they submitted to the operation.[130]

The *Relf* case exposed other incidents of sterilization abuse. In Montgomery, Alabama alone, in addition to the two teenage plaintiffs, eleven other teenage girls were subjected to similar op-

---

125. *See, e.g.*, GENA COREA, THE HIDDEN MALPRACTICE: HOW AMERICAN MEDICINE TREATS WOMEN AS PATIENTS AND PROFESSIONALS 180–81 (1973) (quoting doctors who believed that sterilization of poor women was appropriate to control the population growth of the poor).

126. *Id.*

127. *See* Laurie Nsiah-Jefferson, *Reproductive Laws, Women of Color, and Low-Income Women, in* REPRODUCTIVE LAWS FOR THE 1990s 23, 44–45 (Sherill Cohen & Nadine Taub eds., 1989).

128. 372 F. Supp. 1196 (D.D.C. 1974), *on remand sub nom.* Relf v. Mathews, 403 F. Supp 1235 (D.D.C. 1975), *vacated sub nom.*, Relf v. Weinberger, 565 F.2d 722 (D.C. Cir 1977). The *Relf* case involved two sisters, Minnie Lee, twelve years old, and Mary Alice, fourteen years old, who were sterilized. The operation was ordered by the Department of Health, Education, and Welfare-funded Montgomery Community Action Committee. The girls' mother, who could not read, consented to the operation by placing an "X" on a document whose contents were not described to her. She believed she was merely consenting to Depo-Provera injections, a contraceptive drug which had been previously administered to Minnie Lee and Mary Alice. DAVIS, *supra* note 21, at 216.

The fact that two Black girls, barely in their teens, had previously been injected with a contraceptive bears mention. One can very well ask why such a precaution appeared to be necessary. I am sure that middle- and upper-class white teens were not subjected to contraceptive injections. This only provides further support for the proposition that African-American women were never perceived, even at a young age, as good women. Clearly, they were seen as sexually promiscuous even as girls, and sexual promiscuity was seen as a potential threat to the social order — hence the need to regulate the reproduction of young, Black girls.

129. *Relf*, 372 F. Supp. at 1199.

130. *Id.* One such woman, Nial Ruth Cox, was sterilized at the age of eighteen because officials had threatened to cut off her family's welfare payments. Ms. Cox was assured that her infertility would be temporary. Les Payne, *Forced Sterilization for the Poor?* S.F. CHRON., Feb. 26, 1974, at 19.

erations, and Department of Health, Education, and Welfare (HEW)-funded birth control clinics in other states had been subjecting young girls to sterilization abuse for years.[131] The Eugenics Commission of North Carolina was responsible for 7,686 sterilizations since 1933 and about 5,000 of the sterilized persons were Black.[132] HEW's Population Affairs Office estimated that in 1972 between 100,000 and 200,000 sterilizations were funded by the federal government.[133]

Today, poor African-American women are still most likely to be subjected to sterilization abuse. Although HEW mandated informed consent as a prerequisite to sterilization, presumably to ensure neutrality on the part of the hospital, a 1975 ACLU study revealed that many hospitals were ignoring the regulations and continuing to coerce women of color to consent to sterilization.[134] Presently, 42 U.S.C. section 300a-8 prohibits public officials and authorities from forcing sterilization on welfare beneficiaries. As both Professor Dorothy Roberts and Professor Lisa Ikemoto point out, however, the practical effect of recent government enactments is to do just that.[135] Since subchapters XIX and XX of the Social Security Act authorize matching Medicaid reimbursement for sterilization, but other enactments, such as the Hyde Amendment, restrict the use of federal payments for abortion, ". . . sterilization is the only publicly-funded birth control method readily available to poor women of color."[136]

Although statistics indicate that direct sterilization abuse may no longer occur, indirect means may still be employed to convince women of color to submit to sterilization.[137] In addition, the attitudes which prompted many doctors to coax African-American women to undergo sterilization still exist. Poor women of color are still charged with excessive child-bearing which poses both a financial drain on the system and the threat of children who will become criminals and paupers. "Gena

131. *See Relf*, 372 F. Supp. at 1199.

132. DAVIS, *supra* note 21, at 218.

133. *Id.*

134. *See* Ikemoto, *supra* note 21, at 1232 n.109.

135. *See* Roberts, *supra* note 21, at 1443; Ikemoto, *supra* note 21, at 1232 n.109.

136. Roberts, *supra* note 21, at 1444. Sterilization was the only readily available form of birth control until Norplant.

137. *See* Adele Clarke, *Subtle Forms of Sterilization Abuse: A Reproductive Rights Analysis, in* TEST-TUBE WOMEN: WHAT FUTURE FOR MOTHERHOOD? 188, 200 (Rita Arditti et al. eds., 1984).

Corea reports a conversation with Dr. C., chief of surgery in a northeastern hospital: 'The doctor feels that a girl with lots of kids, on welfare, and not intelligent enough to use birth control, is better off being sterilized.' 'Not intelligent enough to use birth control' is often a code phrase for 'black' or 'poor.' "[138] Another doctor said, "As physicians we have obligations to our individual patients, but we also have obligations to the society of which we are a part . . . . The welfare mess . . . cries out for solutions, one of which is fertility control."[139] Thus, poor African-American women are still subjected to coercive sterilization because they are perceived as having too many children. Implicit in this view is the belief that poor women of color are less deserving of motherhood, and that at some point, their right to have children is subsumed by the right of the State and the American people to be free from the financial and social burden ostensibly caused by their child-bearing.[140]

A thorough examination of history reveals that African-American women consistently have been subjected to various forms of reproductive abuse. These abuses have been justified and grounded in the belief that African-American women deviated from the ideal of true womanhood. Black women were the antithesis of the patriarchal standard of good motherhood. They were perceived as unable to pass on the correct values to their children and to ensure that their children would grow to contribute to the social order. As such, they were seen as undeserving of both reproductive choice and motherhood. These same beliefs underlay the Norplant proposals of today and the notion that it is acceptable to restrict the reproductive freedom of poor, African-American women.

V. African-American Women and Norplant

> [A]t the center of the tangle of pathology is the weakness of the family structure. Once or twice removed, it will be found to be the principal source of most of the aberrant, inadequate,

138. Corea, *supra* note 125, at 180–81.

139. *Id.* at 181.

140. Although this notion could apply to all poor women, there is a strong belief in this country that the majority of poor people in America are people of color. Therefore, when people discuss "poor women" and the problems they present to society, they often mean poor women of color. *See infra* part V.A.

or antisocial behavior that did not establish, but now serves to perpetuate the cycle of poverty and deprivation.[141]

Over time, the ideal standard of womanhood has changed. The idea of women as sexual beings is no longer as shocking to the conscience. Women are no longer expected to stay in the home,[142] and men play a greater role in childcare. Some things, however, remain the same. "Women of color, those who live in poverty, and those made outsiders by virtue of cultural or religious practices are stigmatized by the dominant society and are never presumed good mothers, as are white middle- and upper-class women."[143] White women have to prove that they are unfit mothers; Black women have to prove that they are fit mothers. Moreover, the use of reproductive technology to control the reproduction of Black women persists,[144] and the Norplant proposals promise to continue this control.

In this Part, I argue that the spector of the African-American welfare mother and society's growing fear of the Black underclass have led to attempts to curtail the reproductive freedom of African-American women with Norplant. Like the reformers of the Progressive Era,[145] dominant society devalues poor African-American women as mothers and sees them as a threat to the social order. Norplant is used, not as an option intended to enhance Black women's reproductive freedom or to protect Black children, but as yet another method of curtailing the reproductive liberty of African-American women as a means of social control. Over the past decades, the country has become more and more concerned with "the welfare mess" and with the growth of the Black underclass. Society's latest attempts to regu-

141. Office of Pol'y Planning & Research, U.S. Dep't of Labor, The Negro Family: The Case for National Action 30 (1965).

142. Arguably, a stay-at-home mother is still seen as more dedicated than one who works. For example, a couple of years ago, I was watching a talk show on women who had had difficulties with violent and abusive nannies (one of the unfortunate costs of mothers who work). The audience applauded wildly for a woman who had given up her job as a lawyer to stay home and take care of her children.

143. Ikemoto, *supra* note 21, at 1207.

144. For example, there have been charges that doctors have been using the contraceptive injection Depo-Provera on teenagers and women of color without their consent. Liz Hunt, *U.S. States Use Contraceptive Implant as Social Control*, Irish Times, Aug. 5, 1993, at 8. Also, because of their devalued status as mothers, African-American women and other women of color are more likely to be given court-ordered cesarean sections. *See generally* Ehrenreich, *supra* note 17 (describing how "outsider" women are disproportionately subjected to court-ordered Cesarean sections).

145. *See supra* part III.

late the reproduction of poor, African-American women is inextricably intertwined with these concerns.

The image of the welfare mother is that of an irresponsible, lazy, and sexually promiscuous African-American woman with more children than she can possibly hope to support, "content to sit around and collect welfare, shunning work and passing on her bad values to her offspring."[146] The African-American welfare mother has no ambition to improve her lot, and because of her tendency toward sexual promiscuity, will continue to have children (often to increase her welfare benefits) who will constitute a further drain on the welfare system. In time, these children will grow up and either become pregnant or impregnate, furthering the chain of "generational" welfare dependency.

The African-American welfare mother is never seen as a good mother. She is considered to be deviant both because of her poverty, a condition eugenicists and some modern day theorists see as a personal failing,[147] and because of her lack of connection to men.[148] Moreover, she is expected to pass on these undesirable characteristics to her children. Feminist legal theorist Martha Fineman finds that poor, single mothers, except those who are single due to the death of their spouse or perhaps divorce, are assumed to be bad mothers.[149] "[I]n addition to providing a basis to determine who is undeserving in our culture, the rhetoric constructs single motherhood as dangerous and even deadly, not only to the single mothers and their children, but to society as a whole."[150] Thus, the welfare mother is perceived as posing a threat to society both through her status as a deviant woman and as a mother. This has not always been the case. Providing public aid to single mothers has always been a concern. When the concept was first introduced, professional social workers feared that so-called "mothers' pensions" would weaken the traditional family by encouraging, rather than discouraging single

146. COLLINS, *supra* note 87, at 77.

147. *See* HANDLER & HASENFELD, *supra* note 59, at 9–11 (describing the theories of Charles Murray and other moralists who link poverty to faults in the individual rather than to faults in the social system).

148. *See* Ehrenreich, *supra* note 17, at 512.

149. Fineman, *supra* note 57, at 285–86. Lisa Ikemoto points out that one example of how women of color are perceived to be less than ideal mothers is the misconception that more women of color than white women obtain abortions when, in fact, the opposite is true. Ikemoto, *supra* note 22, at 513 n.64 (citing *Ethnic Pilfering in American Society*, S.F. CHRON., June 11, 1986, at 48).

150. Fineman, *supra* note 56, at 285–86.

motherhood.[151] Many others felt, however, that it would be socially beneficial to allow poor but virtuous mothers to remain home and care for their children.[152] "There was sympathy for the widow, for her impaired ability to be a good mother, which threatened the well-being of the child."[153] "Poor but virtuous" did not necessarily include divorced mothers, mothers who had never been married, or women of color. Although some programs, at least on paper, proposed public aid for these women,[154] by and large, the programs, as administered, were for white widows.[155]

Thus, public aid for white widows was believed to benefit the social order. It was not until the welfare program began to change that welfare mothers came to be viewed as a threat to society. Between 1940 and 1975, the welfare rolls increased from slightly over 1 million to 11 million.[156] The expenditures also rose from $550 million in 1950 to $9.2 billion in 1975.[157] In addition, program participants changed from mostly white widows to include divorced, never married, and nonwhite recipients.[158]

As a result of these changes, welfare reform became federal government policy, and state and local administrators lost their ability to use discretion[159] to exclude those mothers they considered to be unworthy.[160] "[T]he formerly excluded now entered the program, which became more clearly dominated by the undeserving poor — primarily the single African-American mother and her children."[161] As the AFDC recipients began to change,

---

151. HANDLER & HASENFELD, *supra* note 59, at 66.

152. *Id.* at 66–67.

153. *Id.* at 65.

154. *Id.* at 68 ("[S]ubstantial numbers of states did potentially cover children of divorced and deserted mothers; in almost a third of the states, unwed mothers were also included . . . . Of course, statutory coverage was only potential.").

155. *Id.* at 70 ("[N]ationwide, widows constituted 82% of the program. In a 1931 survey, 96% of the families were white, 3% African-American, and 1% other.").

156. *Id.* at 113.

157. *Id.*

158. *Id.* ("By 1975, 40% of AFDC parents were white, 44% African-American, and 12% Hispanic.").

159. During the early part of this century, "[i]t was up to the line officials (judges as well as county agencies) to make the day-to-day distinctions, to separate the worthy mothers from the unworthy." *Id.* at 70.

160. For a more complete discussion of historical changes in welfare and welfare policy, see HANDLER & HASENFELD, *supra* note 59.

161. *See id.* at 123. Handler and Hasenfeld define the "undeserving poor" as those people who society views as being poor as a result of individual fault rather than through circumstances beyond their control, such as physical disability. *Id.* at 15–43.

the program was restructured to respond to the new, deviant clientele. "When African-American divorced, separated, and never-married women began to enter the program, deviant behavior social controls began to be enforced — 'man-in-the-house' rules, 'fit and proper homes,' and so forth."[162] The designation of African-American women as deviant and as bad mothers, from as early as the slavery period, caused welfare mothers to be seen as a threat to the social order. Therefore, concern over welfare was really apprehension about deviant, African-American welfare mothers and the threat they posed to society.[163] This fear was closely intertwined with the growing concern over the Black underclass.

Through their roles as mothers, poor, African-American women are perceived as contributing to America's social problems by their inability to pass on the proper values to their children. African-Americans are among the nation's most visible poor. They dominate the news whenever crime, drugs, and other social ills are discussed. Increasingly, the crime and violence of the inner cities is perceived to threaten middle- and upper-class Americans. In particular, the social unrest in Los Angeles in April of 1992 aroused fears that the discontent and anger of the urban poor would be directed at dominant society. Poor, African-American women have been explicitly blamed for these problems by theorists like Daniel Moynihan in *The Negro Family: The Case for National Action*,[164] and implicitly blamed by Bill Moyers in his television documentary, *The Vanishing Family: Crisis in Black America*.[165]

Attempts to regulate the reproduction of poor African-American women are simply manifestations of the apprehension dominant society feels towards African-American welfare mothers, the Black underclass, and the threat they pose to society. African-American women are still seen as deviant and as

162. *Id.* at 25.

163. Concern over welfare may seem to be motivated primarily by the financial burden welfare imposes on the state. However, AFDC, what most people mean when they refer to welfare, constitutes only about 1% of the federal budget. Marlene Cimons, *Myths Blur the Realities of Welfare*, L.A. TIMES, Jan. 29, 1995, at A1, A18. Therefore, the focus on welfare cannot be contributed solely to financial concerns.

164. *See* U.S. DEPT. OF LABOR, *supra* note 141. In his report, Moynihan invokes the familiar image of the Black matriarch.

165. Bill Moyers, *The Vanishing Family: Crisis in Black America*, (CBS television broadcast, Jan. 25, 1986).

undeserving of motherhood. In this sense, Norplant proposals aimed at poor, African-American women bear striking similarities to the attempts of slaveowners, Progressive Era reformers, and Twentieth Century eugenicists to control the reproduction of African-American women as a means of social control.

A. *Norplant "Bonus" Programs*

Due to dominant society's concern over the African-American welfare mother and the growing Black underclass, various welfare reforms have been proposed.[166] These proposals have included methods intended to "induce" welfare recipients to work (which assumes, of course, that they need inducements, rather than real opportunities to work) and efforts to increase enforcement of child support orders.[167] Proposals for welfare reform have also included various attempts to regulate welfare recipients' sexuality and reproduction, including man-in-the-house rules, under which welfare mothers could lose benefits for cohabitating,[168] and family cap rules, which penalize women on AFDC for having additional children by decreasing monthly benefits or refusing to support additional children.[169] In light of this, the proposed Norplant "bonus" programs should be seen simply as another manifestation of welfare reform rather than as some magnanimous effort on the part of the State to enhance Black women's reproductive options.

Like the reformers of the Progressive Era, the eugenicists of the early Twentieth Century, and those in support of coerced sterilization of women of color during the 1970s, the individuals behind the Norplant "bonus" programs see the regulation of poor, African-American women's reproductive freedom as a solution to social problems, primarily the "welfare mess." The "welfare mess" consists of African-American welfare mothers, generational welfare dependency, and the violent, mostly Black inner city.

---

166. As discussed above, although welfare reform may appear to be neutral, i.e., not directed at any particular race, because of the widely held perception that the majority of those on welfare are African-Americans, welfare reform is, in fact, directed at African-American women. "Ask the average American who is not on welfare to describe the average American who is, and the response will likely be more wrong than it is right: a black unmarried teen-ager in the inner city who has lots of children and no desire to get a job." Cimons, *supra* note 163, at A1, A18.

167. *See* HANDLER & HASENFELD, *supra* note 59, at 201–41.

168. *Id.* at 25.

169. *Id.* at 201–41.

On December 12, 1990, two days after the FDA approved the use of Norplant, the *Philadelphia Inquirer* published an editorial entitled *Poverty and Norplant: Can Contraception Reduce the Under Class?* In the editorial, Donald Kimmelman proposed using Norplant to break the "cycle of inner city poverty — one of America's greatest challenges."[170] Although Kimmelman mentions in a parenthetical that more whites than Blacks live in poverty, he focuses almost exclusively on the use of Norplant to reduce the number of African-American children born into poverty. "The main reason more black children are living in poverty is that the people having the most children are the ones least capable of supporting them.[171] . . . All right, the subject makes us uncomfortable, too. But we're made even more uncomfortable by the impoverishment of black America and its effect on the nation's future."[172] Clearly, Kimmelman, like other Americans,[173] sees African-American welfare mothers as a threat to the social order. Once again, the image of the welfare mother rears its ugly head. In spite of the fact that more whites than Blacks live in poverty and that African-Americans do not make up the majority of people on AFDC,[174] poor, Black women, rather than poor women in general, are the focus of Kimmelman's concern. Kimmelman is concerned with the procreation of African-Americans living in poverty because this community, as opposed to poor whites, is perceived to pose a threat to the rest of society in the form of violence and crime.

Across the country, legislatures also see the "induced" use of Norplant as a solution to the "welfare mess." All fifty states have included the contraceptive in their Medicaid programs to some degree. All states and the District of Columbia now reimburse poor women for the cost of Norplant, and some states have proposed offering cash incentives for the use of Norplant.[175] The Governor of Maryland has even suggested that Norplant should

170. Kimmelman, *supra* note 7.

171. I don't believe that it has been substantiated that Black women living in poverty have more children than other women. However, even if they do, it does not follow that they should lose their right to procreate because of this fact.

172. *See* Kimmelman, *supra* note 6.

173. When I say "Americans," I don't mean just white Americans. Americans of all races, including Blacks, see the African-American welfare mother as a social problem.

174. In spite of popular belief, "[t]he typical welfare recipient is white, has fewer than two children, and often lives in a rural 'mixed' income neighborhood—not a ghetto. . . ." Cimons, *supra* note 163, at A1, A18.

175. *See* Roberts, *supra* note 19, at 20.

be mandatory for women on welfare.[176] Under a plan proposed in Kansas, AFDC recipients would receive a free Norplant implant along with a $500 cash grant.[177] Recipients would also be entitled to a $50 grant for every year they kept the implant.[178]

A Louisiana plan, proposed by David Duke, would offer $100 to women on AFDC who choose to use Norplant. David Duke, a former Grand Wizard of the Ku Klux Klan, is also a supporter of eugenics. "In 1985, Mr. Duke said the 'real answer to the world's problems' was 'promoting the best strains, the best individuals' — ideas he defined as Nazism."[179] Duke's approval of the Louisiana Norplant program adds further support to the argument that Norplant is being offered to welfare mothers not to increase their family planning options, but to decrease their reproductive freedom based on their status as deviant women. Duke believes that by controlling the reproduction of individuals with undesirable characteristics, social problems will be solved. Presumably, considering Duke's ties to the Ku Klux Klan, these individuals include African-Americans and other people of color.

Although Duke's views on social engineering may be extreme, it is arguably simply a question of degree. The notion that social problems can be affected by regulating the reproduction of those believed to cause society's social problems is not that far removed from the belief that the world's problems can be solved by ensuring reproduction of only the "best" characteristics.

Kimmelman states that while the use of Norplant should not be compelled, women could be given an incentive to use it. As Dr. Segal, the developer of Norplant, recognizes, however, "the line between incentive and coercion gets very fuzzy. . . . [W]hen you single out a welfare mother, wave a $500 bill in front of her face and say that the government is going to induce you not to have children, you've gotten into a risky area."[180] All women, including poor women, should have access to affordable and effective birth control. Offering Norplant bonuses to women who can barely support their families on existing welfare grants, however, is a questionable activity. For women desperate to feed,

---

176. Ginzberg, *supra* note 5, at 982.

177. *See* Coale, *supra* note 8, at 195. Representative Kerry Patrick, who introduced the plan, also introduced a plan which would allow courts to require women convicted of cocaine or heroin possession to use Norplant for a year. *Id.*

178. Kan. H.B. 2089 § 1(a), 74th Leg., 2d Sess. (1991).

179. Craig Flournoy, *Duke Says He's Proud of Years as Klan Chief*, DALLAS MORNING NEWS, Jan. 17, 1992, at A1, A16.

180. Goodman, *supra* note 13.

clothe, and house themselves and their children, the temptation to trade their fertility for extra cash may be overwhelming, whether or not they intend to have more children or engage in further sexual activity. As Dr. Segal has pointed out, the distinction between incentive and coercion is often dubious. Ellen Goodman writes, "A profound gap exists between the promise of birth control and the threat of woman control. It's more than a $500 difference."[181]

I commend offering Norplant to women who would be unable to attain it otherwise, but why the unjustified focus on African-American women on AFDC? "Across Appalachia, throughout the Midwest's stricken farm belt, in declining textile regions and heaven knows where else, there are poor women who would welcome the opportunity to use Norplant. Assuring they can get it ought to be the real issue."[182] Proposals to make Norplant, a rather expensive contraceptive device,[183] available to all women have not emerged. This is because legislatures and social commentators, despite claims to the contrary, are not concerned with expanding women's reproductive freedom. Rather, they are determined to restrict the reproductive liberty of certain women.

Because of the fear of generational welfare dependency and the fear that the crime and violence of the inner city will destroy the rest of America, poor, African-American women, as opposed to poor, white women in the Appalachians, are seen as less deserving of motherhood. Dominant society is not concerned with the reproduction of poor, white women in the Midwest because, unlike poor African-American women in the inner city, they appear to pose little threat to the social order. The frightening images of poor, African-Americans, prone to violence and criminality, simply do not exist for poor whites.

B. *Norplant as a Condition of Probation*

In addition to proposals that offer incentives for women on public aid to use Norplant, there has also been a move to use Norplant as a punishment for women convicted of child abuse and drug abuse during pregnancy. For example, the Ohio legislature introduced a bill which would amend the definition of "child

---

181. *Id.*

182. Cynthia Tucker, *The Real Issue in Norplant Controversy is One of Access*, ATLANTA J. & CONST., Jan. 5, 1991, at A15.

183. The price of Norplant ranges from $200 to $500. *Id.*

neglect" to include drug use during pregnancy and which would require women twice convicted of this crime to use Norplant.[184]

One cannot deny the importance of protecting children from child abuse.[185] Neither can one disavow the seriousness of drug abuse during pregnancy.[186] The sudden concern over the welfare of African-American children,[187] however, is curious if not dubious. "When a society has always closed its eyes to the inadequacy of prenatal care available to poor Black women, its current expression of interest in the health of unborn Black children, must be viewed as suspect."[188] In addition, the focus of prosecutors on the use of drugs during pregnancy, to the virtual exclusion of other prenatal dangers, further places dominant society's concern for African-American children in doubt. In spite of the many other dangers women face during pregnancy, legislatures have focused on the use of drugs. "The wealth of evidence regarding maternal and child health conditions in the United States has been ignored in favor of a bizarre and inappropriate obsession with drug use by pregnant women."[189] There has been little if any attention given to the life conditions of poor women and women of color which endanger the lives of unborn children. Aside from environmental hazards women may encounter in the workplace, poor women also experience "limited access to prenatal care due to financial, practical, or cultural barriers, unavailability of decent food because the funds for programs such as the Women, Infants, and Children Food Supplement Program (WIC) have been severely reduced, and lack of adequate shelter and a

184. *See* Roberts, *supra* note 1, at 59.

185. A study by the United States Department of Health and Human Services demonstrates that 1100 children died from abuse in 1986. *See* U.S. DEP'T OF HEALTH & HUMAN SERVS., STUDY OF NATIONAL INCIDENCE AND PREVALENCE OF CHILD ABUSE AND NEGLECT: 1988, at 3–10 (1988). Others estimate that nearly 2000 children are fatally abused each year. *See Child Abuse: What We Know About Prevention Strategies: Hearing Before the House Select Comm. on Children, Youth and Families*, 98th Cong., 2d Sess. 10 (1984) (statement of Anne Cohn, D.P.H. Executive Director, Chicago, IL National Committee for Prevention of Child Abuse).

186. "The statistics most frequently cited are that 375,000 newborns per year, 11% of all births, are exposed to cocaine, marijuana, or heroin *in utero*." Michelle Oberman, *Sex, Drugs, Pregnancy, and the Law: Rethinking the Problems of Pregnant Women Who Use Drugs*, 43 HASTINGS L.J. 505, 507 (1992).

187. Drug use during pregnancy is perceived to affect mostly African-American infants because African-American women are perceived to abuse drugs more often than other women.

188. Roberts, *supra* note 21, at 1446.

189. Oberman, *supra* note 186, at 506–07.

stable environment."[190] If dominant society is so concerned about the welfare of African-American children, why has so little effort been made to address the conditions which continue to endanger their existence?[191]

Professor Dorothy Roberts, among others,[192] believes the focus on crack cocaine use during pregnancy indicates a bias against African-American women and the devaluation of these women as mothers.[193] Crack cocaine is a popular drug among many inner city women.[194] Although it is true that crack use may expose unborn children to serious health problems,[195] it is also true that other more readily available and more commonly used drugs, such as tobacco, alcohol, aspirin, and prescription drugs, may prove just as harmful to the fetus.[196] Studies also show that while Black women test positively for cocaine more frequently than white women, the opposite is true for marijuana.[197] Yet the use of these drugs during pregnancy is rarely, to my knowledge, prosecuted, nor is a great deal of attention paid to their use during pregnancy.[198] "The same proliferation of prosecutions against affluent, white women who abuse alcohol or prescription medication would be unthinkable."[199]

---

190. *Id.* at 507 n.6.

191. It is interesting to note that society is very concerned about the welfare of African-American children when they pose a threat to middle- and upper-class society, but shows very little interest in their welfare when the danger is contained. When upper-class reformers in the Nineteenth Century began to form organizations like the Massachusetts Society for the Prevention of Cruelty to Children, they were more concerned with the threat children in poverty would pose to middle- and upper-class society than with the children themselves. *See* HANDLER & HASENFELD, *supra* note 59, at 58–61.

192. *See, e.g.*, Oberman, *supra* note 186.

193. *See* Roberts, *supra* note 21, at 1434–36.

194. *Id.* at 1428.

195. Cocaine abuse during pregnancy can cause strokes, spontaneous abortion, and abruptio placentae (which necessitates a cesarian section). Newborns exposed to cocaine have lower birth rates, smaller head circumferences, a higher incidence of physical abnormalities, including deformed kidneys and neural tube defects, as well as behavioral problems such as mood dysfunction, organizational defects, and poor attention. Board of Trustees, A.M.A., *Legal Interventions During Pregnancy*, 264 JAMA 2663, 2666 (1990).

196. *See* Oberman, *supra* note 186, at 507 nn.6–7.

197. *See* Ira J. Chasnoff, M.D. et al., *The Prevalence of Illicit Drug or Alcohol Use During Pregnancy and Discrepancies in Mandatory Reporting in Pinellas County, Florida*, 322 NEW ENG. J. MED. 1202, 1204 (1990).

198. Recently, television commercials have begun to caution women about the use of alcohol and cigarettes during pregnancy. Criticism from a television commercial, however, is a far cry from criminal prosecution.

199. Roberts, *supra* note 107, at 1958.

Moreover, studies have shown that African-American women are more likely to be reported for drug use during pregnancy than their white counterparts. In spite of the fact that white women are more likely to abuse drugs than Black women, Black women are ten times more likely to be tested and reported for drug use than white women.[200]

The focus on drug use during pregnancy by African-American women, to the exclusion of others, demonstrates that unlike white women, African-American women are presumed bad mothers. Professor Nancy Ehrenreich recounts the reaction she received from students in her seminar on law, medicine, and reproduction to the information that whites are just as likely to use drugs as people of color. "During one class's discussion of prenatal substance abuse, it was extremely difficult for me to convince the students that whites use illegal drugs as often as people of color do. Although I referred them to specific data regarding comparative drug use . . . , it was still very difficult to get them to seriously consider the possibility that the disproportionate prosecution of women of color for prenatal drug use did not stem from those women's disproportionate engaging in such behavior."[201] Thus, because it is assumed that African-American mothers are more likely to abuse drugs during pregnancy, they are more likely to be reported and prosecuted for prenatal drug use.

One study of the prosecution and sentencing of women for perinatal abuse further supports this point.[202] Young, white, upper-class women, rather then being perceived as inherently bad mothers, were seen as "good girls gone bad."[203] They were seen as having made a terrible mistake from which they had learned a valuable lesson. They usually received mild criminal sanctions like probation.[204] Older white women, low-income women, and all women of color, on the other hand, were perceived much more negatively and received much stiffer sentences.[205] They

200. *See* Dan Holly, *Study: Race Affects Drug-Abuse Testing*, MIAMI HERALD, Sept. 19, 1989, at 2B; Chasnoff et al., *supra* note 197, at 1202.

201. Ehrenreich, *supra* note 17, at 513 n.66.

202. The cases in the study involved women who had given birth alone, outside of the hospital. The babies were either stillborn or died very shortly after birth. The mother disposed of the baby after death, usually in a trash can. Once the baby was found, the mother was charged with murder or some other crime. Anna L. Tsing, *Monster Stories: Women Charged with Perinatal Endangerment*, *in* UNCERTAIN TERMS 282 (Faye Ginsburg & Anna L. Tsing eds., 1990).

203. Ehrenreich, *supra* note 17, at 518.

204. Tsing, *supra* note 202, at 289, 291.

205. *Id.*

were depicted as willfully refusing to give birth in a hospital the way a proper mother would have. "[T]heir acts were perceived as consistent with their general character, rather than as lapses of judgment. They were seen as deviants who could not be changed but only controlled, and their sentences were shockingly severe."[206] Thus, deviant women are perceived as bad mothers, and as a result, undeserving of motherhood. These unwarranted negative images justify the regulation of deviant women's reproductive freedom.

As a result of unequal reporting and unequal emphasis on drugs used primarily by poor African-American women, African-American women are more likely to show up in America's court rooms on charges of drug use during pregnancy. In addition, policies aimed at reducing the number of abused children and the number of women who use drugs during pregnancy will disproportionately affect African-American women. "Because the government is more involved in [the lives of Black women], through their use of public facilities and bureaucracies, they are more susceptible to government monitoring and supervision."[207] It is here that the abuse of Norplant as a means of punishment can take place. In the past, judges have imposed contraception as a means of punishing women.[208] One month after Norplant was approved by the FDA, a judge in California attempted to require Norplant as part of an African-American woman's probation for child abuse.[209] Moreover, some judges have attempted to impose contraception on women even when their crime was unrelated to children or child bearing.[210] "Norplant presents a special temptation to judges because it's so long lasting and doesn't require any cooperation after it's implanted, and can be monitored by a parole officer just by looking at the woman's arm."[211] No appellate court has ever upheld contraception as a condition for probation,[212] but it is unclear how many women have submitted to sterilization or other forms of court-ordered contraception

206. Ehrenreich, *supra* note 17, at 518. One white woman on welfare was sentenced to twenty years in prison after she gave birth to a stillborn baby in a toilet, in spite of the fact that she was unaware of her pregnancy, and her boyfriend called "911" as soon as they realized she was in labor. Tsing, *supra* note 202, at 295.

207. Roberts, supra note 1, at 61.

208. *See* Roberts, *supra* note 1, at 59.

209. *See* Ginzberg, *supra* note 5, at 980.

210. For example, one woman convicted of stealing received Norplant as a condition of probation. Roberts, *supra* note 1, at 59.

211. Lewin, *supra* note 53.

212. *See* Ginzberg, *supra* note 5, at 981.

without appealing the decision. "This kind of thing happens a lot in lower courts and never gets challenged because the defendant's happy not to be in jail."[213] Requiring women to trade their freedom for their fertility in this way sets a dangerous precedent and renders the potential abuses invisible.

Paternalism on the part of judges who see themselves in the role of social crusaders also exposes African-American women to Norplant abuse. In the same way that doctors have been known to use their power to solve what they perceive as a social problem,[214] judges may be tempted to use Norplant as a means of implementing their own social policy. Judges may see themselves as the social reformers of the Twentieth Century with an obligation to enforce and preserve the social order. For example, Judge Howard Broadman, the Visalia, California judge who attempted to require Darlene Johnson, an African-American woman, to use Norplant as a condition of her probation, is perceived by some who work with him to be a well-intentioned but misguided social engineer. "Broadman is criticized for being naive and arrogant enough to think he can make the world better just by acting on his paternalistic impulses. 'Howard really wants to be a social worker, but he's not, he's a judge,' said a Visalia attorney . . . . And even if some of Broadman's orders are in defendants' best interests, said William Mueting, a Tulare County deputy public defender, 'Howard's not their mother. These people should make these decisions on their own.' "[215]

As much as judges and lawyers would like to pretend that justice is blind, judges are affected by the same fears and concerns as the rest of society. Judges too may perceive poor African-American women as improper mothers and feel compelled to restrict their reproductive liberty. Although Broadman denies that the *Johnson* decision involved any social or economic considerations,[216] in a case similar to *Johnson*, he stated, "I want to make [sic] to make it clear that one of the reasons I am making this order is you've got five children. You're thirty years old. None of your children are in your custody or control. Two of them on AFDC. And I'm afraid that if you get pregnant we're

213. Lewin, *supra* note 52.

214. *See supra* notes 138–39 and accompanying text.

215. Bill Ainsworth, *Tulare's Target of Controversy: Judge Howard Broadman's Creative Sentences Have Won Both Plaudits and Condemnation; One Almost Earned Him a Bullet in the Head*, RECORDER, Mar. 28, 1991, at 1.

216. *See 60 Minutes: Norplant* (CBS television broadcast, Nov. 10, 1991).

going to get a cocaine or heroin addicted baby."[217] Thus, negative images that have followed the African-American women all through time motivated Broadman's decision requiring this defendant to use contraceptives as a condition of probation: she had an excessive number of children, her improper care of those children had caused them to be taken from her, two of the children were public charges, and she was expected to use drugs during pregnancy.

The combination of paternalism with the devaluation of African-American women as mothers can be disastrous for poor, Black women. Rather than attempting to address the conditions of pregnant addicts' lives which cause them to abuse drugs during pregnancy, or which cause them to become pregnant in the first place,[218] the legislature and the judiciary have chosen to penalize these women for becoming pregnant. Drug abuse during pregnancy is a serious problem which must be addressed. However, other conditions affecting the health of unborn children should also be addressed. As long as African-American women are disproportionately and unjustifiably subjected to Norplant use as a condition of probation, legislatures, judges, and prosecutors should remain suspect.

C. *Norplant in High Schools*

Access to birth control for young women is a good thing. In light of the realities of teenage sex, affordable and reliable contraception is just as important to them as it is to grown women. Proposals to introduce Norplant to high schools girls, however, raise more than questions of access.

As with other Norplant proposals, supporters have targeted African-American women. The Laurence G. Paquin High School, a predominantly Black school for pregnant girls or girls who have already had babies in Baltimore, Maryland, was the first school to offer Norplant to teens. Paquin High School's principal, Dr. Rosetta Stith, claims that Norplant is not being targeted for anyone, and that her student body is ethnically di-

---

217. People v. Zaring, 10 Cal. Rptr. 2d 263, 267 (1992). The appellate court reviewing the decision in *Zaring* criticized "the apparent imposition of personal social values in the sentencing decision." *Id.* at 270.

218. Many female addicts exchange drugs for sex. Roberts, *supra* note 21, at 1428 n.32. In addition, very little consideration is given to the ability of female addicts to resist sexual advances.

verse.[219] However, all but five of the three hundred girls attending Paquin High School are Black.[220] In addition, according to a report by Esther Oxford, contraceptive counselling at Paquin High School encourages girls to choose Norplant over other contraceptives. "For form's sake, the Pill and the condom will be touched upon. But then the real counselling will begin. By the end of it, the principal confidently predicts, most girls will have agreed to have Norplant."[221]

That Baltimore, Maryland is the first city to begin the use of Norplant in high schools further supports the notion that African-American teenage girls are being targeted for Norplant. Baltimore has had an increasing problem with teen pregnancy: in 1991, one in ten girls ages 15-17 had babies, three times the national average.[222] Moreover, eighty-nine percent of the city's schoolchildren are Black.[223] In spite of the fact that the rate of white teen pregnancy in the U.S. is higher than that for teens in France, Sweden, England, Canada, Wales, and the Netherlands,[224] proposals to offer teens nationwide access to Norplant have not been forthcoming.

By and large, proposals to provide young women with access to Norplant have been based, not on a desire to grant teenage girls greater reproductive freedom, but on goals to decrease the burden on the welfare state and to control the social order. Young, African-American girls who become pregnant are expected to go on welfare and continue generational welfare dependency. According to the Congressional Budget Office, fifty percent of unmarried teenage mothers are on welfare within a year of giving birth, and more than seventy-five percent are on welfare within five years.[225] Those concerned with the rise of welfare dependency often point to the fact that teenage mothers

219. *Sonya Live: Birth Control in Schools* (CNN television broadcast, Feb. 11, 1993).

220. Esther Oxford, *What They Learn at Laurence Paquin School*, INDEPENDENT, Oct. 28, 1993, at 25.

221. *Id.* Norplant may be preferable to other forms of contraception because, unlike the pill, teens cannot forget to take it.

222. *Id.*

223. *Id.*

224. Tucker, *supra* note 182, at A15.

225. Donna St. George, *Pregnant Teens Get Norplant, Advice at Boston School*, TIMES-PICAYUNE, Jan. 27, 1994, at A2. However, according to a study conducted by Mike Males of Occidental College, most unmarried teen mothers are married within five years, most have jobs, and few receive welfare. Alexander Cockburn, *Clinton Blast at Teen Sex Missed Point*, STAR TRIB. (Minneapolis), Feb. 22, 1994, at 12A.

on welfare are the children of women who went on AFDC as teens.[226] Thus, these proposals are based on the desire to prevent more welfare dependency and the social problems welfare is believed to create.

When Americans discuss the ιeenage pregnancy crisis, they are discussing the supposedly skyrocketing pregnancy rate among young African-Americans.[227] African-American girls may be viewed as deviant just as easily as African-American women.[228] In addition, Black teenagers pose as much of a threat to the social order as do Black women. These girls are expected to have babies, go on welfare, continue generational welfare dependency, and perpetuate the "welfare mess." Thus, suggestions to offer Norplant to African-American teens are motivated by the same desire for social order that has motivated the regulation of African-Americans' reproductive liberty for ages.

The use of Norplant by adolescent and teenage girls, rather than other contraceptives, raises other concerns as well. None of the approximately 170 Norplant studies on adults are applicable to teens.[229] Therefore, it is unclear what long term effects, if any, Norplant will have on girls who are still developing. Christine Hohman, a biomedical researcher at Johns Hopkins University, also expresses other concerns about the use of Norplant for teenagers. "Irregular bleeding [the most common side effect of Norplant] is the most worrisome if you are giving Norplant to a population that doesn't regularly see a physician. . . . It can mask things from ectopic pregnancies to ovarian cancer."[230]

The prospect of Norplant use by teenagers requires more consideration. Lisa Ikemoto talks about the "practice of defaulting to science" in which the question of whether or not certain

---

226. For instance, in *The Vanishing Family: Crisis in Black America*, Bill Moyers emphasized the point that the African-American teens featured in his documentary grew up on welfare and that their mothers, and their mothers' mothers also became pregnant as teens. Moyers, *supra* note 165.

227. "We should not be coy as to whom we're talking about — girls in the urban and mainly black underclass." Richard John Neuhaus, *The Wrong Way to Go: What is the Real Problem Among Our Nation's Teenagers? Will Norplant Solve It, or Make It Worse?*, NAT'L. REV., Feb. 1, 1993, at 53.

228. Relf v. Weinberger, 372 F. Supp. 1196 (D.D.C. 1974), clearly supports this point; *see supra* note 128. The fact that two African-American girls, only twelve and fourteen years old, were subjected first to contraceptive injections and then to forced sterilization shows that even at a young age African-American women can be viewed as sexually promiscuous.

229. *For High School Girls, Norplant Debate Hits Home*, N.Y. TIMES, Mar. 7, 1993, § 1, at 28.

230. *Id.*

technologies should be used is never addressed once they are introduced.[231] By assuming a particular technology should be used, physicians create a need for that technology.[232] In addition, Ikemoto addresses a practice she calls "default upon default."[233] This is the process through which "[t]he derived technology or information is implemented by default, not by decision. Simultaneously, its necessity becomes lore, and it is assumed that more of the same is desirable, if not necessary."[234] Through all of these processes, technologies are accepted so quickly that no assessment of the risks is performed nor are other alternatives explored.[235]

Women have been victims of this process countless times in the past.[236] Yet, as each new technology is introduced, it is met with little or no suspicion. I believe that this can be explained by the tendency of Americans to view science as neutral and apolitical.[237] People are likely to believe that scientists are dedicated individuals with no political agendas. However, scientists are susceptible to the effects of stereotypes and social fears.

Although teenage girls should have access to safe and affordable birth control, I am not convinced that proposals to offer Norplant to teenagers are aimed at achieving this end. Instead, these proposals are based on the same motives which have fueled the control of African-American women's reproduction: the fear of African-American women's deviance and the threat that the deviance poses to the social order. In addition, the questionable safety of Norplant use by teenagers makes these proposals even more problematic. Most importantly, the use of contraception as a quick fix for teenage pregnancy needs to be evaluated. Teenage girls may be motivated to engage in sexual activity by numerous factors, some of which have very little to do with a desire for sex.[238] As long as young women engage in sexual intercourse, contraception is needed, both to prevent pregnancy and to pro-

231. Ikemoto, *supra* note 21, at 1286.
232. *Id.* at 1287.
233. *Id.* at 1290.
234. *Id.*
235. *Id.*
236. Take for instance breast implants, medication to prevent miscarriages, and IUDs.
237. Ikemoto, *supra* note 21, at 1289.
238. "At least two-thirds of pregnant teenagers . . . have childhood histories of violence and sexual abuse in their homes." Cockburn, *supra* note 225. In addition, many girls may feel pressure to have sex. One teen, explaining her reasons for wishing she was still a virgin, said, "Because then you don't get hassled by, 'Come sleep

tect them from sexually transmitted diseases.[239] This does not mean, however, that there should not be some focus on why teenage girls engage in sex in the first place. It may very well be that what African-American girls, as well as other teens, need is a boost in self-esteem, rather than a new contraceptive.[240]

Conclusion

> To me, having a baby inside me is the only time I'm really alive. I know I can make something, do something, no matter what color my skin is, and what names people call me. . . . The baby is a good sign, or at least he's *some* sign. If we didn't have that, what would be the difference from death?[241]

The devaluation of motherhood is directly related to the control dominant society exercises over reproduction. Arguably, the patriarchy exercises some degree of control over the reproductive freedom of all women. Once society decides, however, that one group of women does not conform to the ideal of motherhood, that control becomes more overt and more oppressive. For once society devalues the motherhood of a particular group, members of that group become less and less deserving of motherhood, and restrictions on their right to procreate seem more and more appropriate.

For centuries, poor, African-American women have been subjected to this process. Due to their inability to conform to the middle- and upper-class standard of true motherhood, they have been labeled as deviants and bad mothers. As such, their reproductive liberty has been restricted time after time. It is not the fact that they are on AFDC, used drugs during pregnancy, or became pregnant too young that justifies curtailing their reproductive liberty. Rather, it is their previous devaluation as mothers which makes constraint of their reproductive freedom

with me,' or 'Go out with me so I can sleep with you,' or something like that." *Primetime Live: End of Innocence* (ABC television broadcast, Sept. 9, 1993).

239. One of the arguments against Norplant for teens is that Norplant does not prevent the spread of sexually transmitted diseases, like AIDS, and that teenagers using Norplant will stop using condoms.

240. "Young girls who don't get pregnant tend to 'have some belief they are going somewhere,' said one expert who requested anonymity. 'They have some hope for the future, and there are some realistic expectations for them not to do this or put it off for a few years. Find a way to give them this expectation and I think it would be the most powerful way to reduce adolescent pregnancy." Cimons, *supra* note 163.

241. Gerda Lerner, Black Women in White American: A Documentary History 314 (1972) (quoting an anonymous Black woman).

an appropriate solution to the perceived problem, a solution which is often not suggested for other women.

African-American women deserve the right to be mothers. They also deserve the right not to be mothers if they so choose. They need adequate prenatal and perinatal care. They need to be free from the abusive use of contraception, court-ordered cesarean sections, and criminal prosecutions for drug use during pregnancy and perinatal abuse. Most of all, they need a women's movement and a legal community that is responsive to those needs to speak for them when they have no voice. They have been deprived of these rights, not because of what they did, but because of who they are.

'EMPLOYABLE MOTHERS' AND 'SUITABLE WORK': A RE-EVALUATION OF WELFARE AND WAGE-EARNING FOR WOMEN IN THE TWENTIETH-CENTURY UNITED STATES

By Joanne L. Goodwin University of Nevada, Las Vegas

U.S. welfare policy has yet to adequately address a mother's two work roles—care-giving and wage-earning. These two responsibilities have produced conflicting policy responses, sometimes within the same historical period. Citizens and legislators have raised concerns about mothers who worked too much outside their homes; or conversely, mothers who did not work enough to support their families. These contradictions are reflected in the language and practice of welfare policy. During the twentieth century, programs such as mothers' pensions gained public support by promising to subsidize some mothers to raise their children. That language changed after World War II to reflect an expectation that women must enter the workforce to earn, as in the workfare initiatives of the late twentieth century. In contrast to the shift in language over time, the practice of encouraging wage-earning has demonstrated remarkable continuity.

Similarly, the scholarship on gender and welfare has focused on one, not both, of women's responsibilities.[1] Little scholarly attention has been paid to the relationship between wage-earning and welfare for the years preceding World War II. Instead, the standing interpretation for both mothers' pensions and Aid to Dependent Children explains these programs as policies that supported women to stay at home. Whether viewed as a maternalist policy designed to protect motherhood, as presented in Theda Skocpol's recent work; or as a restrictive policy that regulated women's behavior as described by Linda Gordon, the leading scholarship discusses these programs as a relationship between government and families.[2]

In contrast, a secondary theme in welfare literature, the labor regulation thesis, focuses on wage-earning to the exclusion of gender or family issues. According to Frances Fox Piven and Richard Cloward, welfare policy encourages earning regardless of sex, family status, or race and cares little about the protection of family life or the maintenance of traditional gender relations. In fact, both scholars dismiss gender as a central element in welfare's operation. Piven and Cloward's focus on the intersections between markets and welfare policy could have filled this gap in the literature, but their insistence on gender's irrelevance and the primacy of the market over all other factors has kept the interpetations divided between class conflict and gender conflict.[3]

This article argues that policies for impoverished, mother-only families can best be explained by considering the public sector's relationship to both markets and families. This survey of the evolution of such policies illustrates that the expectation to earn has long been a component of public aid for women, even during the early twentieth-century when the greatest support for a maternal endowment existed. The connection between wage-earning and welfare, I argue, redirects us to understand the role of politics in the development of the programs, in the debates over public spending, and its impact on potential beneficiaries of public aid. By following the transitions in the policies from state-level work rules to federal policy, and eventually to the Supreme Court's legitimation of work requirements, this article offers new perspectives on the relationships among gender, race, and welfare over the twentieth century.

The article is organized around three eras of significant transition in public provision. It begins with the origins of public aid to impoverished mothers through the mothers' pension programs of the early twentieth century. The policy that rapidly passed state legislatures, largely due to its appeal to support women to rear their children at home, operated as little more than a wage subsidy. Faced with contests over public spending, the program applied exclusionary criteria, among them a woman's ability to earn. The second period examines the Social Security Act and the development of two provisions for mother-only families: Aid to Dependent Children (1935) and Survivors' Benefits (1939). The Social Security Act retained the language of maternal support but federal policy administrators never enforced this goal. Rather, state and federal governing bodies jockeyed for a workable position between federal regulation and states' rights. The compromise that resulted allowed states to implement work rules. These rules expanded the definition of parental support beyond the male breadwinner to include maternal provision as both wage-earner and care-provider. The "able worker" and "employable mother" criteria worked hand in glove with moral and racial criteria to limit access to benefits. The third section of the article discusses the development of federal work rules in the context of post-war transitions in the economy and civil rights. By the 1960s, welfare policy accommodated state work rules and progressively strengthened earning requirements within federal law—not as a mediator of labor, but as a mediator of political pressure and public spending.

Mothers' Pensions, 1911–1935

Throughout the nineteenth century, two groups made up the largest number of poor relief recipients. Mother-only families were consistently the single largest group using both public and private agencies. The exceptions to this pattern existed during periods of massive unemployment when the second group, unemployed men, turned to poor relief for aid for their families. The long-term need of female-headed families and the intermittent use of relief by male-headed families has been recognized by scholars as the economic consequences of historic social relations. Men, it was assumed, would leave relief as work opportunities returned. Women, however, were more likely to need aid for longer periods due to childrearing obligations and the low wages received for women's work.[4]

The first major transition in provision for impoverished women came with the mothers' pensions laws. Advocates recognized that charity was insufficient, demeaning, sporadic, and inadequate to meet the needs of families in an industrial nation. Social scientists noted the correlation between wage-earning, poverty, and mother-only families. Juvenile court judges and settlement workers wanted to change child welfare policy that took children from their families and placed them in institutions. A gender-conscious mobilization of traditionalist women's groups and progressive settlement house workers called for maternal subsidies to support women to raise their children at home.[5]

The arguments for pensions varied. A traditionalist appeal could be found among members of the National Congress of Mothers. One member, Mrs. G. H. Robertson, urged her coworkers to support the mothers' pension laws to strengthen the bonds of family life. "These friendless mothers are our especial responsibility." Having grown up "to believe in marriage and home-building," Robertson asked her supporters to not "let them find it a tragedy." She called for a national commission and appropriations to aid mother-only families, by which she meant deserted wives, never-married women, and widows with young children.[6] Others on the progressive edge of this movement called for social insurance for mother-only families. They wanted the government to provide insurance to poor mothers that was similar to private insurance benefits. Citing legal precedent for a wife's claim of support when the husband had property, Sophonisba Breckinridge asked if the same premise of a woman's entitlement would not apply to the woman whose husband died and left no property?

> Whether it be termed a pension for the mother, ... or pay for her services as agent of the court, or a grant in aid of family life—it should be available, sufficient in amount, regular in payment, dignifying in its assurance of the community's concern for the well-being of her group.[7]

A diverse group of supporters with a wide range of motivations built support for the new policy and succeeded in passing legislation in a majority of states.

Illinois was the first to pass a state-wide enabling law in 1911, but other states followed quickly. Two years later, eighteen additional states passed mothers' pension legislation. By 1921, forty states and the territories of Alaska and Hawaii had some form of provision for the support of dependent children in their own homes.[8]

Grace Abbott, Director of the U.S. Children's Bureau during the 1920s, explained the rapid spread of mothers' pension legislation in terms of the debate over the value and appropriate location of women's work. Abbott believed the public accepted the new programs because they recognized that "the contribution of the unskilled or semi-skilled mothers in their own homes exceeded their earnings outside of the home." She added that "it was in the public interest to conserve their [women's] child-caring functions." Abbott's assessment emphasized the well-documented evidence on women's poor working conditions, and raised the issue that different causes of poverty could demand different solutions. But she presented an overly optimistic view of the state's interest in providing funds for these programs.[9]

Abbott's statement outlined three problems in creating social policy for mother-only families. The first issue referred to the disparity women faced in the workforce. Occupational segregation and job crowding limited women's work options. Years of research by Abbott's colleagues in the Department of Labor and the nation's settlement houses had documented that the majority of wage-earning women worked in sectors characterized by long hours and low wages. These features made self-sufficiency, let alone support of a family, difficult if not impossible. Abbott's sister, Edith Abbott, had written the pioneering book on the economic status of women workers in the U.S.—a book that highlighted the structural barriers faced by women wage-earners. In addition, Grace Abbott understood very well the opposition women faced from trade unionists in improving these labor conditions for working women. The majority of women could not work their way out of poverty as men theoretically were able to do, and useful social policy would need to recognize these facts.[10]

Second, Abbott recognized that the prevalent cultural view of women's primary duty as a mother had played a significant role in the rapid spread of mothers' pension laws. The National Congress of Mothers and the General Federation of Women's Clubs—two organizations with the largest membership of any women's groups in the early twentieth century—promoted the maternal role over any other. Abbott acknowledged the views of these traditionalist women's groups with her reference to the "public's interest to conserve [women's] child-caring functions." She also knew that social theorists blamed working women for family fragmentation and juvenile delinquency. Any social policy for mother-only families would have to incorporate these ideas into its plan.

The part that Abbott portrayed too rosily—the third problem in creating new policy for mother-only families—was the public's willingness to pay for a subsidy to female-headed families. As Director of the Children's Bureau, she was intimately aware of meager family stipends, requirements for women to earn, and the differential treatment of mothers based on their race, marital status, and in some cases, family size. At the level of implementation, the public discourse which endorsed aid to impoverished mothers met a serious challenge from an equally strong demand for families in poverty, whether male-headed or female-headed, to earn their way to self-sufficiency. A conundrum evolved at the level of implementation. Aid for mother-only families would be endorsed, but a municipality could not possibly support all mothers in need of aid. Consequently, in county after county, administrators of the program developed a variety of screening mechanisms, including one that considered the mother's "ability to earn."[11]

A specific example from Chicago demonstrates the use of work requirements. When first passed in 1911, the mothers' pension law authorized that public funds could be provided to parents for the support of their underage children. That provision included single as well as two-parent families, it listed no limits on the size of grants, nor did it describe other eligibility requirements. The Chicago program became the focus of national attention and debate due to its size and the involvement of its notable reform community. By the next meeting of the state legislature, political pressure had forced a revision of the law and created one of the most restrictive measures at the time (1913). New eligibility rules restricted grants to widows and wives of disabled men; children under the legal work age, which was fourteen years at the time; and citizens.[12]

Wage-earning became intertwined with public aid in two ways—through the law and through denied eligibility. The law stated that a mother "may be absent [from home] for work a definite number of days each week to be specified in the court's order," and women received an examination to determine the degree to which they could earn wages.[13] The history of Mary Legaikas, referred to in the case records of the Juvenile Court as Mrs. C, illustrates a common situation for pensioned mothers.

> In June, 1913, Mrs. C, a Polish woman, applied for a pension for her two children aged 8 and 5 years because she found it impossible to earn enough to support them. Her husband had died of heart disease in 1909, leaving some insurance; but the money had been used for paying funeral bills, debts, and living expenses. The family had been compelled to ask help from the county agent and the united charities a number of times during the four years following the death of the father. A stepson had gone to work at the age of 14, but Mrs. C found him so unmanageable that in 1911 she sent him to his uncle in Tennessee. Mrs. C had been earning only $10 a month by sweeping in a school.[14]

The Juvenile Court's caseworkers estimated this family's budget at thirty-four dollars a month, and granted the two children a pension of ten dollars each. The pension, along with Legaikas' earnings of ten dollars a month, left the family only four dollars short of their need. When Legaikas developed health problems from cleaning the school, caseworkers helped her find a job cleaning in a bank where she earned two more dollars a month. Mary Legaikas' story was recorded as a "success" among pensioned families, but her wage-earning was not exceptional. Approximately one-half of the women who received pensions in Chicago during the first fifteen years of its operation engaged in wage-earning.

The second way in which wage-earning became a component of the mothers' pension program was the practice of denying women aid on the basis of their "potential to earn." Although all able-bodied women were urged to find some type of employment, African American women were believed to be more able to find work. A combination of job segmentation which retained certain classes of jobs for blacks and the perception that "black women had always worked" led to this racialist perception. For example, take the case of an anonymous African American woman who had received a pension for her five children until the 1913 revision made her ineligible. At that point, she turned over the home responsibilities to her fifteen-year old daughter and went out to work. Two years later, when the law became more inclusive, the woman reapplied for a pension, but was denied on the basis that she was "a good wage earner." The court gave precedence to her role as an earner over that of a caretaker for her dependent children, and refused even a subsidy for the children, such as that received by the Legaikas family.[15]

The expectation that women who received mothers' pensions would also work for wages was not unique to Chicago. A Children's Bureau study published in 1923 found that local administration frequently encouraged wage-earning by mothers to meet the gap between family needs and pension stipends. The report noted:

> In each place studied there were some mothers who did not have as much free time to devote to their children as was considered desirable. Some mothers were working because there was no other way to get an adequate income for the family, although the physical strain of work in addition to the care of the house and the children was probably more than they could long endure ... In many instances, however, it was believed that some money-earning occupation on the part of the mother was a wholesome influence in the family life.[16]

Slightly over half—52 percent of the 942 women included in the study—earned to support the family. Twice as many worked outside the home for wages as those who did piecework at home. Furthermore, nearly half (48 percent) of those who worked outside the home worked three or more full-time days per week.[17]

The mothers' pension program in practice diverged quite significantly from the initial proposals of its proponents, and its public benefits were distributed differentially on several grounds. Fiscal and political pressures combined with moral and racialist ideologies to shape the policy in its implementation. Rather than a right or entitlement, mothers' pensions became a needs-based program. From this perspective, it is apparent that using economic criteria like one's ability to earn allowed local administrators broad discretion in providing or witholding benefits. This perspective on the linkage between public aid and work requirements places the failure of women's employment policies in stark relief. The competition between self-sufficiency and care-taking contributed to the weakness of the program at this early stage. To what degree did these ambiguities accompany the program in the transition from locally administered, state-level programs to a federally-supported policy in the Social Security Act?

Aid to Dependent Children, 1935–1962

Historians have looked upon the great safety net of social provision set in place by the Social Security Act as the foundation of the modern welfare state, albeit a limited one. The exclusion of domestic service and agricultural work from social insurance benefits resulted in a two-tier system that protected some workers and not others. Furthermore, many New Deal programs inscribed with family wage ideals limited women's access to social security by emphasizing their primary responsibility within the home, as Linda Gordon has argued. Yet, as this section will illustrate, the *practice* of welfare continued to blur women's dual roles. Most important, the meaning of parental support—as a male contribution to the family—changed. In the decades following passage of Aid to Dependent Children, state and local governments increasingly passed rules that linked welfare with wage-earning.

The Depression all but dismantled the state mothers' pension programs. The policy had existed in 1600 counties in 45 states by 1931, but during the next two years, counties began to drop the program. Declining revenues limited the number of families able to receive the aid, and social services went through a period of considerable fluctuation during the first years of the economic crisis. The Children's Bureau reported that 90 counties had stopped their programs between 1931 and 1933. Some counties retained the program, but added rules that excluded families on the basis of race or family size. Many cities simply kept eligible, but unfunded people on a waiting list. In Chicago, 1,434 families received mothers' pensions in 1935, but the waiting list for the program contained an additional 7,942 families. The public relief department carried 3,870 additional families who could have qualified for mothers' pensions, if additional funds had been available.[18]

During the first few years of the Depression, social service agencies made repeated attempts to convince the federal government of the severity of the economic crisis and the necessity for federal intervention for unemployed families. By 1935, federal funds helped states establish relief and employment programs. That same year, after much debate and compromise, Congress passed a set of provisions designed to provide a safety net for American workers and their families called the Social Security Act. The act had two components: one for social insurance and one for public assistance. Promoters of social insurance met with some success particularly in the provisions for unemployment insurance and old age insurance, but the program had many shortcomings. The social insurance provisions covered limited areas of work and by contemporary estimates excluded about one-half of the working population. In addition, the categories of coverage excluded a majority of women and minority workers. These omissions led to a compromise position to maintain the relief programs; thus the public assistance section of the Social Security Act included categorical aid to the disabled, aged, and dependent children.

Title IV included provisions for Aid to Dependent Children (ADC) and specifically authorized funds for the support of families "who have been deprived of parental support or care by reason of the death, continued absence from the home, or physical or mental incapacity of a parent."[19] As such, supporters maintained their ideal "to enable the mother to stay at home and devote herself to housekeeping and the care of her children."[20] In addition, ADC expanded the benefits of mothers' pensions in two important ways. First, a state's acceptance of the program and its federal funds obligated it to implement the program in every county in the state, share costs with counties, and coordinate the program from one central agency. Mothers' pensions never had this breadth of coverage. Second, ADC expanded eligibility to include deserted, separated and unmarried mothers. Only a few states had previously passed such inclusive eligibility.

On the other hand, ADC passed Congress with a series of compromises that reduced the scope of the Act and left considerable control to local administration. Southern Congressmen had effectively eliminated a provision in the original legislation that would have standardized benefits to all recipients by requiring states to provide "a reasonable subsistence compatible with decency and health." That stipulation in the policy would have authorized an equitable provision of public aid regardless of race, a goal which they perceived to be incompatible with southern social relations. Congressional compromises also allowed localities the flexibility to limit eligibility for ADC relative to their resources. Some states continued to use the "ability to earn" criterion as a mechanism for limiting benefits, and in some cases, aiding local labor markets.[21]

The creation of categorical aid to dependent children complicated policy for female-headed families. Congress divided public provision into two categories during its debates over the Social Security Act. Social insurance covered those determined to be "employable" while an extended welfare system covered those people determined to be "unemployables." This policy decision continued to muddy the definition of women's work and fostered a situation in which mother-only families remained outside provisions in employment programs or social insurance, yet had access to only the minimal resources through ADC. The 1939 amendment to the Social Security Act resolved this situation for one group of mothers. Survivors Insurance for Widows of Workingmen secured benefits for the wives and children of male workers in covered industries. This provision of insurance became the "counterpart of the protection of the wage-earner and his aged wife or widow."[22]

The Social Security Act's contributory system, therefore, institutionalized a division between female-headed families. Wives of industrial workers received a social security insurance payment sufficient to maintain the family. There would be no economic means test, no periodic inspections of the home, no rules governing the behavior of the widow or her children. Justification for the grant came entirely from the male earner's contributions, and it was seen as an extension of the social insurance package for those men working in the specific covered industries. The second group of mother-only families received benefits from the categorical relief program of ADC. If one's husband had worked in agriculture, service, or trades that remained outside the umbrella of social insurance, one received no survivor's benefits. If a woman's husband divorced, deserted or separated from her, or if she never married, she must first apply for benefits and then be subject to regulation and scrutiny.[23]

To what degree did the earlier practices, which expected mothers on aid to supplement family income with their wages, continue during these years? To answer that question we need to turn first to the relief programs that characterized New Deal measures well beyond the passage of the Social Security Act in 1935. Several states took years to put these plans in place. As categorical relief was transferred from the federal emergency programs to state and local governments, some areas failed to raise funds for the aged, blind, and dependent children. In general, work projects under the Federal Emergency Relief Administration (FERA) or the Works Progress Administration (WPA) characterized mother-only families as "unemployable" and designated work relief projects predominantly for male-heads of families. Exceptions existed, however.[24]

The line between government work relief, welfare, and low-wage work could be particularly narrow. In Minneapolis, employers' associations frequently pressured the public welfare department to limit eligibility and reduce stipends. In 1936, hundreds of women who received direct relief had their grants cut. An investigation into the welfare board's action revealed that the department had "made an organized effort to force single girls who are on relief to accept jobs as domestics in homes at starvation wages." The opposition charged the state with forcing women to accept work at sub-standard rates.[25] In Virginia, Washington, D.C., South Dakota and New Jersey, local relief agencies closed welfare cases when local agricultural employment or domestic service was available. FERA supported workers' choices to refuse private sector employment when wages were too low, conditions poor, or the prospect of being paid doubtful. At the same time, the agency maintained that available seasonal employment "such as harvesting of tobacco and cotton" should be taken by the able-bodied in lieu of relief.[26] Urban workers also blended direct relief with wage-earning. A study of city workers in 1934 found over one-half of those who used both relief and earnings worked nearly full-time. Service occupations, particularly servants, hotel and restaurant workers, cleaners, and laundresses exhibited higher incidences of this pattern. The combination of private wages and public relief raised the concern that federal funds served as wage subsidies. These incidents anticipated similar charges by civil rights lawyers thirty years later.[27]

The Woodrum Act of 1939 changed the eligibility for the WPA and explicitly excluded anyone who could meet ADC eligibility. This effectively eliminated work programs as an avenue of support for female-headed families at a time when ADC administration was overwhelmed by demand. Women who received ADC probably continued to supplement it when and wherever possible, despite legislative language that discouraged earning.

The intermingling of race and gender ideas with labor segmentation was also evident. A study of the administration of ADC in southern states explained the underrepresentation of African American women on welfare as a result of local culture and attitudes. A field supervisor explained the discrepancy between need and coverage as "a unanimous feeling on the part of the [welfare] staff and board that there are more work opportunities for Negro women.... The attitude that 'they have always gotten along' and that 'all they'll do is have more children' is definite." Referring again to the local southern welfare board, the study reported that local administrators "see no reason why the employable Negro mother should not continue her usually sketchy seasonal labor or indefinite domestic service rather than receive a public assistance grant."[28] Over the next twenty years, other states would expand their expectations on mothers to earn.

Between 1935 and the early 1960s, state governments contested, challenged, and changed the meaning of "parental support" in Aid to Dependent Children. The original provision intended to aid children whose father had died, deserted, or suffered from a long-term disability. Consequently, the language of the act defined eligible children as those who had "been deprived of parental support or care." To the authors of the Social Security Act, this phrase meant that fathers would provide the income and mothers would provide the care. However, parental support came to mean a mother would assume both roles as the pattern of mixing relief and wage-earning crystallized into policy in some locations during the 1940s and 1950s. Louisiana adopted the first "employable mother" rule in 1943, only eight years after passage of the SSA. The state maintained that all capable women with children over the age of seven years who received public aid could be denied assistance when field work was available. Georgia implemented a similar rule in 1952 but lowered the children's age to three years.[29] The federal Bureau of Public Assistance did not support compulsory earning by mothers, but the agency proved unable to control the practice or maintain desired standards of care. Having been given the leeway to set limits on aid, states used criteria such as family size or structure, residency, and "employable mother" rules to reduce public costs and address local political demands. Local control over eligibility also gave governing bodies the flexibility to address taxpayers' complaints and to retain "customary" practices of race relations. The Bureau of Public Assistance responded with additional regulations, but was unable or politically unwilling to challenge the many and varied state departures from federal rules. It has been estimated that "two-thirds of eligible dependent children" remained uncovered by ADC in 1940.[30]

The dual roles imposed upon poor women can also be seen in the competing directions of federal welfare legislation during the 1950s. Federal initiatives attempted to strike an uneasy balance between "strengthening family life" through reinforcement of women's traditional caretaking roles, while encouraging parents "to attain the maximum of self-support." To move toward the first goal, Congress passed a caretaker grant that made a specific provision for "the mother or other relative" in its 1950 Social Security Act Amendments. The grant brought ADC in line with other categorical aid provisions, such as Old Age

Assistance or Aid to the Blind, and some promoters hoped that this additional stipend would decrease the need for women to earn.[31]

But efforts "to attain the maximum of self-support" may be found in two restrictive measures passed with the same Amendments. The Notification of Law Enforcement Officers had a direct effect because it required welfare officials to notify the police when a deserted child received aid. The Knowland Amendment legitimated a reduction of public aid during labor force shortages. Although it was directed at unemployment insurance, not ADC specifically, the measure paralled those mentioned above in Minneapolis during the Depression, or in the various counties that closed welfare cases when field work became available. It foreshadowed the ability of states to circumvent federal initiatives.[32]

By 1962 when federal legislation first adopted employment policies, it followed the practice of a majority of states. Thirty-three states had work requirements in their ADC regulations, and many included a provision that could deny aid if employment was refused.

Federal Employment Policy in Aid to Families with Dependent Children, 1962–1972

Most commentators refer to the Family Support Act of 1988 as the turning point in work requirements for women on Aid to Families with Dependent Children (AFDC). That statute placed responsibility for the financial support of children upon both parents. It implemented firm measures to collect child support from fathers, and it provided funds for employment and training. The Family Support Act has been characterized as a profound change in welfare policy that created a new dilemma about the role of mothers. Yet, as this section demonstrates, the federal government made a commitment to mandate earning from women on AFDC much earlier and the courts legitimated requirements to earn. These precedents by the legislative and judicial branches validated the authority of localities to control public costs.[33]

Why did federal policy change during the 1960s? The literature offers three explanations. Social scientists most frequently cite post-war demographic patterns, and argue that ever increasing numbers of female-headed families relied on public support which resulted in a welfare crisis. This occurred at the same time that greater numbers of women entered the workforce at all economic levels. A shift in public values about working mothers corresponded with the rise of mother-only families on AFDC and led to the decline of protective, maternalist sentiments. Yet, even though demographics reshaped post-war welfare, the poor had always been asked to contribute to their support. The shift in social values toward wage-earning for mothers applied more to middle-class women than to poor women. The second explanation places the motor of change in labor regulation. Evident throughout U.S. history, labor controls became a justified practice in the post-war years as local administrators expanded or contracted welfare eligibility in accordance with local labor demand and public budgets. Labor segregation accounted for the disproportionate exclusion of certain groups and fostered racial discrimination. As this article discusses, lawyers challenged labor regulation in the states during the 1960s and 1970s. Yet as a comprehensive explanation of change, labor regulation does not explain the absence or outright failure of work policies in some areas. The third explanation argues that welfare attempted to regulate women's sexual behavior particularly through the use of "suitable homes" and "fit mother" criteria, regulations that cut women off grants if their personal behavior deviated from official rules. Welfare policies clearly tried to regulate women's sexual behavior, however the bulk of evidence demonstrates that the policies sought to reduce the size of public aid rather than control household labor or childbearing.[34]

None of these explanations alone explains the shift in federal policy at the time, and they only marginally address the political context of the change. Furthermore, the sketchy historical accounts of "employable mother" rules mistakingly suggest that these state laws were overturned. In fact, the opposite occurred.[35] This section will demonstrate that by the 1960s support for female-headed families had become politically and economically unfeasible. The needs of states to control their expenses featured prominently in both political arguments and court cases. Federal welfare policy placed wage-earning expectations upon both parents and as such accommodated itself to and expanded upon state "employable mother" rules. While exemptions from work rules still existed when family conditions or child care demands necessitated, these care-based ethics also could be suspended by local decisions. The increase in work requirements for all women receiving AFDC in the decades following World War II had its roots in the political as well as economic struggles of welfare policy.

During the 1960s, the central assumption guiding federal welfare policies was continued economic growth. John Kenneth Galbraith, in his influential 1958 book, *The Affluent Society*, predicted that the expanding post-war economy provided the opportunity for increased governmental spending in the public sector—spending which could reduce the degree of poverty in America. Michael Harrington described specific sectors of poverty in the U.S. in *The Other America* published in 1962 and also called for massive increases in federal funding. Influential members of the Kennedy and Johnson administrations lobbied for targeted interventions, and yet, the plans employed relied on fiscal measures. The Council of Economic Advisors designed conservative measures like tax decreases to spur economic growth, and job training—not job creation—to address unemployment.[36]

Social science research, available to policy makers at the time, contradicted the thesis that economic growth alone could reduce the incidence of poverty. Structural barriers to employment emerged during the post-war era that made it more difficult to work one's way out of poverty. The mechanization of agriculture, beginning in the 1940s and continuing for two decades, reduced the need for manual labor in the cotton fields of the South and displaced thousands of African American and white field workers. Thousands of people migrated to the cities of the northeast, upper midwest, and Pacific Coast to fill the wartime demand for labor, but lost those jobs in the post-war demobilization. The aged and children, groups outside the labor force, made up significant sectors of the poor. Policy makers had firm evidence by 1963 that "technological change, shortages of decent paying jobs, and racial discrimination" rested at the core of rising poverty rates, as James Patterson has noted. Yet, the policymakers adopted fiscally conservative anti-poverty strategies and created programs that relied on individual initiative.[37]

William Julius Wilson described the increased rates of poverty for urban racial minorities relative to the effects of migration, economic restructuring, and social isolation. The migration of southern blacks to northern cities during the 1940s and 1950s expanded the population of those cities with disproportionate numbers of unskilled workers, many of whom were young. At first, higher paying industrial jobs existed and later civil rights legislation opened those jobs to blacks, but in about a generation the national economy shifted from one based on manufacturing and industry to one that focused on "higher-order service provision." The inner-city poor typically attained less education, which placed them at a disadvantage for these new jobs. Manufacturing jobs left the industrial northeast and midwest and relocated in other regions of the country or outside the U.S. entirely. The expansion of lower-order service jobs between 1975 and 1985, which could have supplied new jobs to low-skill workers, occurred outside the inner city. Wilson's structural analysis moved the discussion of rising poverty rates away from behavior to focus on larger economic transitions and their relation to poverty rates. His thesis raised questions about the inherent inadequacies of welfare reforms that relied on wage-earning for self-support during periods of declining job availability.[38]

Wilson's study concentrated on male unemployment in inner cities. The increased rates of poverty for female-headed families, which had risen significantly in the post-war decades, would decline, Wilson argued, as the economic prospects of African American men improved and expanded the "marriageable pool." This part of his thesis has been challenged on two specific points. It perpetuated an analysis of women as economically reliant on men, and it consequently obscured the two challenges historically faced by women: their responsibility for child care and their weak position in the wage-economy.

Women were certainly affected by the declining financial fortunes of men, but additional economic trends after the war had an impact. The expanding service sector created a demand for workers in this low-skill, low-wage sector, and women continued to move into the labor market in greater numbers. But the characteristics of these jobs—wage ceilings, absence of benefits (particularly health care and child care), and little correlation between productivity and wage increases—made it difficult, if not impossible, for women supporting families to get ahead.

Federal welfare reform proposals began to discuss openly work incentives for women receiving AFDC in 1962. Over the next thirty years, continuing to date with the Clinton administration's welfare reform proposals, the central premise of these reforms has been to increase women's wage-earning contribution. The 1962 Social Security Amendments allowed states to add unemployed fathers to AFDC benefits, and in so doing provided the first language for work requirements in federal AFDC rules. The Kennedy administration wanted to emphasize services that would enhance both parents' efforts to become self-supporting. It provided funds for the reimbursement of child care expenses to encourage women to work, but the amount allocated remained relatively small. Disincentives remained in the program, in the form of a 100 percent wage deduction from welfare benefits.

By the mid-1960s, the growing number of families receiving AFDC, the rise in the number of never-married mothers receiving assistance, and the increased costs to localities created what academic and popular commentators referred to as a "welfare crisis." States and cities cut welfare budgets to reduce caseloads, but despite these attempts budgets continued to expand. The War on Poverty programs included employment programs, but they solicited primarily unemployed men and teenagers. Neither the employment programs nor the work requirements for mothers proved effective in moving people into self-supporting jobs. By 1967, public support for social programs waned in the context of an expensive war in Vietnam, continued social protest, and unrest in the nation's cities. Politicians found that their constitutents were a receptive audience for calls to reduce welfare costs. Their arguments focused on personal behaviors—illegitimacy and a lack of the work ethic—as the source of the enlarged social programs. Wilbur Mills, Representative from the state of Arkansas and chair of the House Ways and Means Committee, played a major role in the campaign to restrict benefits to AFDC and require wage-earning.

The 1967 Social Security Act Amendments attempted to correct the ineffectiveness of the earlier provisions. Those amendments, referred to as the Work Incentive Program (WIN), created a plan for comprehensive services, job training, and employment. Congress authorized funds for job training and child care. WIN's incentive allowed a woman to keep part of her earnings in addition to her AFDC benefit, thus eliminating the 100 percent tax on earnings. Women could keep the first thirty dollars and one-third of any additional amount earned. The measure included stiff penalties, however, and benefits could be cancelled for anyone who refused to participate or refused to accept employment. WIN also allowed exemptions for mothers with children sixteen years or younger in deference to caretaking responsibilities, yet by the end of 1971, Congress passed the Talmadge Amendments which expanded the work component. They required all recipients to register for jobs and lowered the age exclusion to six years of age. States received a mandate to submit fifteen percent of their AFDC families for jobs or training.[39]

The discussion surrounding work requirements in 1967 and 1971 illustrated clearly that lipservice to motherhood aside, the priority in the law was to reduce AFDC rolls and the corresponding public costs of those programs. The House Ways and Means committee report on the SSA in 1967 expressed its concern "over the rapidly increasing costs to the taxpayers" and the expanding numbers receiving public aid. The report linked the program's expansion to greater numbers of female-headed families and an increase in "out of wedlock" children receiving aid. The report also made it clear that the committee members were aware of an "increasingly critical public attention" to AFDC. Yet, WIN had limited success placing AFDC recipients in jobs. Too few people received referrals to work programs, less than two percent obtained jobs, and the majority of those jobs were at low wages. Furthermore, WIN prioritized recipients for training and placement. Fathers received preferential treatment and were thought to be the most able to re-enter the market. Single mothers and married women followed. While it was politically and fiscally imperative to increase work requirements, the federal government left the details to the states.[40]

Federal policy did not seek to macro-manage labor supply, but at the same time it did not enforce rules that would prohibit states from passing work rules. The following discussion of two seminal court cases reveals that in some locations, state and local public welfare administrators managed the size of their programs using "employable mother" criteria to exclude otherwise eligible families. As mentioned earlier, thirty-three states had work requirements in their welfare codes before 1962. A close examination of these two cases: one in Georgia that provided the first constitutional test on employment as an eligibility requirement, and one in New York that eventually led to the Supreme Court decision, illustrates the emerging legal position on the relationship between women's wage-earning and receipt of AFDC. In both cases, the courts reaffirmed a state's right to make wage-earning a condition of aid.[41]

During the late 1960s and early 1970s, civil rights and anti-poverty lawyers presented cases before state courts and the U.S. Supreme Court in an effort to eliminate racial discrimination in welfare policy and to expand the rights of the poor. The Supreme Court decided in favor of expanded legal protection for welfare recipients and overturned several state laws between 1968 and 1973. These cases ended the "substitute father" laws that had required financial support from non-related men, and residency requirements. They also entitled recipients to due process through hearings regarding their case. But state and Supreme Court decisions that involved issues of employment and the control of public costs fared less favorably. The majority of the Supreme Court believed that welfare represented an experiment in "cooperative federalism" and as such required a delicate balance between state and federal powers. Success in these relations depended upon "considerable latitude" granted to the states. Nowhere is that more apparent than in the "employable mother" rules. The first constitutional test of the "employable mother" rule occurred in Gèorgia in 1968 as part of a larger civil rights legal campaign.[42]

Georgia legislators had historically resisted public provision for female-headed families. The state had the dubious distinction of being one of two states that had not enacted any form of mothers' pensions as late as 1931. By 1938, the state temporarily lost federal matching funds for ADC because it had established a racial quota system. Over the next two decades, African Americans consistently received less access to ADC than whites relative to their population and need. In 1951, the Georgia state legislature passed the first law in the country that denied aid to "more than one illegitimate child of a mother." The following year, the law was repealled on constitutional grounds, but the state legislature implemented a plan that would achieve similar goals. The plan comprised several measures that would establish paternity for all children, enforce "substitute father" liability, enact suitable home rules, and require wage-earning. The "employable mother" rule is the only limitation among these that has not been discussed in the literature.[43]

Georgia's "employable mother" rule required all able-bodied mothers to work when "suitable work" was available, and when they were not needed in the home. This policy had many characteristics of earlier practices in various regions of the country; yet, the Georgia case was distinctive for its subtext of gender and racial expectations regarding employability. Georgia defined "suitable work" as employment in line with a woman's "training and experience." It allowed for individual aptitude, but it also classified types of work appropriate to women workers, for example, domestic service. The rule further defined work as that which kept "with the prevailing wage scale and working conditions in the community." The County Boards of Welfare could use their discretion about "the habits and customs prevalent in the community" to determine if a woman was able to engage in part-time or full-time work. The language of the rule and the flexibility it granted to local boards to make decisions based on existing community "habits and customs" created an opportunity for discriminatory practices in the context of Georgia's previous history of aid to dependent families and the period's racial politics.[44]

In the fall of 1966 and spring of 1967, several African American welfare recipients working with C. B. King, an Albany, Georgia attorney known for his litigation of civil rights cases, and Jack Greenberg, Charles Jones, and Stephen Ralston of the NAACP Legal Defense Fund, challenged Georgia's regulation. The lawyers for the plaintiffs argued that county welfare boards cut black families off AFDC when agricultural work or domestic service was available. The boards assumed the work to be full-time employment, and eliminated any supplemental aid from public funds to those families. The rule violated their constitutional rights, compelled women to leave their young children, maintained "a stable underpaid agricultural labor force," and fostered racial discrimination.[45]

The plaintiffs' case focused on three parts of the "employable mother" rule. First, the rule made work mandatory for any mother who received aid. It stated that she "must accept available 'suitable' employment" if it provided necessary "safeguards" for her children. Second, the rule allowed county welfare boards located in areas with seasonal employment to "designate such periods as periods of full-time employment," and to deny aid to AFDC applicants and close current cases during those months. The fact of actual employment was secondary to the premise that work was available. Third, the rule distinguished between full-time and part-time employment and denied supplementary aid to full-time workers regardless of their need. The defendants in the case, officials of the county and state welfare departments, argued that Title IV of the SSA called for parental responsibility for the support and care of the child. Mothers, they argued, had the same standard of responsibility as fathers. This position collapsed any division of labor between parents and placed the burden of both on the single mother.

The Fifth Circuit Court of Appeals found Georgia's Public Welfare "employable mother" rule valid and within constitutional protections. Women must accept employment when it was available. The case did result in a change of policy on the second and third points, however. Before the case went to court, the state department of welfare revised two parts of the rule. Actual employment replaced presumed seasonal employment and the distinction between full-time and part-time workers was modified. Shortly before the Court of Appeals finding, the plaintiffs added charges against the federal and regional representatives of the U. S. Department of Health, Education, and Welfare who had approved the plan, and filed the case in U.S. District Court. The District Court agreed with the Court of Appeals and supported the work requirement in Georgia's plan. The judges found "there is no federally protected right of [a] mother to refuse employment while receiving assistance and remaining at home with her children." But the District Court did agree that Georgia had violated the Fourteenth Amendment's equal protection clause by denying aid to full-time workers and thus "discriminating on [the] basis of source of income."[46]

The plaintiffs' lawyers succeeded in eliminating discriminatory practices in the receipt of AFDC. The decision allowed women to combine public aid with full-time wages, a benefit previously denied. This had particular importance considering the low wages of the work. But the plaintiffs had hoped to challenge broader premises of welfare law—the "right not to work outside the home" and the segregation of employment patterns.

The issue of work requirements in AFDC reached the Supreme Court in 1973. The case involved public assistance recipients who had challenged New York State's requirement that people must accept jobs as a condition of eligibility. This case followed the implementation of WIN, and the plaintiffs sought to have the federal provisions supercede the state work rules. WIN offered greater support services and broader exemptions from employment. The District Court agreed, but the state appealed the decision.

The U.S. Supreme Court upheld New York State's work program, but the majority and minority reports offer an illustrative comparison of the issues surrounding employable mothers. At the center of the majority decision was the state's right to control work policies as a way to manage costs. Justice Powell, writing for the majority, rejected the claim that WIN superseded state plans. If Congress wanted that to be the case, it could have stated so directly, he argued. Furthermore, a system that gave the federal government sole jurisdiction over work programs presented several problems. The federal government would have "the exclusive manner of applying the carrot and stick" in employment policies. To give the federal government such powers, Powell continued, "could impair the capacity of the state government to deal effectively with the critical problem of mounting welfare costs and the increasing financial dependency of many of its citizens." The opinion affirmed the state's authority to set guidelines requiring self-support, limit expenses to those "most in need," and adopt mechanisms "to cope with the fiscal hardships enveloping many state and local governments." The decision set forward the Supreme Court's belief in the necessity of flexibility in allowing states to manage the fiscal politics of their welfare programs.[47]

Justice Thurgood Marshall wrote the dissent with Justice William Brennan joining. The dissent articulated two major objections: the Justices had misinterpreted Congress' intent in establishing a federal work program, and they granted more extensive powers to states than the Court's earlier decisions had authorized. On the first point, Marshall argued that Congress had specifically designed a program of training, education and employment in the 1967 WIN program. It represented a "carefully coordinated system" built upon a foundation of support services, designed to shape skills which could lead to real jobs and self-support. Congress recognized that the success of such a plan demanded supervision and evaluation, and made such provisions, Marshall continued. It did not envision a system of state make-work programs, such as New York State's, nor that employment would become a condition of welfare. On the second point, Marshall chastised the Justices for their interpretation of state's rights. States had been granted flexibility in the SSA in two areas: the allocation of resources and the establishment of standards of need. The majority opinion erred when it allowed states to propose "additional conditions of eligiblity" such as work requirements, Marshall argued.[48]

Marshall's dissent also cautioned his peers and the public about the problems inherent in work programs. He acknowledged that the original intention of

ADC had failed in practice, and mothers had often left home to earn additional income. But he questioned the purpose, efficacy, and potential abuses of work programs. What kind of job could be required—work for which one was trained and from which one could derive self-support, or any job designated by the welfare authority? The latter had proved to be the case in state experiences and as ex-director of the NAACP Legal Defense Fund, Marshall was well aware of the way local boards had operated. Yet, in the context of growing impatience with increased public expenditures, rising unemployment, and inflation, the American public expressed overt hostility toward welfare, and the courts reinforced the authority of states to mandate earning by mothers.[49]

Two points may be drawn from this overview of twentieth century policies for impoverished mother-only families. First, no analysis of public aid to mother-only families will ever fully evaluate the development of policy if it only focuses on the relation between states and families. It must add markets to that analysis. Women had to balance wage-earning with welfare receipt even though policy did not provide many resources to do so. The findings presented here challenge the standard interpretations that public aid removed women from the workforce. In fact, the studies from the early twentieth century show a rate of employment among women receiving mothers' pensions at nearly 50 percent, much higher than scholars have previously thought. The evidence also indicates that the practices of the 1910s and 1920s continued through the following decades, although more case studies are needed for the 1930s and 1940s. The New Deal social policies that privileged male workers and identified mothers as "unemployables" because of their home responsibilities contributed to a false dichotimization of poor women's experience. So too did the attitudes of organized labor and social insurance proponents that failed to incorporate mother-only family issues into plans for universal social security. The flexibility granted to local welfare boards allowed officials to deny full or partial grants to women on the basis of their "ability to earn." Local administrators determined eligibility, not through standardized budgets or uniform rules, but in consideration of fiscal management and, in some areas, labor needs. Furthermore, an examination of wage-earning and welfare explains links between race and welfare in specific local economies. Some women, like southern African Americans, had historically been marked as "more employable," and the "employable mother" rules required wage-earning in racially segregated areas of the labor force. The continued use of work rules allowed state policies to operate in tandem with these racially constructed labor practices.

During the 1960s, the issue of wage-earning became more visible at the national level. Civil rights lawyers challenged laws at the same time that city and state officials charged welfare recipients with fraud and castigated them for lacking initiative. Within this context, a federal policy emerged that insisted upon wage-earning requirements for mothers. Its history raises implicit concerns about the success of employment solutions in welfare reform. While welfare has changed in many ways over the century, the barriers women face supporting themselves—locating child care and working in low-wage sectors—have remained fairly constant. The provision of supplemental services may have the greatest long-range impact on self-sufficiency in families, but the additional costs of these programs continue to raise political opposition.

The second point to be drawn from this overview is the significant presence of fiscal politics in shaping programs at the local and federal levels. Politicians, policy administrators, and constitutents repeatedly attempted to limit the size and cost of their programs with varying success. By 1981, fiscal politics and budget reduction became so central to federal welfare reform that the Omnibus Budget Reconciliation Act (OBRA) eliminated the "work incentives" built into the policy since the early 1960s. OBRA eliminated deductions for work-related expenses including daycare and gave no incentive for earned wages. The requirements to earn stayed in place, however. At the state level, experimental programs in work training and education have seen periodic cutbacks if not complete reduction of their budgets. Ironically, at a time in American politics when "family values" dominated political discourse, welfare policy expressed those values by eclipsing women's care-giving role in the family. If welfare policy continues to emphasize wage-earning, politicians and planners will need to convince the public of the long-term benefits that additional expenditures in child care, training, and education will bring. They will need to address, finally, the specific challenges presented by the poverty of mother-only families.

Department of History
LasVegas, NV 89154-5020

ENDNOTES

This research received financial assistance from the Nevada Humanities Committee and the UNLV Graduate College. I would like to thank Terry McDonald, James Patterson, Colin Loader and the anonymous readers of this journal for their helpful comments on earlier versions of this article. David Anderson provided helpful research at an early stage of this project.

1. The research on gender and the welfare state has developed into a complex and theoretically varied body of literature. For the most fully developed argument on maternalist social policy, see Theda Skocpol, *Protecting Soldiers and Mothers: The Political Origins of Social Policy in the United States* (Cambridge, MA, 1992); and Seth Koven and Sonya Michel, eds. *Mothers of a New World: Maternalist Politics and the Origins of Welfare States* (New York, 1993). The most helpful overview of Aid to Dependent Children may be found in Linda Gordon, *Pitied But Not Entitled: Single Mothers and the History of Welfare* (New York, 1994). Recent work on gender and welfare states in Europe offers an interesting comparison to the U.S. on this point. For example, see Alisa Klaus, *Every Child a Lion: The Origins of Maternal and Infant Health Policy in the United States and France, 1890–1920* (Ithaca, 1993); Susan Pedersen, *Family, Dependence, and the Origins of the Welfare State: Britain and France, 1914–1945* (New York, 1993); and Miriam Cohen and Michael Hanagan, "The Politics of Gender and the Making of the Welfare State, 1900–1940: A Comparative Perspective," *Journal of Social History* 24:3 (Spring 1991): 469–484.

2. Linda Gordon and Theda Skocpol discuss their different understandings of gender and welfare in *Contention* 2:3 (Spring 1993): 157, 185.

3. Frances Fox Piven and Richard Cloward, *Regulating the Poor: The Functions of Public Welfare* (New York, 1971), and "Welfare Doesn't Shore Up Traditional Family Roles: A Reply to Linda Gordon," *Social Research* 55:4 (Winter 1988): 631–647. Michael Katz, *The Undeserving Poor: From the War on Poverty to the War on Welfare* (New York, 1989). For analyses of work and welfare since the 1960s, see Mimi Abramovitz, *Regulating the*

Lives of Women: Social Welfare Policy from Colonial Times to the Present (Boston, 1988), 338–342. Roberta Spalter-Roth and Heidi Hartmann have written several pieces on the inadequacies of contemporary work policies as anti-poverty plans. For example, see Roberta Spalter-Roth, Heidi Hartmann, and Linda Andrews, *Combining Work and Welfare: An Alternative Anti-Poverty Strategy*, A Report to the Ford Foundation from the Institute for Women's Policy Research (Washington, D.C., 1992).

4. Excellent overviews of the history of welfare may be found in James Patterson, *America's Struggle Against Poverty, 1900–1935* (Cambridge, MA, 1986); and Michael Katz, *In the Shadow of the Poorhouse: A Social History of Welfare in America* (New York, 1986). For particular studies on women in poverty, see Linda Gordon, *Heroes of Their Own Lives: The Politics and History of Family Violence, Boston 1880–1960* (New York, 1988), and Beverly Stadum, *Poor Women and Their Families: Hard Working Charity Cases, 1900–1930* (Albany, 1992).

5. For the origins of mothers' pensions, see Mark H. Leff, "Consensus for Reform: The Mothers' Pension Movement in the Progressive Era," *Social Service Review* 47 (1973): 397–417; Theda Skocpol, *Protecting Soldiers and Mothers*.

6. Quote is from Mrs. G. H. Robertson, "The State's Duty to Fatherless Children," *Child Welfare Magazine* (January 1912): 156–160. A good source on the contemporary debate surrounding mothers' pensions may be found in *Proceedings*, National Conference of Charities and Corrections (Fort Wayne, IN, 1912).

7. Sophonisba P. Breckinridge, "Neglected Widowhood in the Juvenile Court," *American Journal of Sociology* (July 1910): 67.

8. U.S. Department of Labor, Children's Bureau, *Mothers' Aid, 1931*, Publication No. 220 (Washington, D.C., 1933), 2.

9. Grace Abbott, *The Child and the State, Select Documents*, vol. II (Chicago, 1938), 229, cited in Winifred Bell, *Aid to Dependent Children* (New York, 1965), 6.

10. Edith Abbott, *Women in Industry: A Study in American Economic History* (New York, 1910).

11. The problem of insufficient funding and work requirements was discussed in several Children's Bureau reports. For example, see U.S. Children's Bureau, *Minimum Standards for Child Welfare Adopted by the Washington and Regional Conference on Child Welfare*, 1919, Publication No. 62 (Washington, D.C., 1920), 10; U.S. Children's Bureau, *Proceedings of the Conference on Mothers' Pensions, Providence, Rhode Island, June 28, 1922*, Publication No. 109 (Washington, D.C., 1922), 11, 24; U.S. Children's Bureau, *Standards of Public Aid to Children in their Own Homes*, by Florence Nesbitt, Publication No. 118 (Washington, D.C., 1923), 17–20. For a discussion of settlements' racially-based views on working mothers, see Ruth Crocker, *Social Work and Social Order: The Settlement Movement in Two Industrial Cities, 1889–1930* (Urbana, 1992).

12. *Laws of Illinois*, Forty-Seventh General Assembly, 1911, 126; *Laws of Illinois*, Forty-Eighth General Assembly, 1913, 129.

13. *Laws of Illinois*, Senate Bill No. 300, 48th General Assembly, 1913, 129.

14. U.S. Department of Labor, Children's Bureau, *The Administration of the Aid-to-Mothers' Law in Illinois*, by Edith Abbott and Sophonisba Breckinridge, Publication No. 82 (Washington, D.C., 1921), 35–36.

15. U.S. Department of Labor, Children's Bureau, *The Administration of the Aid-to Mothers' Law in Illinois*, 109–110. The majority of denied pensions were refused on the as-

sessment of adequate means of support. Forty-nine percent were deemed economically sufficient, 37 percent were legally ineligible, and 12 percent were either uncooperative or unfit. Economic sufficiency also provided the single largest reason for removing women from the program once they had been accepted. Cook County, IL, Board of County Commissioners, *Charity Service Reports*, County Agent Report (1911–1917) and Juvenile Court, Mothers' Pension Department (1918–1927).

16. The study took data from large, medium, and small cities and one rural county. U.S. Department of Labor, Children's Bureau, *Standards of Public Aid to Children in Their Own Homes*, 17.

17. U.S. Department of Labor, Children's Bureau, *Standards of Public Aid to Children in Their Own Homes*, 20.

18. Grace Abbott, "Recent Trends in Mothers' Aid," *Social Service Review* 8:2 (June 1934): 208; Grace Abbott, *From Relief to Social Security: The Development of the New Public Welfare Services and their Administration* (Chicago, 1941), 183.

19. Grace Abbott, *The Child and the State*, vol. 2 (Chicago, 1938), 309–312.

20. U.S. Committee on Economic Security, *Social Security in America: The Factual Background of the Social Security Act as Summarized from Staff Reports to the Committee on Economic Security* (Washington, D.C., 1937), 223.

21. For an overview of ADC, see Winifred Bell, *Aid to Dependent Children*; and Linda Gordon, *Pitied But Not Entitled*. Grace Abbott describes the circumstances surrounding the construction of the provision in Grace Abbott, *From Relief to Social Security*, 279.

22. "Final Report of the Advisory Council on Social Security," 30, Grace and Edith Abbott Papers, Special Collections, Joseph Regenstein Library, University of Chicago, Chicago.

23. Linda Gordon, "Social Insurance and Public Assistance: The Influence of Gender in Welfare Thought in the United States, 1890–1935," *The American Historical Review* 97 (February 1992): 19–54.

24. Some women did receive work relief, but their participation was restricted by quotas, restrictions on the types of jobs, such as sewing, and a wage scale that remained lower than men's wages. Elizabeth Faue, *Community of Suffering and Struggle: Women, Men, and the Labor Movement in Minneapolis, 1915–1945* (Chapel Hill, 1991), 132–133; Nancy E. Rose, "Gender, Race, and the Welfare State: Government Work Programs from the 1930s to the Present," *Feminist Studies* 19:2 (Summer 1993): 319–342.

Also, see Claudia Goldin, "The Changing Economic Role of Women," *Journal of Interdisciplinary History* (Spring 1983): 707–733; Ruth Milkman, "Women's Work and the Economic Crises, Some Lessons from the Great Depression," in *A Heritage of Her Own*, Nancy Cott and Elizabeth Pleck, eds. (New York, 1979), 507–537. Grace Abbott citing a Children's Bureau study of the 1921–1922 depression and its effect on families, in Grace Abbott, *From Relief to Social Security*, 125.

25. Faue, *Community of Suffering and Struggle*, 119–120, 154–155.

26. Federal Emergency Relief Administration, *Monthly Report, July 1 to July 30, 1935* (Washington, D.C., 1935), 49.

27. Federal Emergency Relief Administration, *Monthly Report, June 1 to June 30, 1935* (Washington, D.C., 1935), 9–11.

28. Mary S. Larabee, "Unmarried Parenthood Under the Social Security Act," *Proceedings of the National Conference of Social Work* (Washington, D.C., 1939), 447–449.

29. Bell, *Aid to Dependent Children*, chp. 6. For the Georgia case, *Anderson v. Burson*, 300 F. Supp. 401 (1968).

30. Scholars disagree on the explanations for the Bureau's ineffectiveness. Piven and Cloward argued that the principle of local responsibility which retained significant control at the local level was deliberate and functional to local labor demands. Piven and Cloward, *Regulating the Poor*, 128–130. Patterson found the federal bureau unable to act alone in the absence of "potent lobbies" or organizations of the poor. Patterson, *America's Struggle Against Poverty*, 68, 87. Yvonne Zylan's political analysis of ADC in the post-war years suggests that the "backlash" on ADC programs, particularly in the states, might have been tempered had organized women mounted some political opposition. Unlike organized women's support for mothers' pensions in the early twentieth century, Zylan has found no evidence for support of the ADC program at mid-century. Yvonne Zylan, "Constructing the Patriarchal Welfare State: Aid to Dependent Children and the Politics of Gender, 1945–1960." Unpublished paper in the author's possession.

31. *New York Times*, 18 August 1950, p. 14, c. 2. Abramovitz believes states pressured the federal government to add the caretaker grant to help relieve some of the costs, *Regulating the Lives of Women*, 316, 322; Patterson, *America's Struggle Against Poverty*, 88.

32. The Knowland Amendment effectively allowed states to cut workers' benefits during a labor shortage, even a strike. President Truman and organized labor spoke out against the measure as one that would seriously challenge a worker's right to reject employment at sub-standard wages. *New York Times*, 12 August 1950, p. 1, c. 4; 29 August 1950, p. 20, c.6.

33. Aid to Dependent Children was renamed Aid to Families with Dependent Children in the 1962 revisions. Family Support Act, *Statutes at Large*, 100th Cong., 2d Sess., 102 Stat. 2343 (1988).

34. Social scientists have explained the shift toward increased federal work expectations as part of a transition that replaced social workers' influence over policy with that of economists. Irwin Garfinkle and Sara McLanahan, *Single Mothers and their Children: A New American Dilemma* (Washington, D.C., 1986); Katz, *The Undeserving Poor*. For labor regulation theory, see Piven and Cloward, *Regulating the Poor*. For sex regulation theory, see Mimi Abramovitz, *Regulating the Lives of Women*.

35. The 'employable mother' rule has been included among the state welfare laws overturned by the courts. Instead, as this article discusses, the courts reaffirmed a state's right to require work. Piven and Cloward, *Regulating the Poor*, 308; and Patterson, *America's Struggle Against Poverty*, 179.

36. John Kenneth Galbraith, *The Affluent Society* (Boston, 1958); Michael Harrington, *The Other America* (New York, 1962); Katz, *The Underserving Poor*, 91–94.

37. For example, see Nicholas Lemann, *The Promised Land: The Great Black Migration and How it Changed America* (New York, 1991), 345. The impact specifically on African American women is discussed in Jacqueline Jones, *Labor of Love, Labor of Sorrow: Black Women, Work, and the Family from Slavery to the Present* (New York, 1985), 235–268, 301–310. Also, see James Patterson, *America's Struggle Against Poverty*, 136.

38. William Julius Wilson, *The Truly Disadvantaged: The Inner City, the Underclass, and Public Policy* (Chicago, 1987), chp. 2, especially 33–41; and chp. 4, especially 160–161.

39. Wilson, *The Truly Disadvantaged*, 39; Patterson, *America's Struggle Against Poverty*, 175, 179.

40. *Report of the House Committee on Ways and Means*, Social Security Amendments of 1967, H.R. Rep. No. 544, 90th Cong., 1st Sess. 95–96. On the limits of WIN, see Patterson, *America's Struggle Against Poverty*, 176; Abramovitz, *Regulating the Lives of Women*, 332.

41. *Anderson v. Burson*, 300 F. Supp. 401 (1968); *New York State Department of Social Services v. Dublino*, 93 S.Ct. 2507 (1973). For a legal review that applies a patriarchal analysis, see Sylvia A. Law, "Women, Work, Welfare, and the Preservation of Patriarchy," *University of Pennsylvania Law Review* 131:6 (May 1983): 1249–1339.

42. Jack Greenberg, *Crusaders in the Courts: How a Dedicated Band of Lawyers Fought for the Civil Rights Revolution* (New York, 1994). The court cases are respectively, *King v. Smith*, 88 Sup. Ct. 2128 (1968); *Shapiro v. Thompson*, 89 Sup. Ct. 1322 (1969); and *Goldberg v. Kelly*, 397 Sup. Ct. 254 (1970).

43. U.S. Department of Labor, Children's Bureau, *Mothers' Aid, 1931*. South Carolina was the other state. Act No. 445, *Georgia Laws* 1951, pp. 692–693, cited in Bell, *Aid to Dependent Children*, 71, 35; Georgia, Department of Public Welfare, *Public Assistance Manual*, Pt. III, Sec V, p. 15 (1952).

44. Regulations, *Georgia Manual of Public Welfare Administration*, Part III, Section V-C(3) (b) (2), cited in Brief for Plaintiffs (October 12, 1966), p. 11, *Anderson v. Burson* RG 21, U.S. District Court, No. District of Georgia, Atlanta Division, Box 880.

45. According to Stephen Ralston, King contacted the NAACP Legal Defense Fund. Personal correspondence, Stephen Ralston, Senior Staff Attorney, NAACP Legal Defense and Educational Fund, Inc., December 24, 1993. Complaint, June 5, 1967, *Anderson v. Burson* RG 21, U.S. District Court, No. District of Georgia, Atlanta Division, Box 880.

46. *New York Times*, 8 October 1967, p. 38 c. 3; *Anderson v. Burson*, 300 F. Supp. 401 (1968).

47. *New York State Department of Social Services v. Dublino*, 93 S.Ct. 2507 (1973), 2512–2513.

48. *New York State Department of Social Services v. Dublino*, 93 S.Ct. 2507 (1973), 2522.

49. *New York State Department of Social Services v. Dublino*, 93 S.Ct. 2507 (1973), 2520.

Fathers' Rights, Mothers' Wrongs? Reflections on Unwed Fathers' Rights and Sex Equality

MARY L. SHANLEY

This article examines arguments concerning the right of an unwed biological father to consent to the adoption of his offspring, and to take custody of the child even against the mother's wishes. The understanding of gender-neutrality that supposedly supports many such arguments is false, and risks diminishing women's decision-making authority under the guise of sex equality. Laws governing unwed parent's rights must emphasize the centrality of parental responsibility in establishing parental rights.

The recent case of *Baby Girl Clausen*, involving a custody dispute between the biological parents, Cara Clausen and Daniel Schmidt of Iowa, and the adoptive parents, Roberta and Jan DeBoer of Michigan, over who should be recognized as the legal parents of baby Jessica, focused national attention on the issue of what rights biological unwed fathers may have to custody of their infant offspring. When Cara Clausen, at the time unmarried, gave birth to a baby girl, she gave her irrevocable consent to the child's adoption two days after its birth, as did the man she named as the child's father on the birth certificate. Within weeks, however, Clausen regretted her decision and informed Daniel Schmidt that he was the baby's father. Schmidt responded by filing a petition to establish paternity, and initiating legal action to block the adoption. Schmidt contended that a biological father has a right to custody of his child unless it is shown that he is "unfit" to be a parent. After some two years of litigation, Michigan declared it did not have jurisdiction in the matter and Iowa proceeded to enforce its decree that Schmidt's parental rights had never properly been terminated and the child had to be returned to his physical custody.[1]

Many of those who commented on the case asked whether Daniel Schmidt's alleged rights should be enforced in the face of the trauma Jessica, now a

Hypatia vol. 10, no. 1 (Winter 1995) © by Mary L. Shanley

toddler, would suffer in being removed from the only family she had known. While this matter merits serious attention, I focus on a different issue raised by the case, namely the basis and nature of an unwed biological father's right to veto an adoption decision of an unwed mother. This question was obscured in the case of *Baby Girl Clausen* because Daniel Schmidt initiated his action with the full cooperation of Cara Clausen, who had come to regret her decision. But it was *his* rights that the Iowa courts upheld, holding that an unwed biological father had a right to preclude an adoption initiated by the biological mother or the state.

There have been a significant number of cases in which an unwed biological father has sought to reverse the biological mother's decision to allow a child to be adopted, statutory provisions stipulating under what conditions an unwed biological father's consent is necessary vary greatly from state to state, and legal thinking on the matter is quite unsettled.[2] Many courts continue to apply the traditional rule that they must consider the "best interest" of the child in making any decision about custody. Supporters of biological fathers' rights, by contrast, argue that when the biological mother does not wish to retain custody, the biological father's claim automatically takes precedence over that of some "stranger" or potential adoptive parent (Boyd 1990; Buchanan 1984; Eveleigh 1989; Hamilton 1987-88; Raab 1984; Serviss 1991; Steward 1990; Zinman 1992).[3] Interestingly, some advocates of women's rights have also criticized the best interest standard as too subject to the biases of individual judges but have argued that women's unique role in human gestation and childbirth, as well as various aspects of their social and economic vulnerability, dictate that an unwed biological mother must be able to make the decision to have her child adopted without interference by the father or the state (Becker 1989; Erickson 1984; Erickson 1990, 39 n. 22; Rothman 1989). According to this view, neither the biological mother nor the state has an obligation to seek the biological father's consent to the adoption decision, or even to inform him of his paternity. From such a perspective, statutes that require biological mothers, but not biological fathers, to consent to the adoption of their newborn infant do not deny men equal protection. The debate between advocates of these two perspectives takes us to the difficult issue of what, indeed, should be the grounding of *anyone's* claim to parental rights.

From my perspective, neither the "fathers' rights" nor the "maternal autonomy" position provides a fully satisfactory basis for thinking about the custodial claims of unwed biological parents. I am persuaded by considerations advanced by advocates in both camps that the best interest standard is unsuitable for cases involving newborns surrendered by their mothers for adoption, in part because that standard does not adequately recognize the claims of biological paternity, and in part because it is difficult to guard adequately against the biases of individual judges. But almost all arguments for unwed fathers' rights are based on a notion of gender-neutrality that is misleading, not only because of women's biological experience of pregnancy but also because of the inequality inherent in the social structures in which sexual and reproductive activity currently take place. Many arguments in favor of a mother's right to decide on the custody of her child, by contrast, expose the ways in which gender-neutral rules applied to situations of social and economic inequality may in practice perpetuate male privilege. These insights might suggest that one good way to compensate for the present social and economic inequalities would be to give women complete or at least preponderant decision-making authority about reproductive matters until present social and economic inequalities based on sex diminish, but such a policy would run the risk of reinforcing the gender stereotype that women, not men, are the natural and proper nurturers of children. Law and social policy in the area of parental rights must walk a very fine line between adopting false gender-neutrality by treating men and women identically on the one hand, and reinforcing gender stereotypes on the other (on this dilemma more generally, see Eisenstein 1988).

The theoretical question of who should have parental rights and on what grounds is complicated by the practical consideration of what will prove to be the best means of moving toward both greater sexual equality and acceptance of diverse family forms. (I am, for example, concerned that any principles I develop here be compatible with enabling lesbian life-partners to parent a child free from threats by a known sperm donor to seek parental rights after agreeing prior to conception not to do so.)[4] I am also influenced by my belief in the desirability of sexual equality in both the public and private spheres, which requires not only access to jobs and public activity for women, but also the assumption of concrete, day-to-day, "hands on" responsibility for child-rearing by men (See Johnson 1994).

I argue here that a liberal polity interested in protecting the possibility for intimate association and family life for all its members should articulate norms that ground parental claims in a mixture of genetic relationship, assumption of responsibility, and provision of care to the child (including gestation). In the case of a newborn, this means that the biological father must take concrete steps to demonstrate his commitment to the child prior to the biological mother's relinquishment of the child for adoption, and courts must have the authority to judge both his efforts and the mother's objections to his claim. Only some such standard, I believe, recognizes the complexity of the sexual, genetic, biological, economic, and social relationships between adults and among adults and children that are involved in human reproductive activity.

Beyond the legal issues raised by unwed biological fathers' claims to custody, considering the rights of unwed biological parents also raises issues about individual rights in situations dealing with family relationships. By looking behind abstract assertions of individual rights to examine the dependency, reciprocity, and responsibility involved in family relationships, I suggest that the traditional liberal understanding of autonomous individuals must be revised to take account of the fact that persons are not fundamentally isolated and discrete but constructed in and by their relationships to others.[5]

UNWED PARENTS' CUSTODY RIGHTS IN COMMON LAW AND CONSTITUTIONAL LAW

The Patriarchal Construction Of The Common Law Regarding Custody

A frequent theme in much of the literature advocating unwed fathers' rights is that unmarried fathers have been treated unfairly as a result of widespread social hostility toward them. Arguing that an unwed biological father has a right to be informed of the birth of his child under any circumstances, John Hamilton complains that "until a few years ago, unwed fathers were ignored or received virtually no protection from either the United States Constitution or the statutes of most states. Indeed, courts and legislatures traditionally have been openly hostile to the recognition of parental rights of unwed fathers" (Hamilton 1987-88, 949-50). Elizabeth Stanton argues that "custody was vested in the mother because the law presumed that she was better suited to raise children," a presumption Stanton regards as both unwarranted and hurtful to men.[6] Commentaries that assume that laws placing custody of nonmarital children in their biological mothers' hands reflected hostility to biological fathers display a profound misunderstanding of the patriarchal roots of family law, as well as a stunning indifference to the devastating social and economic consequences of unwed motherhood for women.

The common law, which largely regulated legal aspects of family relationships in America well into the nineteenth century was profoundly patriarchal; legal definitions of who is a father and the extent of paternal responsibilities governed not only a man's relationships with his children but also with women both inside and outside his family. Under the common law a man had complete custodial authority over any children born of his wife, and no legal relationship at all to children he sired out of wedlock. The child of his wife took his surname; the nonmarital child did not. The marital child had a right to financial support from him; the nonmarital child did not. The marital child had the right to inherit from him if he died without a will; the nonmarital child did not (Becker 1989, 498).

The husband's authority over the marital child was an extension of his authority over his wife. Under the common law doctrine of coverture, a wife's legal personality was subsumed in that of her husband during marriage. A wife could not enter into contracts, sue or be sued, or engage in other legal transactions without being joined by her husband. He owned outright her movable property and had control of (although he could not alienate) her real estate. Torts she committed in his presence were chargeable to him, not to her. A married woman also had no right to refuse her husband sexual access—marital rape was not recognized as an offense. So complete was the husband's custodial authority that during his lifetime he had the power to convey his parental rights to a third person without the mother's consent, and he could name someone other than the mother to be the child's guardian after his death.[7]

Thus, under the common law, a man's legal relationship to his offspring was governed by his relationship to their mother. If the woman was his wife, a child was "his," so much so that he exercised exclusive custodial authority. If the mother was not his wife, however, the child was "filius nullius," the child of no one. Obviously, these rules affected both lineage and property. They allowed a man to lay claim to legitimate heirs (for without marriage, who would know for certain who the father of a child might be?) and to avoid squandering his estate supporting other children. While the father was shielded from financial responsibility for his "spurious" offspring, a woman who bore children outside of marriage was "ruined"; unmarried mothers' desperate attempts at suicide and infanticide dot the pages of social histories and nineteenth-century novels. Although the nonmarital child could inherit from no one, Poor Laws assigned mothers financial responsibility for their offspring and gave them custodial rights as long as they kept their children off public support. A woman's responsibility for her nonmarital children punished her for sex outside of marriage and increased women's incentives to join themselves to men through marriage.

It is no wonder that women's rights advocates from the mid-nineteenth century on protested against the patriarchal assumptions and the sexual double-standard implicit in this configuration of rules governing marital and nonmarital procreation. During the nineteenth century, in part due to women's rights advocacy, legislatures began to replace common law rules with statutes that granted wives equal custodial rights with their husbands. By the early twentieth century, when parents divorced, judges began to prefer mothers as custodians of marital children of "tender years" (usually under seven or ten years of age).[8] Eventually the standard of the "best interest of the child"—which did not automatically prefer either spouse and which purported to recognize the needs of the child as paramount—replaced any presumption explicitly favoring the custodial claim of either married parent if they divorced.

With respect to the custody of nonmarital children the law changed more slowly, and the impetus came mainly from children's rights advocates who wanted to get rid of the legal disabilities of "illegitimacy," such as the inability to collect survivors' benefits, receive child support, and inherit from the father (*Weber v. Aetna Casualty & Surety Co.*, 406 U.S. 164 [1972]; *Gomez v. Perez*,

409 U.S. 535 [1973]; and *Trimble v. Gordon*, 430 U.S. 762 [1977]). Thus the common law protections of fathers against the claims of nonmarital children and their mothers have been largely dissolved, and paternal responsibility for out-of-wedlock children established. But the common law history provides no basis for paternal *rights* to such children. Some advocates of unwed biological fathers' rights have argued, however, that while unrecognized in the common law, the ability to raise one's biological child is a fundamental interest protected by the United States Constitution, and that under the Constitution men and women must have equal rights to claim custody of their nonmarital offspring.

The Emergence Of Unwed Fathers' Constitutional Rights

Over the past twenty years, five decisions of the Supreme Court—*Stanley v. Illinois*, *Quilloin v. Wolcott*, *Caban v. Mohammed*, *Lehr v. Robertson*, and *Michael H. v. Gerald D.*—laid down some guidelines for thinking about unwed fathers' rights. These decisions have established that at least in those instances where an unmarried biological father has established a relationship with his child by an unmarried woman, the father's right to continue the relationship may be constitutionally protected. Although these decisions do not by any means resolve all the dilemmas surrounding the custody of infants born to unmarried biological parents, they provide a useful starting point for thinking about those issues.

The Court first considered the custodial rights of unmarried biological fathers in 1972 in *Stanley v. Illinois* (405 U.S. 645 [1972]). Mr. Stanley had lived with his three biological children and their mother, to whom he was not married, intermittently for eighteen years. When the mother died, the state of Illinois declared the children wards of the state and placed them with court-appointed guardians. This was done without a hearing as to Stanley's fitness as a parent. Stanley protested, arguing that Illinois law denied him equal protection of the laws since neither unwed mothers nor married fathers or mothers could be deprived of custody of their children unless they were shown to be unfit. The state argued that Stanley's fitness or unfitness was irrelevant, because an unwed father was not a "parent" whose existing relationship with his children must be considered; an unwed father was presumed unfit because he had not married the mother. The Supreme Court rejected Illinois's argument, and stated that "the private interest here, that of a man in the children he has sired and raised, undeniably warrants deference and, absent a powerful countervailing interest, protection" (*Stanley v. Illinois*, 405 U.S. 645 [1972] at 651). Failure to provide a hearing on parental fitness for an unwed father violated both the due process and the equal protection clauses of the Fourteenth Amendment.

In cases after *Stanley* the Court drew distinctions between biological fathers who, like Stanley, had been involved in raising their biological children and those who had not assumed day-to-day practical responsibility for them. In *Quilloin v. Walcott* (434 U.S. 246 [1978]), Leon Quilloin sought to prevent the adoption of his eleven-year-old biological child by the child's stepfather, Mr. Wolcott. The Court upheld a Georgia statute that stipulated that a biological mother alone could consent to the adoption of her child; the consent of the unwed biological father was required only if he had legitimated the child. The Court said that no due process violation occurred when contact between the biological father and child had been only sporadic.

The next year, hearkening back to *Stanley*, the Court said in *Caban v. Mohammed* (441 U.S. 380 [1979]) that a New York statute that required unwed biological mothers, but not biological fathers, to consent to the adoption of their children was unconstitutional when an unwed biological father's relationship to his child was "fully comparable" to that of the mother. Caban, like Stanley, had been involved with raising the children, having lived with their mother and supported them for several years. The case arose when the mother's new husband sought to adopt her children. Mr. Caban, who had previously lived with the children and their mother for several years, argued that the law which required only the biological mother's consent to adoption violated the equal protection clause. Caban also claimed that biological fathers had a due process right or liberty interest "to maintain a parental relationship with their children absent a finding they are unfit as parents" (*Caban v. Mohammed*, 441 U.S. 380 [1979] at 385). New York argued that the distinction between biological mother and biological father was justified because of the fundamental difference between maternal and paternal relationships with their children. The Supreme Court agreed with Caban, holding that "maternal and paternal roles are not invariably different in importance," but it explicitly reserved any opinion about whether a distinction such as New York had made would be valid with regard to newborn adoptions (441 U.S. 380 at 392, n. 11).

Lehr v. Robertson (463 U.S. 248 [1983]) also concerned a biological father's effort to block the adoption of his child by her stepfather, but unlike Mr. Stanley and Mr. Caban, and like Mr. Quilloin, Jonathan Lehr had almost no contact with his biological daughter, Jessica. Lehr claimed, however, that he had a liberty interest in an actual *or potential* relationship with Jessica and that the State's failure to provide him notice of her pending adoption violated due process. He also asserted that the New York statute violated equal protection because it required the consent of the biological mother, but not the biological father, for an adoption. The case was complicated by the fact that the majority and minority disagreed over whether Lehr had ignored the child until recently, or, as he claimed, had repeatedly attempted to establish contact with her from the time of her birth but was thwarted by her mother. The majority held that Lehr had not made sufficient contact to establish a parental right, declaring that a biological father's constitutional rights are a function of the tangible and affirmative responsibilities he accepts toward his offspring. The biological connection alone, the Court held, is not sufficient to *guarantee* an unwed father a voice in the adoption decision, although it affords him an *opportunity* to be heard: The "biological connection . . . offers the natural father an opportunity that no other male possesses to develop a relationship with his offspring. If he grasps that opportunity and accepts some measure of responsibility for the child's future, he may enjoy the blessings of the parent-child relationship," but if he fails to grasp the opportunity, "the Equal Protection Clause does not prevent a State from according the two parents different legal rights" (*Lehr v. Robertson*, 463 U.S. 248 [1983] at 268). In a dissent, Justice White, joined by Justices Marshall and Blackmun, disagreed not only about the factual question of whether Lehr had attempted to contact Jessica but also with the framework used by the majority to determine the existence of the unwed biological father's liberty interest. Justice White asserted that "the 'biological connection' is itself a relationship that creates a protected interest" (463 U.S. 248 at 272). In Justice White's view, the majority decision is more demanding than the Constitution requires and fails to recognize that the biological tie gives an unwed biological father a constitutionally protected interest that warrants, in this case, that Lehr be afforded a hearing to establish legal paternity and to be heard with regard to Jessica's adoption.

The rule that the Supreme Court seemed to be developing in these cases, namely, that an unwed biological father who had established a substantial relationship with his child had a constitutionally protected interest in maintaining that relationship, was sidelined in the Court's most recent decision dealing with an unwed biological father's rights. In *Michael H. v. Gerald D.* (491 U.S. 110 [1989]), the Court found that a California statute creating an irrebuttable presumption that a woman's husband was the father of a child she bore was constitutional. A biological father, Michael H., who had lived intermittently with and provided care to his biological daughter and her mother even though the mother was married to and intermittently lived with her husband as well, argued that he had a right to a hearing to establish his paternity when the husband and wife sought to cut off his contact with the child.

The case produced no fewer than five opinions from a deeply fractured Court. The plurality decision, written by Justice Scalia and joined in full only by Chief Justice Rehnquist, rejected Michael's claim (491 U.S. 110 at 113-32). Justice Scalia contended both that the state had an interest in preserving the "unitary family" and that neither Michael nor his genetic daughter had a constitutionally protected liberty interest in maintaining their relationship.[9] Justices O'Connor and Kennedy agreed with Justice Scalia's conclusion but not his reasoning.[10] In a concurring opinion, Justice Stevens asserted, without much apparent basis, that Michael could have obtained visitation rights as an "other person having an interest in the welfare of the child" and so did not need further protection (491 U.S. 110 at 133). Two dissenting opinions supported Michael's right to a hearing but used quite different grounds to do so. Justice White, joined by Justice Brennan, reiterated the view he expressed in *Lehr* that biology itself creates a presumptive parental right. Justice Brennan, joined by Justices Marshall and Blackmun, argued that the *combination* of biology and nurture establishes the liberty interest Michael claimed (491 U.S. 110 at 142-43).

It is important to note that despite the fact that they arrived at opposite conclusions, both Justice Scalia's and Justice White's opinions adopted male-centered models of the basis of parental rights. Justice Scalia looked to the law to protect the paternal rights of married men by basing those rights on a man's legal relationship with the child's mother. Justice White grounded paternal rights in a biological tie established by blood tests, regardless of a man's legal ties to the mother. Both opinions made it unnecessary for the law to ascertain the *mother's* wishes or intentions with respect to paternal claims to her child. By contrast, and although they too reached opposite conclusions, both Justice Stevens and Justice Brennan considered the interests and actions of the mother to be relevant to establishing paternal claims. For Justice Stevens, the inherent biological and sociological differences between care of the fetus by a woman and a man justified different treatment of parental rights at the time of birth. Justice Brennan suggested that once a relationship between father and child exists, the mother cannot then exclude an otherwise fit father from being heard with respect to his paternal rights. Justice Brennan's opinion left open the question of to what extent Michael's situation was like or unlike that of an unwed biological father of a child born to an unmarried woman.

These decisions do not tell us whether under the Constitution an unwed biological father has a right to veto the adoption of a newborn even if he has had no opportunity to establish the kind of relationship and provide the kind of care that the Court has declared protects parental rights.[11] Must the law provide an unmarried biological father an opportunity to demonstrate his commitment if the biological mother does not allow him access to the child? Advocates of fathers' rights insist that unwed biological fathers do have such a constitutional right and that when the biological mother has decided to relinquish her parental rights, a biological father, unless shown to be "unfit," is entitled to assume custody of his offspring without a hearing on the best interest of the child.

Fit Fathers and Competent Mothers: Who Shall Decide When Biological Parents Disagree?

An Unwed Father's Right To "Grasp The Opportunity" To Become A Parent: The Pursuit Of Abstract Equality

In recent years some biological fathers have claimed that, because of the mother's lack of cooperation, they have not found any way to meet the Court's demand, articulated in *Lehr*, that a biological father who wants to retain his parental rights and "enjoy the blessings of the parent-child relationship" act to "grasp the opportunity" to develop a relationship with his offspring by assuming some "responsibility for the child's future" (463 U.S. 248 at 262). Should a biological father have the opportunity to veto an adoption regardless of the wishes of the mother? Should adoption proceedings be precluded until the father has been heard? What considerations should guide us as we try to evaluate such issues? To answer these questions we need to think about both the basis of claims for custodial rights and the relative claims of biological mothers and fathers outside of marriage.

The argument that the biological father must be given custody when the biological mother chooses not to raise the child is grounded first of all in the conviction that parenthood is a significant good in the lives of men as well as women. Fathers might wish to raise their children for the same reasons mothers do—sharing intimacy and love, nurturing a child to adulthood, seeing one's genetic inheritance survive into the next generation, and passing on ethnic and religious traditions. A commitment to gender neutrality led most states to abandon an automatic maternal preference if mother and father, married or unmarried, each sought custody, and the same commitment would suggest that the law presume that the law must require the consent of *both* parents before the child can be adopted.

Various commentators support an unwed biological father's right to veto the adoption of his child on the grounds that fathers have a fully comparable interest to that of mothers in exercising parental rights and responsibilities. Claudia Serviss says all parents have "a constitutionally-protected opportunity interest in developing a parent-child relationship" (Serviss 1991, 788). John Hamilton argues that all unwed fathers have a right to be notified by the state of the existence of their offspring and be heard before any adoption can proceed, and that the state may therefore require the biological mother to identify the biological father (Hamilton 1987-88), while Daniel Zinman insists that the state must allow a biological father to take custody of his child when the biological mother has relinquished her rights, unless he is shown to be unfit (Zinman 1992). In a recent California case, *In the Matter of Kelsey S.*, the California Supreme Court appeared to agree that a gender-neutral standard should prevail. It held unconstitutional a statute that gave unwed mothers and legally recognized or "presumed" fathers a greater say in preadoption proceedings to terminate parental rights than it gave to unwed biological fathers. The court declared that the statute rested on a "sex-based distinction" that bore no relationship to any legitimate state interest once the child was outside the mother's body and she had decided to relinquish custody.[12]

The presumption of fitness for biological parents also avoids the dangers of subjective judgment and cultural prejudice that seem unavoidable in attempts to determine the child's best interest (Bartlett 1988, 303). One supporter of an unwed father's right to custody argues that a best interest determination "is subject to abuse and may lead to paternalistic infringement on the parent-child relationship in the name of the child's welfare. Given the long waiting list of adoptive parents that exists today and the traditional preference for rearing a child in a two-parent home, a best interest test is a no-win situation for the unwed father of a newborn with whom he has not yet had the opportunity to develop an emotional tie."[13] In 1987, the Georgia Supreme Court explicitly rejected the best interest test in favor of a fitness standard on the grounds that it was presumptively unfair to compare the putative father with adoptive parents with whom the child had never lived. It held that "If [the father] has not abandoned his opportunity interest, the standard which must be used to determine his rights to legitimate the child is his fitness as a parent to have custody of the child. If he is fit he must prevail" (*In re Baby Girl Eason*, 257 Ga. 292 at 297, 358 S.E. 2d 459 at 463 [1987], qtd. in Zinman 1992, 993-94).

Use of the best interest test in cases of an infant who has lived with no adult care-giver for any appreciable period of time should be changed not only in the interests of sex equality but of family diversity as well. The best interest standard invites the court to make judgments about the relative merit of a whole array of "lifestyle" issues that are not subject to scrutiny when an unwed biological father does not contest a biological mother's wish to retain custody of her child. A "fitness" standard applied to unwed biological fathers would avoid the possibility that an adoption decision might rest on a judge's preference that a child be raised in a two-parent household rather than by a single male, or a judge's prediction that middle-class professionals will give a child more "advantages" than the child would receive in a working-class home. If no action (or failure to act) by the biological father shows that he should not be entrusted with custody, then value judgments concerning different lifestyles and household arrangements should be precluded from inappropriately influencing the custody decision concerning the placement of a newborn infant.

To argue that the courts should abandon the best interest standard when an unwed biological father wishes to raise an infant does not, however, imply that the only relevant consideration is the biological father's "fitness." The question of whether an unwed biological father shall have a right to custody of his newborn infant is not about paternal rights *tout court*, but also or alternatively about an unmarried woman's authority to decide who shall take custody of her newborn child. For the biological father to assume custody, the biological mother's expressed wishes concerning the child's placement will of necessity be overridden. Arguments that a biological father should be able to veto the adoption decision of the biological mother and assume custody unless proven to be unfit run up against counter arguments that the courts should defer to an unwed biological mother with respect to placement of her child. It is to these considerations that I now turn.

An Unwed Mother's Claim To Decisional Autonomy: Taking Context Seriously

Arguments in favor of the "fitness" standard for unwed biological fathers falsely assume that once a biological mother has surrendered the child for adoption she has no further relevant wishes with respect to custody. Defenders of an unwed biological father's right to veto an adoption often contrast what they portray as his laudable desire to assume custody and to care for the child with the biological mother's uncaring decision not to raise the child herself. The image of the "bad mother," and the assumption that if the mother chooses not to raise the child, she must be indifferent to its fate, hover just beneath the surface of such depictions. The notion that once a mother decides to relinquish her child for adoption she can have no further relevant concerns distorts and denigrates both her experience of pregnancy and the nature of her decision.[14] But relinquishment of the newborn for adoption may reflect any of a wide array of circumstances: lack of money or job prospects, youth or immaturity, feelings of inadequacy or isolation. While some women may be indifferent to the placement of their children, in most cases relinquishment is not a sign that the biological mother does not care for the child; in most cases women agonize over the adoption decision and try to make certain to do what is best for their offspring (Sweeney 1990, Erickson 1984, 459 and n. 65; Erickson 1990, 38 n. 44).

A woman's decision to place her biological child for adoption also does not mean that she is indifferent about the question of who raises the child. The argument that an unwed biological father should be preferred to adoptive parents because they are "strangers" to the child inappropriately ignores the biological mother's preference that the child be adopted through an agency or private placement rather than placed with a guardian or in the father's custody. If the mother has had very little contact with the father beyond the act of intercourse that led to her pregnancy, the father may be as much a *social* "stranger" to her and the child as the adoptive parents, and his claim rests on genetics alone. Contrasting the biological father's rights to those of strangers obscures the fact that the fundamental or precipitating disagreement about custody is not between the adoptive parents and the biological father, but between the two biological parents. And if the mother has known the father over a considerable period of time, her unwillingness to make him the custodial parent needs to be examined to see why she feels as she does, just as it would be if the parents were married.

Are there any reasons to weigh the biological mother's wishes about who shall (or shall not) take custody of the child more heavily than those of the biological father? At the time of birth the relationship of biological father and mother to the child is neither biologically nor socially symmetrical. She has borne the child for nine months, activity for which there is no precise male analog; indeed, no one else can perform functions analogous to those of gestation.[15] The biological mother's "expectant" state has affected both her own physiological experience and the ways in which others view and interact with her.[16] The Supreme Court has recognized the significance of this asymmetry between mother and father during pregnancy by holding that a wife is not required to notify her husband or obtain his consent before getting an abortion (*Planned Parenthood of Southeastern Pennsylvania v. Casey*, 112 S. Ct. 2791 [1992]). To what extent should asymmetry of biological function during gestation affect the right to make custodial decisions concerning a newborn?

Some theorists argue that the fact that only the woman is engaged in the physical gestation of the human fetus should make a decisive difference in the rights to be accorded unwed mothers and fathers in deciding on the custody of their offspring. According to sociologist Barbara Katz Rothman, parenting is a social relationship and parental rights are established by care-giving. In her view, the biological difference between mother and father is crucial and conclusive in establishing their respective claims for custody of newborns: "Infants belong to their mothers at birth because of the unique nurturant relationship that has existed between them up to that moment. That is, birth mothers have full parental rights, including rights of custody, of the babies they bore" (Rothman 1989, 254). By the same token, other persons with a genetic tie to the child do not have such rights: "We will not recognize genetic claims to parenthood, neither as traditional 'paternity' claims nor as genetic maternity in cases of ovum donation" (254). Rothman would have the gestational mother's absolute claim last for six weeks after giving birth, and so the adoption decision would rest solely in the mother's hands during that period. After six weeks, "Custody would go to the nurturing parent in case of dispute" (255). Rothman emphasizes that her preference for the gestational mother rests on her understanding of pregnancy as "a social as well as a physical relationship," and that "*any* mother is engaged in a social interaction with her fetus as the pregnancy progresses" (97). Neither the physical interdependence nor the social relationship between mother and fetus can be fully shared by any other adult, no matter how attentive. Actual care-giving, not genetic connection, creates familial bonds and, in this case, Rothman argues, custodial rights.

Others also have argued that parental rights usually are not symmetrical and that the social or biological bonds (or both) between mothers and children should give mothers the authority to decide who should have custody of their offspring. Nancy Erickson argues that the liberty interest that a parent has "to control the care, custody, and upbringing of the child" pertains only to the mother (not the father) of a newborn. At birth the mother is "not only the 'primary caretaker parent,' she is the only caretaker parent" because of her role during pregnancy (Erickson 1984, 461-62). Thinking about custody of older children of parents who divorce, Mary Becker argues that mothers are so frequently the primary care-givers of their children that it makes sense to adopt an automatic "maternal deference" standard rather than hold a hearing to try to determine what arrangement would be in the child's best interest: "When the parents cannot agree on a custody outcome, the judge should defer to the mother's decision on custody provided that she is fit, using the 'fitness' standard applicable when the state is arguing for temporary or permanent separation of parents and children in intact families" (Becker 1992, 971). Becker is not terribly worried that giving primacy to the mother's wishes might in some instances permit a woman to deprive a caring father of custody: "A maternal deference standard would recognize that mothers, as a group, have greater competence and standing to decide what is best for their children . . . than judges, fathers, or adversarial experts. . . . Mothers will sometimes make wrong decisions, but in the aggregate they are likely to make better decisions than the other possible decision makers" (972). Becker's reasoning applied to custodial decisions affecting newborns suggests that courts should defer to a biological mother, both because the woman has provided direct nurture to the fetus during pregnancy, and because, on average, biological mothers' decisions are likely to be as good as or better than those of anyone else.

Martha Fineman, similarly very critical of the best interest standard, would replace it with a "primary care-giver" standard (Fineman 1991). Fineman argues that the best interest of the child standard frequently disadvantages mothers by looking to the likely future financial resources of father and mother. It would be more appropriate (both in terms of fairness to the parents and of the child's emotional well-being), Fineman asserts, to look instead at who has actually given the child physical and emotional care up to the present. In most, but not all, instances, this will be the mother. Although Fineman does not discuss custody of newborns, if courts were to apply the primary care-giver standard to the kinds of disputes I am discussing, it would suggest that the mother who has borne and given birth should make the custody decision concerning the infant.

Many arguments for giving an unwed biological father custody of an infant child whom a biological mother wishes to have adopted not only ignore the physical and social experiences of pregnancy but also invite no inquiry at all into the conditions under which the woman became pregnant. Just as inquiry into the biological father's actions during the mother's pregnancy is permissible to encourage paternal nurturance and to counter notions of male ownership of children, so attention to the circumstances under which conception took place is reasonable to ensure that the child was not conceived as the result of abusive behavior toward the mother.[17] In trying to determine which parent's wishes concerning adoption should prevail, it would not be unreasonable for the law to regard an unmarried biological father who had been in a long-term relationship with the mother or shared living expenses with her and their offspring differently than one who engaged in casual or coercive sex (perhaps a "date rape" that the woman did not prosecute) or deceived the woman (perhaps saying he was single when he was in fact married) or willfully ignored the fact that the girl was under the age of consent.

While they refute the patriarchal premise of many fathers' rights arguments that a father has a right to custody of his biological offspring unless he is proven to be "unfit," many arguments that mothers should have the exclusive right to decide to place their offspring for adoption run the risk of treating some men unjustly and of locking both women and men into traditional gender roles. If parental claims are properly grounded in the first instance in a *combination* of biological ties and nurturance, then while a father's genetic link per se does not give him parental rights, it becomes a reason *to look to see* if he has attempted to assume responsibility for the child, and has done so without interfering with the mother's well-being. If, and only if, he has acted accordingly, should a court recognize his claim to custody.

RETHINKING THE BASES OF PARENTAL RIGHTS: RESPONSIBILITY, RELATIONSHIP, AND CARE

If unwed biological fathers should have some custodial claim to their children but not the extreme claim qualified only by "fitness," what standards should define the extent of their rights? The law needs to adopt stringent criteria for assessing the biological father's intention to take responsibility for and act as a parent to his child even prior to birth. Such criteria will require us to shift our thinking and mode of argumentation away from an emphasis on parents as owners to parents as stewards, from parental rights to parental responsibilities, and from parents viewed as individuals to parents as persons-in-relationship with a child.[18]

Many discussions of the "rights" of biological mothers and fathers reveal the inherent tension in liberal theory and legal practice between protecting individuals and their freedoms and protecting and fostering those relationships which in fundamental ways constitute every individual.[19] The language of parental rights emphasizes the parent's status as an autonomous rights-bearer, and invoking individual rights has proved useful in minimizing the role of the state in people's procreative and child-rearing decisions. For example, begetting, bearing, and raising children are for many people part of the good or fulfilling life that the liberal state is obligated to protect. No one seriously proposes that children should simply be assigned at birth to the best possible or next available parents without regard to who begot and bore them. Courts have recognized the importance of intergenerational ties for many people and protected the liberty to procreate and parent a child not only in custody cases like *Stanley* and *Caban*, but in decisions prohibiting forced sterilization, such as *Skinner v. Oklahoma* (316 U.S. 535 [1942]).[20] And since biological parents have a variety of incentives to care for their children to the best of their ability, assigning custody to them simultaneously protects children's rights as well as those of adults and sets important bounds to the exercise of state power.[21]

Yet in other contexts, use of the language of parental rights inappropriately focuses on the individual parent rather than on the relationships that are inherent in being a "parent." Katharine Bartlett has advocated recasting many legal disputes that involve parents and children in such a way that the language used does not pit one "right" against another, but emphasizes the view that parenthood implies deep and sustained human connection and must be grounded in adult responsibility for children: "The law should force parents to state their claims . . . not from the competing, individual perspectives of either parent or even of the child, but from the perspective of each parent-child relationship. And in evaluating (and thereby giving meaning to) that relationship, the law should focus on parental responsibility rather than reciprocal 'rights.' " Bartlett suggests that language based more explicitly on open-ended responsibility toward children would capture the nature of the parent-child relationship better than discussions framed in terms of parental rights (Bartlett 1988, 295).[22]

When someone is considered in the role of parent, he or she cannot be viewed apart from the child that makes him or her a parent; an "autonomous" (in the sense of unfettered or atomistic) individual is precisely what a parent is *not*. A "parental right" should not be viewed as pertaining to an individual per se, but only to an individual-in-relationship with a dependent child. It is therefore entirely appropriate for the law to require that efforts be made to establish a relationship before a parental right can be recognized.

Asking a court to determine whether a man or woman has made efforts to establish a parental relationship with a newborn is, however, fraught with difficulties that involve the different physical relationship of biological father and mother to the fetus during pregnancy, the social relationships between biological father and mother, and the need to minimize both intrusiveness by the courts and subjectivity in their judgments. Indeed, part of the attraction of both the paternal fitness test and the maternal deference standard is that each of these provides a fixed criterion for determining an unwed biological father's custodial claim. Unfortunately, however, the efficiency and clarity of each of these criteria are purchased at the cost of reducing legal discourse about family relationships to an assertion of either fathers' or mothers' rights.

My proposal that an unwed biological father have an opportunity to establish through his behavior his intention to parent his offspring tries to minimize the legal effects of biological asymmetry without ignoring altogether the relevance of sexual difference. I assume that an unwed biological mother has demonstrated a parental relationship with her newborn by virtue of having carried the fetus to term, while an unwed biological father may be required to show actual involvement with prenatal life if he wishes to have custody of the child. The model or norm of "parent" in this case, therefore, is established not by the male who awaits the appearance of the child after birth, but by the pregnant woman (Eisenstein 1988, 79-116).

Some people might object that assuming maternal care simply by the fact of pregnancy is invalid, especially in cases in which the mother has taken drugs or engaged in other behavior that might have a harmful effect on the fetus. As Cynthia Daniels has pointed out, the image of the pregnant drug addict is deeply disturbing, "representing as it does the paradox of a woman simultaneously engaged in the destruction of life (addiction) and the perpetuation of life (pregnancy)" (Daniels 1993, 98). In such cases, should the mother forfeit either her claim to custody or her claim to make the decision to place the child for adoption?[23]

It is tempting to blame, and to seek to punish, the pregnant addict when confronted by the needs of children who are physically or mentally impaired as a result of their exposure to harmful substances during gestation. It is important, however, to ascertain just what the drug-dependent mother is guilty of, and whether punishment or taking away her right to be heard concerning custody of her child is an appropriate response to her behavior.[24] In cases of maternal drug use, it is sometimes difficult to distinguish whether damage to a newborn resulted from drugs taken by the mother or from other factors such as drugs taken by the father, environmental pollutants (particularly high in poor neighborhoods), and malnutrition.[25] What is the degree of a woman's culpability if, like many drug-addicted pregnant women, she sought treatment for her addiction, but was turned away?[26] Even if in a particular case it could be ascertained that fetal damage was uniquely caused by drugs the mother took, it does not follow that she was so indifferent to the well-being of her child that she should be deprived of her right to be heard with respect to placing the child for adoption. The care the pregnant woman has given the fetus through bearing it to term and the harm her actions have caused it cannot be separated; both involve the biology and chemistry of gestation, the passage of materials across the placenta through the bloodstream. To see the pregnant drug addict as a child abuser rather than a person who is herself in need of medical treatment is to ignore the inseparability of mother and fetus during pregnancy (see Young 1994). The mother's decision to place the child for adoption will often be made in an effort to protect the child from harm now that it can be cared for outside her body.

The different biological roles of men and women in human reproduction make it imperative that law and public policy "recognize that a father and mother must be permitted to demonstrate commitment to their child in different ways."[27] What actions might a court accept as indications that an unwed biological father had made every effort to act as a parent to the child? Recent legislative efforts in New York State show that this question is not easily answered. In 1990, in *In re Raquel Marie X.* (76 NY 2d387 [1990]), the New York Court of Appeals struck down a statute that stipulated that only a father who had established a home with the mother for six months prior to her relinquishment of the child for adoption could veto the mother's adoption decision. The court held that the provision imposed "an absolute condition . . . only tangentially related to the parental relationship" and allowed a woman who would not live with a man the power unilaterally to cut off his constitutionally protected interest in parenting his child (76 NY2d at 405, 559 N.E. 2d at 426, 559 N.Y.S. 2d at 863). It instructed the legislature to find some other way to gauge a father's commitment to his unborn child's welfare, and set forth certain standards that lower courts were to follow in the meantime when judging an unwed father's parental commitment. "The father must be willing to assume full custody, not merely attempt to prevent the adoption, and he must promptly manifest parental responsibility both before and after the child's birth."[28] In assessing the father's demonstration of responsibility, judges should look at such matters as "public acknowledgment of paternity, payment of pregnancy and birth expenses, steps taken to establish legal responsibility for the child, and other factors evincing a commitment to the child" (76 NY2d at 428, 559 N.E. 2d at 428, 559 N.Y.S. 2d at 865).

Although courts in New York have used these guidelines in resolving a number of cases involving unwed fathers' efforts to block mothers' adoption decisions in the years since *In re Raquel Marie X.*,[29] four years have gone by without the New York legislature passing a new statute governing an unwed father's right to veto an adoption. Three recent bills introduced in the legislature (two in the Assembly, one in the Senate) have differed strikingly in their underlying approaches to unwed fathers' rights. Both the variety of provisions in these bills and the fact that none has been enacted into law reflect a widely shared uncertainty over what considerations should be brought to bear to determine the nature and extent of an unwed biological father's custodial rights. I discuss these bills in the order of what each requires for a father to establish his right to consent to the adoption of his offspring, from the weakest to the strongest stipulations.

The first of two bills (A. 8028) introduced in the Assembly during the 1993 and 1994 legislative sessions, listed a number of actions an unwed father of an infant under six months might take to establish his right to consent to the adoption. The bill would make his consent necessary if he openly lived with the child or the child's mother prior to the placement of the child for adoption; *or* held himself out to be the father of such child during such period; *or* paid or offered to pay a reasonable sum, consistent with his means, for the medical expenses of pregnancy and childbirth; *or* initiated judicial proceedings to obtain custody of the child; *or* married the child's mother.[30] Since the father needs to have taken only one of these actions and may have initiated judicial proceedings after the child was born, this bill applies a simple "fitness" test and requires no showing of interest prior to the child's birth.

By contrast with the minimal expectations put on unwed fathers by A. 8028, the Senate bill (S. 3776) introduced in the Senate during the 1991 and 1992 legislative sessions requires that a father have demonstrated his commitment to his offspring in a number of ways both prior to and after the birth of the child.[31] It does so by replacing most "or"s in the Assembly bill with the conjunctive "and," thus insisting that a biological father have supported the mother or baby financially, held himself out as the father, and taken steps to initiate legal proceedings to assume custody of the child.

The second Assembly bill (A. 8319) introduced during the 1993 and 1994 legislative sessions would exact an even stronger commitment from the father. It stipulates that the father have paid or offered to pay a reasonable part of the medical expenses of pregnancy and childbirth and the child's living expenses, and that he have initiated judicial proceedings to establish paternity and to obtain sole custody of the child within clearly specified time limits. The bill further stipulates that " 'ability to assume sole custody' shall mean ability to assume guardianship and custody of the child and become the primary caretaker of the child for the foreseeable future."[32] This bill clearly means to grant the right to consent to an adoption only to unwed fathers who demonstrate that they have been and will be actively engaged in the care and upbringing of their offspring; the stipulations rest on an image of father as caretaker and nurturer, not simply as progenitor.

A. 8319 goes a long way to enact the spirit of the principles set forth in this essay, but a fully adequate statute would go further. A court should be required to hear a mother's objections, if she has any, to a father's assuming custody of the child, both because the birth of a child has resulted from a web of social interactions and relationships, and because the mother's relinquishment of the child for adoption should be viewed as the last in a series of actions meant to provide care for the child, not as an act of abandonment that gives her no interest in the child's placement. In cases in which the mother objects to the father's assumption of custody, a court should listen to the *reasons* the mother opposes placing the child in the biological father's custody. Because parental rights must be grounded in the provision of care and the assumption of responsibility, if an unwed mother could demonstrate that her pregnancy was a result of force, coercion, or deception, or that she had been under the age of consent when intercourse occurred, the father would be held to be "unfit." This still might not meet the possibility that a man who desired children might impregnate a woman whom he knew would neither abort nor raise a child, provide care and financial support throughout the pregnancy, and petition for paternity and custody—a kind of inexpensive "surrogacy." I am convinced that such intentional instrumental use of any woman's body is morally abhorrent, but I am not certain how to ensure that such a man could not assume custody of his offspring under such circumstances.

Finally, a statute should provide that a pregnant woman who wishes to make plans for her child should be able to ascertain early in the pregnancy whether or not the father will step forward later to oppose the adoption. The law should provide that she be able to notify him in writing of the pregnancy and preclude him from a veto if he fails to act soon after receipt of such notification. Similarly, if a father is found to be entitled to veto an adoption, a mother should be able to negate her consent to the child's adoption and be put back in the same position she was in prior to her consent, that is, as one of two unwed parents each of whom seeks custody (Erickson 1991).

One purpose of spelling out what actions the father needs to take to establish his claim would be to ascertain as early as possible during the pregnancy or after birth whether or not he wished custody, so that infants could be definitively freed for adoption. Where the mother objected to the father's assuming custody, a hearing would be necessary. A hearing would, of course, take more time than assigning custody based on a rule that any "fit" biological father prevail or that a mother be able to make the decision to place her child for adoption unimpeded by the biological father. But a hearing to ascertain whether an unwed biological father has grasped the opportunity to parent his newborn should not cause more delay than a best interest hearing. Such a hearing would be to ascertain facts about the unwed father's behavior and the mother's considered opinion concerning custody, not to try to project what custodial arrangement might be in the child's best interest.

These considerations leave unresolved the thorny issue of how the law should deal with cases in which a biological mother lies to the biological father about his paternity or otherwise hides her pregnancy, making it impossible for him to take any action to signal his willingness to take care of his offspring. In 1992 the New York Court of Appeals addressed the question of what effect a lack of knowledge of a woman's pregnancy should have on a biological father's right to seek custody after learning of the child's existence. In *Matter of Robert O. v. Russell K.*, an unwed biological father sought to overturn the adoption of his son on the grounds that either the mother or the State had a duty to ensure that he knew of the child's birth, and that their failure to inform him denied him his constitutional rights. The New York court acknowledged that "the unwed father of an infant placed for adoption immediately at birth faces a unique dilemma should he desire to establish his parental rights." His opportunity to "shoulder the responsibility of parenthood may disappear before he has a chance to grasp it." But although the father, Robert O., acted as soon as he knew of the child's existence, the adoption had been finalized ten months previously. "Promptness," said the court, "is measured in terms of the child's life, not by the onset of the father's awareness." Robert O., having failed to determine in a timely fashion whether the woman with whom he had lived was pregnant, lost the right he would have had to an opportunity to manifest his "willingness to be a parent" (*Matter of Robert O. v. Russell K.*, 80 NY2d 252 [1992] at 262). The responsibility to know of a child's existence should fall on the man who would assume responsibility for raising the child. By contrast, one defender of unwed fathers' rights proposes a jail sentence of up to two years for a woman who refuses to name the father of her child when surrendering the infant for adoption! (Hamilton 1987-88, 1103 n. 406).[33] A biological father aware of a woman's pregnancy should be required to act prior to birth and soon after he suspects his paternity; a biological father who is actively kept ignorant might be allowed to step forward for some specified period after birth (probably not less than eight weeks nor longer than six months), but thereafter the importance of establishing a firm parent-child relationship would preclude his advancing a parental claim. The child's need for such a relationship should also lead to requirements that courts hear and decide disputes concerning the adoption of infants expeditiously.

Although the reflections set forth in this essay suggest various reforms in the laws governing the custody of nonmarital children, they do not in and of themselves answer the question of whether the case of *Baby Girl Clausen* was decided correctly. I find that very hard to do because neither side grounded its position in the kinds of principles I have put forward here. The Iowa statute that Daniel Schmidt invoked to claim that the adoption could not be finalized required the biological father's consent, but no showing that he demonstrate his commitment to the child prior to (or even subsequent to) birth. The father's mere opposition to the adoption was a sufficient basis upon which to grant him custody. The DeBoers, for their part, based their claim that they should be allowed to adopt Jessica on the best interest standard. Placing the child with the Schmidts reinforced the notion that a biological tie between man and child automatically creates a custodial claim. On the other hand, favoring the DeBoers would not only have reinforced the best interest standard but might have been viewed as rewarding them for prolonging legal proceedings after Schmidt raised his claim.

The outcome consonant with the principles advanced here would have granted a hearing to Schmidt, recognizing that while his biological tie alone did not guarantee him custodial rights, the fact that he acted immediately after learning that he was Jessica's biological father and within four weeks after her birth established grounds for a hearing and, provided that Cara Clausen did not object, for custody. Had Clausen objected, the hearing would not have attempted to determine whether the child's best interest would be better served by granting custody to Schmidt or the DeBoers, but whether Schmidt's actions were sufficient to establish a claim to custody. To establish his right to consent to the adoption, he would have to demonstrate that he had good reason to believe that the child Cara Clausen was carrying was not his offspring, that he acted immediately and decisively to assume full custody after learning that he was Jessica's biological father, and that he had done so within the statutory limit for advancing such a claim. It seems to me likely both that Schmidt's claim would have been recognized and that the likelihood of a ruling in his favor would have been much clearer to the DeBoers and their lawyer than was the case under the law then in effect.

The main lesson to be drawn from cases like *Baby Girl Clausen* is that it is imperative that states formulate adoption laws that will reflect the principle that parental rights are established in the first instance by a combination of biology and the provision of care, a principle already articulated by the Supreme Court. Another lesson may be that in certain instances it would make sense to allow some form of legal recognition to the fact that a child may have multiple "parents": genetic parents (sperm and egg donors), biological parents (the gestational mother and genetic father), stepparents, adoptive parents, social parents (that is, those who actually provide care), and legal guardians.[34] The possibility of some such recognition might avoid some cases in which unwed biological fathers who have not married the mothers of their offspring seek to block the adoption of the infant. Some of these cases seem motivated not so much by the man's desire to raise the child as by his fear of losing all opportunity to know child he has sired. There may be ways of dealing with this fear short of blocking the adoption. Adoption registries that allow adopted children and birth parents to contact one another by mutual consent when the child has reached his or her majority seem to have been helpful to biological parents, adoptive parents, and children alike. They allow for the simultaneous recognition of the importance of both biological and social parenting, and in doing so undercut the suggestion that something about adoption is shameful and best kept hidden. Such registries also take into account the perspective of children who want to know their biological forebears, without weakening either the legal rights and responsibilities of the social (adoptive) parents or the primacy of the emotional bonds between adoptive parents and children.

Beyond these legal changes, cases such as *Matter of Robert O. v. Russell K.* and *Baby Girl Clausen* should also lead us to try to understand the circumstances that might lead an unwed mother to lie about or conceal the paternity of her child, such as fear of violence or harassment, or shame over an unwanted sexual relationship. Working toward justice in family relationships requires struggling to eliminate the social conditions that give rise to such fear and shame, and also requires making sure that all citizens have access to the resources that allow family relationships to survive and flourish, so that no biological parent will be forced by economic factors to relinquish custody of children they would prefer to raise themselves had they the resources to do so.

CONCLUSION

This analysis of disputes over paternal custody of nonmarital newborns makes it abundantly clear that the language of individual rights, so central to liberal political theory, and to the due process and equal protection guarantees of the U.S. Constitution, is not well-suited to dealing with complex issues of parent-child relationships. Although notions of maternal or paternal rights are not useless (for example, they allow us to think about limits to state intervention), they tend to focus attention on an adult *individual*, whereas parental issues involve two adults and a child, and the relationships among them.[35] Legal and social discourse alike must put the lived relationship between parents and between parent and child, not the rights of individuals alone, at the center of the analysis of parental claims. In particular the language of a father's "right" to custody of his infant child based on his genetic tie obscures the complexity of the relationships involved in human reproductive activity.

Because parenting involves being in a relationship with another dependent person, a parental "right" cannot properly be conceived of as something independent of the relationship. An individual can exercise a parental right, but the existence or the nature of the right cannot be explained by reference to that individual alone. Only by taking account of the interpersonal dependency, reciprocity, and responsibility involved in family relationships will we be able to approach a world dedicated to achieving both lived equality between men and women and committed parents for every child.

NOTES

I thank Julie Bartkowiak, Joan Callahan, Ann Congleton, Stephen Ellmann, Nancy Erickson, Leslie Goldstein, Mona Harrington, Alice Hearst, Wolfgang Hirczy, Martha Minow, Uma Narayan, Susan Okin, and Joan Posner for helping me think about the issues raised in this essay. I began work on this article while a Fellow at the Center for Human Values at Princeton University, for whose support I am very grateful.

This essay was originally presented as a paper at the Conference on Feminist Ethics and Social Policy, University of Pittsburgh, November 5-7, 1993. It appears in somewhat different form in Callahan (1995). An expanded and revised version appears with the title "Unwed Fathers' Rights, Adoption, and Sex Equality: Gender-Neutrality and the Perpetuation of Patriarchy" (Shanley 1995).

1. *In the Interest of* B.G.C., Supreme Court of Iowa, No. 207/91-476, 92-49, September 23, 1992, and *In the Matter of Baby Girl Clausen*, Michigan Court of Appeals, No. 161102, March 29, 1993.

2. A summary of the different statutory provisions in all fifty states regarding who must give consent to an adoption and under what conditions is found in *Adoption Laws: Answers to the Most-Asked Questions* (Rockville, MD: National Adoption Information Clearinghouse, n.d.).

3. Hirczy (1992) and Hamilton (1987-88) argue that the law should insist that the paternity of every child be established at birth, a necessary prerequisite for an unwed father's assertion of paternal rights.

4. New York City Family Court denied parental rights to a man who donated sperm to a lesbian couple in *Thomas S. v. Robin Y.* (599 N.Y.S. 377 [1993]), on the grounds that the child's mothers were her "parents" and that the sperm donor's genetic tie and occasional visits with the child did not in themselves establish a parental relationship. Denial of parental rights to sperm donors does not address the question of whether a child has a right to know who his or her biological parents are.

In this article I leave aside cases of contract parenthood, artificial insemination by donor, and embryo transfer. All of these practices raise issues that involve the distinctions between genetic, biological, and "intentional" parents, and all involve more than two adults who claim the status of "parent." I discuss the claims of a married sperm donor who together with his wife tries to become a parent by hiring a gestational ("surrogate") mother in Shanley (1993). I argue there that the enforcement of pregnancy contracts against the gestational mother's will would dismiss the relevance of a woman's biological and social experiences of relationship to a fetus during pregnancy and undermine the principle that parental rights are rooted in the assumption of concrete responsibility for a fetus or child.

5. On reconceptualizing the liberal individual, see Held (1993, esp. 192-214). On the concept of autonomy in liberal theory, see Nedelsky (1989, 7-36) and (1990, 162-89). On the limitations of the legal concept of individual rights see Glendon (1991).

6. Stanton (1987, 92). Stanton quotes Note, "Custody Rights of Unwed Fathers" (*Pacific Law Journal* 4 [1973]: 923): "Various reasons have been proposed for considering the unwed mother a more natural guardian than the father. These include: (1) the mother is more easily identified than the father; (2) she is biologically better suited to care for and nurture the child; and (3) the natural bonds of love and affection for the child are stronger in the mother than in any other person."

7. Blackstone ([1783] 1978, 1: 453). On coverture in general see Shanley (1989). On laws governing custody in the early United States, see Grossberg (1985) and Hartog (1993).

8. Also important was the rise in both social and judicial attention to childhood and its particular needs. See Zainaldin (1979, 1038-89); Grossberg (1985); and Hartog (1993).

9. Scalia argued that the proper methodology of discerning what interests are protected by the due process clause is to look at "the most specific level at which a relevant tradition protecting, or denying protection to, the asserted right can be identified" (491 U.S. 110 at 127-28, n. 6).

10. O'Connor and Kennedy disagreed with Scalia's attempt to identify "the most specific level" of a "tradition" to discover what interests are protected by the due process clause (491 U.S. 110 at 132).

11. Jonathan Lehr raised this issue but it was not definitively answered in *Lehr v. Robertson*, since the majority of the Supreme Court rejected evidence Lehr offered to support his contention that his early efforts to know his daughter had been rebuffed by her mother.

The Supreme Court may be asked to decide whether an unwed biological father has a right to veto the adoption of a newborn if he has had no opportunity to establish a relationship during its early days if it hears an appeal from the decision of the Illinois Supreme Court, *In re Petition of John Doe and Jane Doe, Husband and Wife, to Adopt Baby Boy Janikova* ([No. 76063] 1994 WestLaw 265086 [Ill.]).

12. See Goldberg (1992).

13. "Recent Developments: Family Law—Unwed Fathers' Rights—New York Court of Appeals Mandates Veto Power over Newborn's Adoption for Unwed Father Who Demonstrates Parental Responsibility—*In re Raquel Marie X*." *Harvard Law Review* 104 (January 1991): 807. Fineman (1988, 770) and Eveleigh (1989) also express strong reservations about the best interest test in custody adjudication.

14. For a discussion of the many ways legal and social discourses label women as "bad mothers" see Roberts (1992). By contrast, Eveleigh (1989) writes as if the biological mother who has relinquished custody has no further relevant interest in her child; when a biological father seeks custody of his newborn offspring, the relevant interests, she says, are those of the state, the father, and the child.

15. The uniqueness of pregnancy has implications for the custodial claims not only of unwed fathers but also for those of a lesbian partner who had planned to co-parent a child, as well as those of a genetic mother who might turn to a "surrogate" to bear a child on her behalf.

16. On the social construction of the experience of pregnancy and childbirth see Martin (1987) and Rothman (1982).

17. It is not necessary to accept Catharine MacKinnon's view that fully consensual sexual relations between men and women are virtually impossible to achieve in our society in order to acknowledge that the context in which sexual relations take place—including great disparities of social and economic resources—can make sexual relationships more or less consensual or coercive. See MacKinnon (1989, 174) and (1987, 88). For an opposing view see Roiphe (1993), which sees many feminist challenges to social structures of inequality that affect relationships between men and women as contributions to an ideology not of female empowerment but of female weakness and need for protection.

18. Smith (1983) argues against the "property model" of parenting and family relationships.

19. Excellent discussions of the ways in which classical liberal theory pays insufficient attention to the ways in which individuals are constituted in and by their relationships to others are found in Held (1993); Nedelsky (1989) and (1990); Ruddick (1989); and Tronto (1993).

Issues involving children raise in a particularly acute manner the tension between protecting people as individuals and protecting family associations or family ties. On the dilemmas inherent in using privacy language to afford protection to both individuals and families see Karst (1980).

20. Although see also cases that permitted sterilization of the mentally retarded, *Buck v. Bell*, 274 U.S. 200 (1927) and *Sterilization of Moore*, 289 N.C. 95, 221 S.E. 2d 307 (1976).

21. See Okin (1989) and Olsen (1985) for a clear analysis of the impossibility of complete state neutrality toward the family.

22. For an interesting critique of Bartlett that is flawed by its failure to recognize the deeply individualistic as well as the patriarchal characteristics of contemporary family law, see Dolgin (1990).

The idea that parental rights depend on the fulfillment of parental duties receives one of its classic expressions in John Locke's argument against the patriarchal ideas of Sir Robert Filmer. Filmer held that the very act of begetting a child gave a father sovereignty over his offspring. By contrast, Locke asserts that not procreation but only providing care for a child bestows parental authority and that by nature (although not necessarily by law) mothers share such authority. But despite Locke's rejection of patriarchal reasoning with respect to both polity and family, patriarchal notions crept back into much liberal political theory and law because these used the language of the rights of autonomous individuals to talk about or characterize the relationship between parent and child. To some extent, when nineteenth-century feminists began their assault on the common law doctrine of coverture, they used Lockean notions of individual freedom and equality to argue that mothers should have equal rights over marital children. Even these reforms, however, did not subvert the notion of parental autonomy or of parental rights to custody. See Locke ([1690] 1980, 2.6.58). On patriarchalism see Schochet (1975); on Locke's antipatriarchalism see Shanley (1979).

23. I leave aside the question of whether the mother should be subject to prosecution for fetal abuse, although I do not think she should be.

24. This is not to deny responsibility to drug-dependent pregnant women. As Cynthia Daniels writes, "Clearly, race, class, and gender inequality limit an individual's access to a full range of options, but they do not negate all choice. The behavior of poor addicted women may be 'determined' in the historical sense that they are born into conditions which they did not create and cannot control, and which set strict limits on their lives, but this does not mean that they lose all power to shape their own destinies" (Daniels 1993, 125).

Dorothy Roberts has written with great insight about the ways in which issues of equality and privacy arise in prosecutions of drug-addicted women of color who have babies (Roberts 1991).

25. Studies have shown that the consequences for offspring of poor women who used alcohol and drugs during pregnancy were much more severe than for offspring of upper-income women, chiefly because of differences in nutrition (Nesrin Bingol et al. 1987. "The Influence of Socioeconomic Factors on the Occurrence of Fetal Alcohol Syndrome" *Advances in Alcohol and Substance Abuse* 6(4): 117; cited in Daniels 1993, 125).

26. Two-thirds of the major hospitals in fifteen cities surveyed by the House Select Committee on Children, Youth, and Families in 1989 reported that they had no place to which to refer drug-addicted pregnant women for treatment (Karol L. Kumpfer. 1991. "Treatment Programs for Drug-Abusing Women" *The Future of Children* 1[1]: 52; cited in Daniels 1993, 126). In the same year, there were 135 treatment beds available for the more than four thousand pregnant drug-dependent women in the state of Florida (*Johnson v. State of Florida*, Petitioners' Initial Brief, Appeal from the District Court of Appeal, Fifth District 77, 831 Sup. Ct. of Fla. [1990], p. 31; cited in Daniels 1993, 126). Many drug treatment programs do not accept pregnant women for fear that they will be sued if the woman loses the pregnancy as a result of treatment (Chavkin 1991, 1556).

27. "Recent Developments: Family Law—Unwed Fathers' Rights—. . . *In re Raquel Marie X.* . . ," p. 805 (footnote omitted).

28. "Recent Developments: Family Law . . .—*In re Raquel Marie X.* . . ," p. 803.

29. For example, *Matter of Kiran Chandini S.*, 166 A.D. 2d 599, 560 N.Y.S. 2d 886 (1990); In the *Matter of John E. v. John Doe*, 564 N.Y.S. 2d 439 (1990); *Matter of Stephen C.*, 566 N.Y.S. 2d 178 (1991); *Matter of Robert O. v. Russell K.*, 80 NY2d 252 (1992).

30. New York State Legislature, Assembly, A. 8028, May 17, 1993; introduced by Member of the Assembly Vito Lopez and referred to the Committee on the Judiciary.

31. New York State Legislature, Senate, S. 3776-B, March 11, 1991; introduced by Senator Mary Goodhue.

32. New York State Legislature, Assembly, A. 8319A, June 4, 1993; introduced by Member of the Assembly Vito Lopez (at the request of the Governor) and referred to the Committee on Children and Families.

If the father needs to work to support the child, he clearly cannot provide uninterrupted childcare. The stipulation that he be the primary caretaker of the child is meant to preclude men who intend to turn the care of their offspring over to other family members from blocking adoptions. This raises very hard issues about how the law might recognize the roles of members of extended families in raising children, and this aspect of any proposed bill needs further deliberation.

33. Less drastically, Justice Titone, concurring in *Robert O. v. Russell K.*, also felt it unreasonable to require that a man who wishes to assert paternal rights know of the pregnancy of his sexual partner.

34. See interesting suggestions for ways in which the law might recognize more than two parents in Bartlett (1984). See also Bartholet (1993).

35. I leave aside cases of contract parenthood, artificial insemination by donor, and embryo transfer, for example, in all of which issues involving distinctions between genetic, biological, and "intentional" parents arise, and in all of which there may be more than two adults claiming the status of "parent." See above note 4.

REFERENCES

Adoption laws: Answers to the most-asked questions. N.d. Rockville, MD: National Adoption Information Clearinghouse.

Bartholet, Elizabeth. 1993. *Family bonds: Adoption and the politics of parenting*. Boston: Houghton Mifflin.

Bartlett, Katharine. 1984. Rethinking parenthood as an exclusive status: The need for legal alternatives when the premise of the nuclear family has failed. *Virginia Law Review* 70(5): 879-963.

———. 1988. Re-expressing parenthood. *Yale Law Journal* 98(2): 293-340.

Becker, Mary. 1989. The rights of unwed parents: Feminist approaches. *Social Service Review* 63(4): 496-518.

———. 1992. Maternal feelings: Myth, taboo, and child custody. *Review of Law and Women's Studies* 1: 901-92.

Blackstone, William. [1783] 1978. *Commentaries on the laws of England*. 4 vols. 9th ed. Ed. Berkowitz and Throne. Oxford: Clarendon Press.

Boyd, Jeffrey S. 1990. The unwed father's custody claim in California: When does the parental preference doctrine apply? *Pepperdine Law Review* 17: 969-1010.

Buchanan, Elizabeth. 1984. The constitutional rights of unwed fathers before and after *Lehr v. Robertson*. *Ohio State Law Journal* 45: 311-82.

Buck v. Bell. 1927. 274 U.S. 200.

Caban v. Mohammed. 1979. 441 U.S. 380.

Callahan, Joan C., ed. 1995. *Reproduction, ethics and the law: Feminist perspectives*. Bloomington: Indiana University Press. Forthcoming.

Chavkin, Wendy. 1991. Mandatory treatment for drug use during pregnancy. *Journal of the American Medical Association* 266(11): 1556-61.

Daniels, Cynthia R. 1993. *At women's expense: State power and the politics of fetal rights*. Cambridge: Harvard University Press.

Dolgin, Janet L. 1990. Status and contract in feminist legal theory of the family: A reply to Bartlett. *Women's Rights Law Reporter* 12: 103-13.

Eisenstein, Zillah. 1988. *The female body and the law*. Berkeley: University of California Press.

Erickson, Nancy S. 1984. The feminist dilemma over unwed parents' custody rights: The mother's rights must take priority. *Journal of Law and Inequality* 2: 447-72.

———. 1990. Neither abortion nor adoption: Women without options. Paper presented at the American Association of Law Schools (AALS), San Francisco, 6 January.

———. 1991. Proposal for a model law on unwed fathers' adoption rights. Unpublished paper. Brooklyn, N.Y.

Eveleigh, Laurel J. 1989. Certainly not child's play: A serious game of hide and seek with the rights of unwed fathers. *Syracuse Law Review* 40: 1055-88.

Fineman, Martha Albertson. 1988. Dominant discourse, professional language, and legal change in child custody decisionmaking. *Harvard Law Review* 101(4): 727-74.

———. 1991. *The illusion of equality: The rhetoric and reality of divorce reform*. Chicago: University of Chicago Press.

Glendon, Mary Ann. 1991. *Rights talk: The impoverishment of political discourse*. New York: Free Press.

Goldberg, Stephanie B. 1992. Having my baby. *ABA Journal* 78: 84-86.

Gomez v. Perez. 1973. 409 U.S. 535.

Grossberg, Michael. 1985. *Governing the hearth: Law and the family in nineteenth-century America*. Chapel Hill: University of North Carolina Press.

Hamilton, John R. 1987-88. The unwed father and the right to know of his child's existence. *Kentucky Law Journal* 76: 949-1009.

Hartog, Hendrick. 1993. Breaking the marital bond. Princeton University. Unpublished manuscript.

Held, Virginia. 1993. *Feminist morality: Transforming culture, society, and politics*. Chicago: University of Chicago Press.

Hirczy, Wolfgang. 1992. The politics of illegitimacy: A cross-national comparison. Paper presented at the Annual Meeting of the American Political Science Association, Chicago, 3-6 September.

In the Interest of B.G.C. September 23, 1992. Supreme Court of Iowa, No. 207/91-476, 92-49.

In the Matter of Baby Girl Clausen. March 29, 1993. Michigan Court of Appeals, No. 161102.

In re Raquel Marie X. 1990. 76 NY2d 387.

In re Baby Girl Eason. 1987. 257 Ga. 292, 358 S.E. 2d 459.

Johnson, Sally. 1994. Helping fathers become parents. *New York Times*, 24 February, metropolitan edition, C1.

Karst, Kenneth L. 1980. The freedom of intimate association. *Yale Law Journal* 89(4): 624-92.

Lehr v. Robertson. 1983. 463 U.S. 248.

Locke, John. [1690] 1980. *Second treatise of government*. Ed. C. B. Macpherson. Indianapolis: Hackett.

MacKinnon, Catharine A. 1987. *Feminism unmodified: Discourses on life and law*. Cambridge: Harvard University Press.

———. 1989. *Toward a feminist theory of the state*. Cambridge: Harvard University Press.

Martin, Emily. 1987. *The woman in the body: A cultural analysis of reproduction*. Boston: Beacon Press.

Matter of Robert O. v. Russell K. 1992. 80 NY2d 252.

Michael H. v. Gerald D. 1989. 491 U.S. 110.

Nedelsky, Jennifer. 1990. Law, boundaries, and the bounded self. *Representations* 30: 162-189.

———. 1989. Reconceiving autonomy: Sources, thoughts, and possibilities. *Yale Journal of Law and Feminism* 1(1): 7-36.

New York State Assembly. A. 8028. May 17, 1993.

New York State Assembly. A. 8319A, June 4, 1993.

New York State Senate. S. 3776-B. March 11, 1991.

Okin, Susan M. 1989. *Justice, gender, and the family*. New York: Basic Books.

Olsen, Frances. 1985. The myth of state intervention in the family. *University of Michigan Journal of Law Reform* 18(4): 835-64.

Planned Parenthood of Southeastern Pennsylvania v. Casey. 1992. 112 S. Ct. 2791.

Raab, Jennifer J. 1984. *Lehr v. Robertson*: Unwed fathers and adoption—How much process is due? *Harvard Women's Law Journal* 7: 265-87.

Recent developments: Family law—Unwed fathers' rights—New York Court of Appeals mandates veto power over newborn's adoption for unwed father who demonstrates parental responsibility—In re Raquel Marie X. . . . 1991. *Harvard Law Review* 104(3): 800-807.

Roberts, Dorothy E. 1991. Punishing drug addicts who have babies: Women of color, equality, and the right of privacy. *Harvard Law Review* 104(7): 1419-82.

———. 1992. Racism and patriarchy in the meaning of motherhood. Paper presented at the Feminism and Legal Theory Project Workshop on Motherhood, Columbia Law School, New York, 4-5 December.

Roiphe, Katie. 1993. *The morning after: Sex, fear, and feminism on campus*. Boston: Little Brown and Co.

Rothman, Barbara Katz. 1982. *In labor: Women and power in the birthplace*. New York: Norton.

———. 1989. *Recreating motherhood: Ideology and technology in a patriarchal society*. New York: W. W. Norton.

Ruddick, Sara. 1989. *Maternal thinking: Towards a politics of peace*. Boston: Beacon.

Schochet, Gordon. 1975. *Patriarchalism in political thought*. New York: Basic Books.

Serviss, Claudia. 1991. *Lehr v. Robertson*'s "Grasp the opportunity": For California's natural fathers, custody may be beyond their grasp. *Western State University Law Review* 18: 771-90.

Shanley, Mary L. 1979. Marriage contract and social contract in seventeenth-century English political thought. *Western Political Quarterly* 32(1): 79-91.

———. 1989. *Feminism, marriage, and the law in Victorian England*. Princeton: Princeton University Press.

———. 1993. "Surrogate mothering" and women's freedom: A critique of contracts for human reproduction. *Signs* 18(3): 618-39.

———. 1995. Unwed fathers' rights, adoption, and sex equality: Gender-neutrality and the perpetuation of patriarchy. *Columbia Law Review* 95(1). Forthcoming.

Smith, Janet Farrell. 1983. Parenting and property. In *Mothering: Essays in feminist theory*, ed. Joyce Trebilcot. Totowa, N.J.: Rowman & Allanheld.

Stanley v. Illinois. 1972. 405 U.S. 645.

Stanton, Elizabeth Rose. 1990. The rights of the biological father: From adoption and custody to surrogate motherhood. *Vermont Law Review* 12: 87-121.

Sterilization of Moore. 1976. 289 N.C. 95, 221 S.E. 2d 307.

Steward, Rebecca L. 1990. Constitutional rights of unwed fathers: Is equal protection equal for unwed fathers? *Southwestern University Law Review* 19: 1087-1111.

Sweeney, Maureen A. 1990. Between sorrow and happy endings: A new paradigm of adoption. *Yale Journal of Law and Feminism*, 329-70.

Thomas S. v. Robin Y. 1993. 599 N.Y.S. 377.

Trimble v. Gordon. 1977. 430 U.S. 762.

Tronto, Joan C. 1993. *Moral boundaries: A political argument for an ethic of care*. New York: Routledge.

Weber v. Aetna Casualty & Surety Co. 1972. 406 U.S. 164.

Young, Iris Marion. 1994. Punishment, treatment, empowerment: Three approaches to policy for pregnant addicts. *Feminist Studies* 20(1): 33-58.

Zainaldin, Jamil S. 1979. The emergence of a modern American family law: Child custody, adoption, and the courts, 1796-1851. *Northwestern University Law Review* 73: 1038-1089.

Zinman, Daniel C. 1992. Father knows best: The unwed father's right to raise his infant surrendered for adoption. *Fordham Law Review* 60: 971-1001.

Acknowledgments

Grossberg, Michael. "Who Gets the Child? Custody, Guardianship, and the Rise of a Judicial Patriarchy in Nineteenth-Century America." *Feminist Studies* 9 (1983): 235–60. Reprinted with the permission of the publisher, FEMINIST STUDIES, Inc., c/o Women's Studies Program, University of Maryland, College Park, MD 20742.

Sheppard, Annamay T. "Lesbian Mothers II: Long Night's Journey Into Day." *Women's Rights Law Reporter* 8 (1985): 219–46. Reprinted with the permission of the *Women's Rights Law Reporter*.

Rothman, Barbara Katz. "Women as Fathers: Motherhood and Child Care Under a Modified Patriarchy" *Gender and Society* 3 (1989): 89–104. Reprinted with the permission of Sage Publications.

Binion, Gayle. "*Webster* v. *Reproductive Health Services*: Devaluing the Right to Choose." *Women and Politics* 11 (1991): 41–59. Reprinted with the permission of Haworth Press, Inc. Copyright 1991.

Fineman, Martha L. "Images of Mothers in Poverty Discourses." *Duke Law Journal* (1991): 274–95. Copyright (1991). Reprinted with the permission of the *Duke Law Journal*. This article is based on her book *The Neutered Mother, the Sexual Family, and Other Twentieth Century Tragedies* (New York: Routledge, 1995).

Rutherford, Charlotte. "Reproductive Freedoms and African American Women." *Yale Journal of Law and Feminism* 4 (1992): 255–90. Reprinted with the permission of the *Yale Journal of Law and Feminism*.

Blankenship, Kim M., Beth Rushing, Suzanne A. Onorato, and Renée White. "Reproductive Technologies and the U.S. Courts." *Gender and Society* 7 (1993): 8–31. Reprinted with the permission of Sage Publications.

Rhode, Deborah L. "Adolescent Pregnancy and Public Policy." *Political Science Quarterly* 108 (1993–94): 635–69. Reprinted with the permission of the author and The Academy of Political Science.

Mezey, Susan Gluck, Raymond Tatalovich and Michael Walsh. "Keeping Abortion Clinics Open: The Importance of *Ragsdale* v. *Turnock* in the Post-*Casey* Era." *Policy Studies Review* 13 (1994): 111–26. Reprinted with the permission of Transaction Publishers.

Nolan, Laurence C. "The Unconstitutional Conditions Doctrine and Mandating Norplant for Women on Welfare Discourse." *American University Journal of Gender and the Law* 3 (1994): 15–37. Reprinted with the permission of the Washington College of Law.

Apel, Susan B. "Communitarianism and Feminism: The Case Against the Preference for the Two-Parent Family." *Wisconsin Women's Law Journal* 10 (1995): 1–26. Reprinted with the permission of the University of Wisconson-Madison School of Law.

Burrell, Darci Elaine. "The Norplant Solution: Norplant and the Control of African-American Motherhood." *UCLA Women's Law Journal* 5 (1995): 401–44. Reprinted with the permission of the *UCLA Women's Law Journal.*

Goodwin, Joanne L. "'Employable Mothers' and 'Suitable Work': A Re-Evaluation of Welfare and Wage-Earning for Women in the Twentieth-Century United States." *Journal of Social History* 29 (1995): 253–74. Reprinted with the permission of the *Journal of Social History,* Carnegie Mellon University.

Shanley, Mary L. "Fathers' Rights, Mothers' Wrongs? Reflections on Unwed Fathers' Rights and Sex Equality." *Hypatia: A Journal of Feminist Philosophy* 10 (1995): 74–103. Reprinted with the permission of Indiana University Press.